Microbial Production

Hideharu Anazawa • Sakayu Shimizu
Editors

Microbial Production

From Genome Design to Cell Engineering

Springer

Editors
Hideharu Anazawa, Ph.D.
Director
Japan Bioindustry Association
Grande Bldg. 8F, 2-26-9 Hatchobori, Chuo-ku
Tokyo 104-0032, Japan
anazawahdhr@jba.or.jp

Sakayu Shimizu, Ph.D.
Professor
Department of Bioscience and Biotechnology
Graduate School of Enviromental Science
Kyoto Gakuen University
Nanjo-Ohtani, Sogabe, Kameoka
Kyoto 621-8555, Japan
sshimizu@kyotogakuen.ac.jp

ISBN 978-4-431-54606-1 ISBN 978-4-431-54607-8 (eBook)
DOI 10.1007/978-4-431-54607-8
Springer Tokyo Heidelberg New York Dordrecht London

Library of Congress Control Number: 2013956366

Printed on acid-free paper

Springer is part of Springer Science+Business Media (www.springer.com)

Preface

It has long been considered essential to introduce energy-saving and environmentally friendly bioprocesses to incorporate resource-saving concepts in production systems. Production of useful substances using microbial and enzymatic reactions, for example, is a truly environmentally friendly process and should be actively explored if there is even a small possibility for the process to replace a chemical-industrial one based on conventional petrochemical reactions.

Historically speaking, the microbial production of useful substances has shown expansion of its fundamental and technological platforms and evolved in a unique manner, mainly through fermentative or enzymatic transformation of bioactive compounds such as antibiotics, amino acids, nucleic acid-related compounds, and vitamins. There have been some relatively recent developments in technologically and industrially new areas, such as the production of chiral chemicals using chemo-enzymatic methods, the production of commodity chemicals (e.g., acrylamide, ethanol, isopropanol, and *n*-butanol), and single-cell oil production.

Many of the technologies originated in Japan and have made prominent contributions to mankind. One of the bases of these developments has been established through extensive screening using the rich and diverse microbial resources of Japan, a country that has been one of the major players in the establishment and development of scientific and technological platforms.

As already mentioned, a bioprocess, especially a microbial one, is essentially environmentally friendly. However, there are many unresolved issues related to energy savings and resource depletion. Is CO_2 reduction really possible by introducing biosystems in place of petrochemical systems? Are biosystems really clean? At this time, unfortunately, we still do not have enough data, concrete evidence, and in-depth discussions about these issues. What are always referred to are the cases of the nitrile hydratase process for acrylamide production and the lactonase process for pantothenate production. In each instance, it is evident that the overall process is simple and rapid, and requires less energy (30 % CO_2 reduction compared with conventional chemical processes). Undoubtedly, this tendency can be found in many of the processes already in use, but to our regret, no relevant data have been presented to society.

According to a report from the Department of Trade and Industries in the UK, Japan's strength in this area of biotechnology lies in the fact that chemical industries have been actively promoting the industrialization of bioprocesses with the use of their rich microbial resources and have incorporated their technologies into their industrial structures. However, I believe that these facts may not, in themselves, be obvious in Japan, or may already be self-evident and allow no room for further debate, which could be why there are not many active discussions about these matters now. Am I the only person who has the impression that all relevant political actions at the national level supporting this biotechnology are also inadequate?

Many of the chapters collected here are based on the results of the work for the decade-long METI/NEDO project, the so-called Minimum Genome Factory, in which I was involved as a project leader.

Kyoto, Japan Sakayu Shimizu

Contents

Part I
Minimum Genome Factory

Chapter 1
Creation of Novel Technologies for Extracellular Protein Production Toward the Development of *Bacillus subtilis* Genome Factories

Katsutoshi Ara, Kenji Manabe, Shenghao Liu, Yasushi Kageyama, Tadahiro Ozawa, Masatoshi Tohata, Keiji Endo, Kazuhisa Sawada, Nozomu Shibata, Akihito Kawahara, Kazuhiro Saito, Hiroshi Kodama, Yoshiharu Kimura, Katsuya Ozaki, Yoshinori Takema, Hiroshi Kakeshita, Kouji Nakamura, Kunio Yamane, Takeko Kodama, Junichi Sekiguchi, Takuya Morimoto, Ryosuke Kadoya, Shigehiko Kanaya, Yasutaro Fujita, Fujio Kawamura, and Naotake Ogasawara

Abstract *Bacillus subtilis* has been widely used for the industrial production of useful proteins because of its high protein secretion ability and safety. We focused on genome reduction as a new concept for enhancing production of recombinant enzymes in *B. subtilis* cells based on detailed analysis of the genome mechanism. First, we reported that a novel *B. subtilis* strain, MGB874, depleted 20.7 % of the genomic sequence of the wild type by rationally designed deletions to create simplified cells for protein production. When compared with wild-type cells, the productivity of cellulase and protease from transformed plasmids harboring the corresponding genes was markedly enhanced. These results indicate that a bacterial factory specializing in the production of substances can be constructed by deleting the genomic regions unimportant for growth and substance production from *B. subtilis*. Second, deletion of the *rocDEF-rocR* region, which is involved in arginine degradation, was found to contribute to the improvement of enzyme production in strain MGB874.

K. Ara (✉)
Integrated Medical Research Laboratories, Kao (China) Research & Development Center Co., Ltd, Shanghai, China

Biological Science Laboratories, Kao Corp, Tochigi, Japan
e-mail: ara.katsutoshi@kao.sh.cn

K. Manabe • S. Liu • Y. Kageyama • T. Ozawa • M. Tohata • K. Endo • K. Sawada
N. Shibata • A. Kawahara • K. Saito • H. Kodama • Y. Kimura • K. Ozaki • Y. Takema
Biological Science Laboratories, Kao Corp, Tochigi, Japan

H. Kakeshita
Biological Science Laboratories, Kao Corp, Tochigi, Japan

Graduate School of Life and Environmental Sciences, University of Tsukuba, Tsukuba, Japan

K. Nakamura • K. Yamane
Graduate School of Life and Environmental Sciences, University of Tsukuba, Tsukuba, Japan

H. Anazawa and S. Shimizu (eds.), *Microbial Production: From Genome Design to Cell Engineering*, DOI 10.1007/978-4-431-54607-8_1, © Springer Japan 2014

The present study indicated that our results demonstrated the effectiveness of a synthetic genomic approach with reduction of genome size to generate novel and useful bacteria for industrial uses. Furthermore, the design of the changes in the transcriptional regulatory network of the nitrogen metabolic pathway in *B. subtilis* cells could facilitate the generation of improved industrial protein production.

Keywords *Bacillus subtilis* • Recombinant protein productivity • Refined genome factory

1.1 Introduction

Bacillus subtilis (*B. subtilis*), a gram-positive soporiferous bacillus, has been widely used for the industrial production of useful proteins because of its high protein secretion ability and safety (Simonen and Palva 1993). Beginning about 1990, mainly European and Japanese research groups implemented a project to sequence the entire *B. subtilis* strain 168 genome, and reported the sequence of the entire 4,215-kbp genome in 1997 (Kunst et al. 1997). At that time, they reported 4,101 genes in the entire genome and identified or inferred the functions of 58 % of these genes. A subsequent project to analyze the functions of the unknown genes identified the functions of about half of the functionally unknown genes. In addition, the individual disruption of 4,101 genes of *B. subtilis* showed that 271 genes (essential genes) were absolutely essential for growth, and the majority was involved in DNA replication, gene transcription and translation, cell structure formation, and cell division (Schumann et al. 2000; Kobayashi et al. 2003).

Deleting genes unnecessary for the production and secretion of useful proteins from the *B. subtilis* 168 strain and introducing necessary genes to improve the *B. subtilis* genome, we aimed to create a host microorganism cell (MGF) that can efficiently

T. Kodama
Biological Science Laboratories, Kao Corp, Tochigi, Japan

Faculty of Textile Science and Technology, University of Shinshu, Matsumoto, Japan

J. Sekiguchi
Faculty of Textile Science and Technology, University of Shinshu, Matsumoto, Japan

T. Morimoto • R. Kadoya
Graduate School of Information Science, Nara Institute of Science and Technology, Ikoma, Japan

Biological Science Laboratories, Kao Corp, Tochigi, Japan

S. Kanaya • N. Ogasawara
Graduate School of Information Science, Nara Institute of Science and Technology, Ikoma, Japan

Y. Fujita
Faculty of Life Science and Biotechnology, Fukuyama University, Fukuyama, Japan

F. Kawamura
Laboratory of Molecular Genetics, Rikkyo University, Tokyo, Japan

produce commercial enzymes promising for application in a wide range of production processes. The number of essential genes required for the growth of eukaryotic bacteria is expected to be about 300–500, regardless of the genus and species, and the majority of genomes consist of nonessential genes, which can probably be deleted from the chromosome without influencing growth. Westers et al. deleted two prophage regions (SP* and PBSX), three prophage-like regions (prophage 1, prophage 3, and skin), and the largest operon *pks* from the *B. subtilis* genome, thereby constructing a sextuple-deletion strain lacking 332 genes, accounting for 7.7 % of the entire genome (Westers et al. 2003). In this sextuple-deletion strain, the deletion of regions was found not to influence the growth of *B. subtilis*, protein secretion, transformation competence, or sporulation ability, indicating that the genome of *B. subtilis* can be artificially reduced by deleting nonessential gene regions.

We expected that it would be possible to construct a database for determination of the minimum set of genes in *B. subtilis* by efficiently deleting these regions. In addition, it is interesting to know whether the deletion of a large genomic region influences the production of useful enzymes and cell growth. We previously attempted to delete regions of more than 10 kbp present in essential genes, and successfully constructed a strain (MGF874) with a reduced genome, with deletions of 866 genes accounting for 20.7 % (about 874 kbp) of the *B. subtilis* genome, which was similar in morphology and growth to the wild strain. Moreover, a genome-reducing strain (*B. subtilis* strain RGB1334) was constructed from *B. subtilis* MGB874 by deletion of 27 gene regions, which resulted in a 20 % increase of cellulose productivity in the later growth phase when compared to *B. subtilis* MGB874 (Ara et al. 2007; Morimoto et al. 2008; Kageyama et al. 2009). These results indicate that a bacterial factory specializing in the production of substances can be constructed by deleting the genomic regions unimportant for growth and substance production from *B. subtilis*. Furthermore, based on detailed analysis of the genome mechanism (Kobayashi et al. 2012), the design of the amino-acid metabolism of a *B. subtilis* (Manabe et al. 2011, 2012), improvement of secretion equipment (Kakeshita et al. 2010, 2011a, b, c), a high translation system of a target gene (Tagami et al. 2012), and cell surface engineering (see Chap. 15; Kodama et al. 2007a; Kodama et al. 2007b; Kodama et al. 2011), etc., were implemented.

1.2 Genome Reduction in *B. subtilis*

Strategies for genome reduction, which represents a relatively new field in synthetic genomics, have been used with *Escherichia coli* and *B. subtilis* to investigate microbial genomic architecture and improve their characteristics (Westers et al. 2003; Posfai et al. 2006; Mizoguchi et al. 2007, 2008; Fabret et al. 2002). For effective production of enzymes, we have been performing a study aiming at the creation of *B. subtilis* MGF, by deleting genes that are not essential under normal cultivating conditions, and enhancing essential genes. Single deletion of each of a number of genes and multiple deletions of dozens of genomic regions were carried out efficiently by using different deletion techniques.

1.2.1 Genome Deletion Technology

A genetic tool to introduce marker-free deletions is essential for multiple manipulations of genomes (Liu et al. 2007). We reduced the *B. subtilis* genome by step-by-step deletion, using the *upp* (encoding uracil-phosphoribosyltransferase) cassette and 5-fluorouracil (5-FU) selection (Fabret et al. 2002), to select cells that had lost a drug-resistant cassette used to introduce primary deletions through intramolecular homologous recombination at repeated sequences flanking the cassette (Morimoto et al. 2008) (Fig. 1.1a). We also developed a system using the AraR repressor to improve the efficiency at which marker-free mutants can be obtained (Liu et al. 2008). In this method, we replaced the native *araR* gene with a promoter-less neomycin resistance gene (*neo*) fused to the arabinose operon (*ara*) promoter, and the selection marker cassette containing a chloramphenicol resistance gene and the *araR* gene encoding the repressor for the *ara* operon was then integrated into the target site. Transformants became neomycin sensitive after integration of the marker cassette because of the repression of *neo* expression by AraR, and marker cassette-free cells were obtained by selection for neomycin resistance. However, the methods just described require the use of specific genetic backgrounds, such as inactivation of the native *upp* gene for 5-FU selection, or replacement of the *araR* gene with the *Para-neo* construct in the latter system (Liu et al. 2008). We developed a simple and efficient method to create marker-free deletion mutants of *B. subtilis* through transformation with recombinant PCR products, using the *E. coli mazF* gene encoding an endoribonuclease that cleaves free mRNAs as a counter-selection tool (Morimoto et al. 2009, 2011a, b).

The *mazF*-encoding cassette is fused with the flanking sequences of the target region using splicing by overlap extension-polymerase chain reaction (SOE-PCR). Upstream and downstream sequences (fragments A and B) of the flanking region to be deleted are amplified from the genomic DNA of the *B. subtilis* strain to be manipulated. The *mazF* cassette is amplified from the genomic DNA of *B. subtilis* strains that contain a drug resistance gene and the *mazF* gene under the control of an isopropyl β-D-1-thiogzalactopyranoside (IPTG)-inducible *spac* promoter (Fig. 1.1b). An internal sequence (fragment C) in the target region is also amplified. These PCR products are fused by recombinant PCR in the order A–B–*mazF*-cassette–C (as illustrated in Fig. 1.1) and integrated into the target region through homologous recombination between fragment A and C loci. The resulting recombinants are selected for drug resistance in the absence of IPTG. Thereafter, the primary transformant is cultivated in the presence of IPTG (i.e., *mazF* toxin-inducing conditions), and clones in which the *mazF* cassette has been excised by intramolecular homologous recombination at region B are selected (Fig. 1.1b).

1.2.2 Multiple Deletion Design for MGF

To construct the multiple deletion series mutants, we rationally designed to maintain cellular function for recombinant protein production. Among 4,106 genes in the

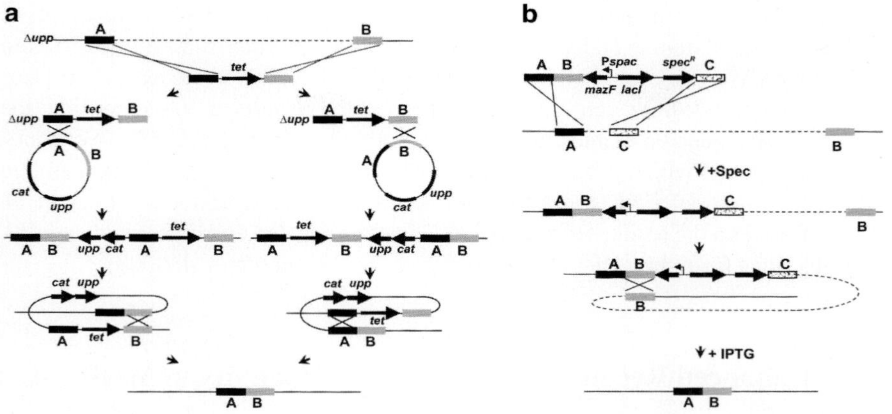

Fig. 1.1 Method to introduce marker-free deletions in the *Bacillus subtilis* genome. **a** A derivative of *B. subtilis* 168, 168 **Δ***upp*, in which the *upp* gene encoding uracil-phosphoribosyl transferase is inactivated by replacement with the erythromycin resistance gene (*erm*), was used as the starting strain for generation of the deletion mutant series. The entire length of the tetracycline-resistant gene (*tet*) cassette with its 5′-regulatory region was amplified from the pBEST307 plasmid. At least 500 bp of sequences flanking both sides of the region to be deleted were amplified by PCR, and joined upstream and downstream of the *tet* cassette by ligation using overlapping sequences in primers. The *B. subtilis* 168 **Δ***upp* strain and its derivatives were transformed with the resultant fragment to obtain a strain in which the target sequence was replaced with the *tet* gene. Next, to obtain marker-less mutants, fragments upstream and downstream of the target sequence were amplified, ligated, and cloned into the pBRcat/upp plasmid harboring the *upp* and chloramphenicol resistance (*cat*) gene. The resultant plasmid was integrated into the genome of the primary transformant with selection for tetracycline and chloramphenicol resistance. The resultant strain became 5-fluorouracil (5-FU) sensitive as a result of introduction of the functional *upp* gene, and mutants without the plasmid sequence were selected on LB plates containing 10 μM 5-FU. **b** Development of a new simple method to introduce marker-free deletions. Upstream (fragment A), downstream (fragment B) of the target region for deletion, and the first homologous recombination sequence (fragment C) were amplified by PCR from the chromosomal DNA of *B. subtilis*. By using the chromosomal DNA of the strain TMO310 (168, *aprE*::*spec*^{R}, *lacI*, P*spac-mazF*) or TMO311 (168, *aprE*::*Km*^{R}, *lacI*, P*spac-mazF*) as a template, a DNA fragment was amplified by PCR as MazF cassette. Using the thus-obtained PCR fragments, i.e., the MazF cassette, fragment A, fragment B, and fragment C, recombinant PCR was performed. As a result, a DNA fragment that has fragment A, fragment B, MazF cassette, and fragment C, aligned in that order, was obtained. *B. subtilis* 168 was transformed with this PCR product according to the competent cell transformation method. The transformants were selected on a LB medium plate containing 100 μg/ml spectinomycin or 5 μg/ml kanamycin overnight at 37 °C. The transformant having spectinomycin resistance or kanamycin resistance was cultured overnight in LB liquid medium. After dilution of the culture, the culture solution was applied on a LB medium plate supplemented with 1 mM isopropyl **β**-D-1-thiogalactopyranoside (IPTG). These colonies growing on IPTG-containing LB medium plate are the *B. subtilis* transformants in which the target region for deletion and MazF cassette were both popped out from the host chromosome DNA

genome of the *B. subtilis* 168 strain, 271 genes were reported to be essential for cellular maintenance and growth (Kobayashi et al. 2003). We initially identified contiguous genome sequences greater than 10 kb that did not code for RNA or essential proteins (Morimoto et al. 2008). In addition, we excluded all known and possible genes involved in primary metabolism to maintain growth in minimal medium, as well as those related to DNA metabolism to avoid genome instability (Morimoto et al. 2008). We also evaluated the production of the cellulase protease and amylase using about 3,000 single gene deletion mutants of the *B. subtilis* 168 to consider the effects of each gene on extracellular protein production (Ara et al. 2007). We identified 116 genes of which deletion enhanced the secretory protein production and 96 genes of which deletion suppressed the secretory protein production (Ara et al. 2007). Based on the result, we designed the deletion regions and performed step-by-step deletions from the genome of wild-type cells (Morimoto et al. 2008).

1.3 Enhanced Recombinant Protein Productivity in MGB874

By step-by-step deletions in regions where single deletions did not affect cell growth, we construct a reduced genome *B. subtilis* strain. Accordingly, sequential deletions in 23 nonessential gene clusters were introduced in *B. subtilis* 168 to generate MGB874, depleted of 874 kb (20.7 %) of the sequence including 865 genes from the original *B. subtilis* 168 genome (Fig. 1.2a). Although the growth rate of MGB874 cells was reduced compared with the wild-type 168 strain, cell morphology and chromosome distribution were normal (Fig. 1.2b). The reduced growth rate indicates that certain nonannotated and deleted genes contribute to the metabolic capacity of

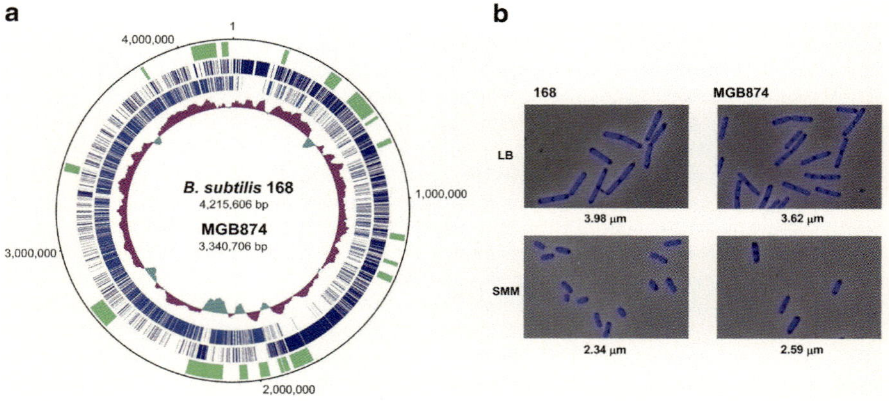

Fig. 1.2 Design and cell morphologies of MGB874. **a** Outer concentric ring: genome coordinate (bases) of the *B. subtilis* 168 genome. Ring 2 (*green*): positions of deleted sequences in MGB874, including prophages and prophage-like regions (SPβ, PBSX, skin, pro1-7), and polyketide and plipastatin synthesis operons (*pks*, *pps*). Rings 3 and 4 (*blue*): protein-coding regions in clockwise (Ring 3) and counterclockwise (Ring 4) orientations. Ring 5: GC contents. **b** Cell morphology, chromosome distribution, and mean values of cell lengths of wild-type 168 and MGB874 cells. MGB874 and 168 cells were cultured at 37 °C in LB or SMM medium, and images were obtained during the exponential growth phase after staining with 4,6-diamindino-2-phenylindole (DAPI). The average cell length is indicated (~200 cells analyzed)

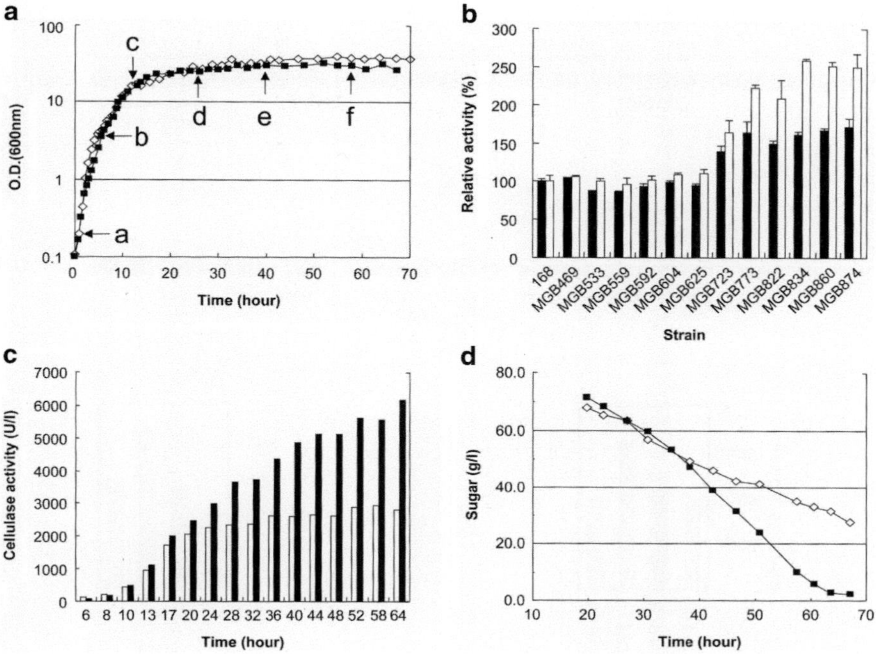

Fig. 1.3 Productivity of extracellular enzymes by the multiple-deletion series strains. **a** Growth profiles of wild-type 168 (*open diamonds*) and MGB874 (*closed squares*). *Arrows* with *a–e* indicate times of cell collection for transcriptome analysis. **b** Relative activities of cellulase Egl237 (*black bars*) and M-protease (*white bars*) in growth medium of the multiple-deletion series, compared to those of the wild-type 168 strain after 75 h cultivation in 2xL-Mal medium, are indicated with error bars (average of three experiments). **c** Cellulase production as a function of cell growth. Extracellular cellulase activities of wild-type 168 (*white bars*) and MGB874 (*black bars*) cultures (0.4 μl). **d** Consumption of sugars in the growth medium. Wild-type 168 (*open diamonds*) and MGB874 (*closed squares*) cultures were collected at the indicated times

B. subtilis cells under normal growth conditions. It is also possible that this phenotype is caused by unexpected synergetic effects of the deletions of genes.

To assess the productivity of MGB874 cells in terms of exogenous protein secretion, we examined the production of thermostable alkaline cellulase, Egl237 (Hakamada et al. 2000), and alkaline protease (M-protease) (Kobayashi et al. 1995) from a multicopy plasmid. We introduced these plasmids into the multiple deletion series strains and measured the protease and cellulase activities after 75 h culture in 2xL-Mal, a model medium for industrial protein production, respectively (Morimoto et al. 2008) (Fig. 1.3a). Unexpectedly, the production of both enzymes increased in proportion to genome deletion lengths, with maximum levels estimated in the MGB874 strain (Fig. 1.3b). The activities of cellulase and protease in the culture medium of MGB874 cells were about 1.7- and 2.5 fold higher than those from wild-type cells, respectively (Fig. 1.3b). Increase in cell mass was similar for MGB874 and wild-type cells, and cellulase production was arrested in wild-type 168 cells

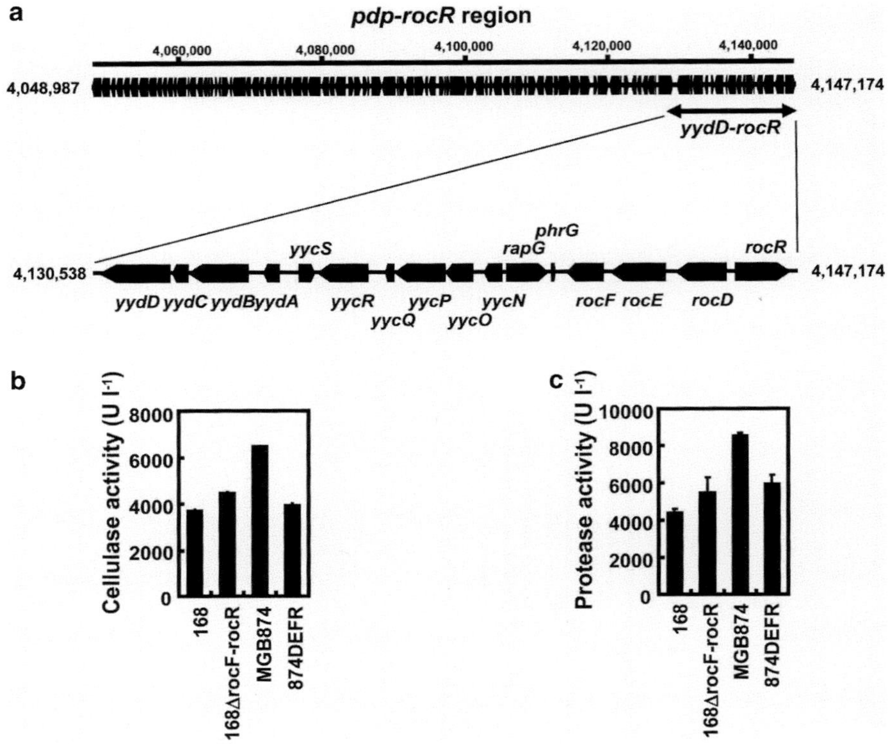

Fig. 1.4 Determination of the deletion contributing to enhanced enzyme production. The *pdp-rocR* region was deleted from MGB625 to construct MGB723. Systematic inactivations were carried out to narrow the range of region responsible for elevating the enzyme production level. **a** The *pdp-rocR* region (98,188 bp) corresponding to 4,048,987–4,147,174 bp in the genome of *B. subtilis* 168. The *pdp-rocR* region (16,636 bp) corresponding to 4,130,538–4,147,174 bp in the genome of *B. subtilis* 168. **b** Cellulase activities in the growth media of *B. subtilis* derivative strains harboring pHYS237 for Egl-237 production. **c** Protease activities in the growth media of *B. subtilis* derivative strains harboring pHP237-K16 for M-protease production. Cells were precultured in LB medium containing 15 μg ml^{-1} Tet at 30 °C for 15 h and were inoculated into 2xL-Mal medium. Cells further cultured at 30 °C for 75 h were separated from the growth medium by centrifugation and the supernatants were used for determinations of cellulase and protease activities

from entering the stationary phase (Fig. 1.3c). In contrast, the cellulase level in MGB874 cells continued to increase throughout the culture period to about twice that obtained from the wild type (Fig. 1.3c). Furthermore, maltose consumption in the culture medium was enhanced in MGB874 cells, indicating that the efficiency of carbon source utilization also is improved as a result of genome reduction (Fig. 1.3d).

We attempted to determine which gene deletions elevated the enzyme production level in strain MGB874. In a reduced-genome strain of MGB723 (Morimoto et al. 2008), which was constructed from MGB625 (Morimoto et al. 2008) by the deletion of the 98-kb genome region including *pdp-rocR* genes (Fig. 1.4a), the production of cellulase and protease were improved greatly (Fig. 1.4b). First, we demonstrated that the deletion of *rocR* and *rocDEF* operon in this region, related to the arginine metabolic pathway, was at least partly responsible for the enhanced enzyme production in strain MGB874 (Manabe et al. 2011). Furthermore, we determined that deletion of the monocistronic *rocR*, coding-positive regulator for a major glutamate dehydrogenase RocG, markedly increased cellulase production, whereas deletion of the *rocDEF* operon only slightly increased cellulase production by strain MGB625 (Manabe et al. 2011). Therefore, we next evaluated cellulase production in a mutant strain named 874DEFR, which was constructed by reintroducing *rocDEF* and *rocR* into MGB874, and found that cellulase activity in the growth medium was considerably lower than that detected for MGB874, and was almost equivalent to that of wild-type strain 168 (Fig. 1.4b). Conversely, cellulase production in the mutant, in which the *rocDEF-rocR* region was deleted in the wild type, was approximately 1.2 fold higher than that by parental strain 168, although cellulase production by strain MGB874 was approximately 1.6 fold higher than that by the strain 874DEFR (Fig. 1.4b). These data suggest that other deleted genes in the *rocDEF-rocR* region or the extensive genome reduction itself enhanced the positive effect of the deletion of *rocDEF-rocR* on cellulase production in reduced-genome strain MGB874. To verify whether deletion of the *rocDEF-rocR* region influenced the production of other enzymes, the levels of the alkaline protease, M-protease, were evaluated. As a result, the production of M-protease was also improved by deletion of the *rocDEF-rocR* region (Fig. 1.4c).

Moreover, we investigated the mechanism of the enhanced enzyme production. RocG is also serves as a regulatory protein that inhibits GltC (Commichau et al. 2007), a transcriptional regulator specific to the *gltAB* operon that encodes large and small subunits of glutamate synthase (GOGAT) (Bohannon et al. 1989; Picossi et al. 2007). Further studies revealed that metabolic changes were caused by *rocR* depletion in MGB874 cells. We measured extracellular and intracellular amino-acid levels of MGB874 compared to wild-type cells (Manabe et al. 2011). In the culture broth of strains 168 and 874DEFR, both arginine and glutamate were depleted before cells entered the stationary phase (Manabe et al. 2011). In contrast, in the culture medium of strain MGB874, the arginine level decreased gradually throughout the culture period following inactivation of the arginine degradation pathway (Manabe et al. 2011). Glutamine was below the limit of detection in the culture broths of strains 168 and MGB874. On the other hand, the glutamate pool size in strain MGB874 cells was significantly larger compared to that in strain 168 and 874DEFR cells (Manabe et al. 2011). The metabolic changes in MGB874 cells are illustrated in Fig. 1.5, with reference to these results, and the data of transcriptome analyses revealed that the genetic expression pattern in the stationary phase of MGB874 cells was characteristic compared with wild-type cells (Manabe et al. 2011).

In addition, we examined the transcriptional level of Egl-237 cellulase from a plasmid-encoded gene to determine the high productivities of recombinant cellulase in strains MGB874. The plasmid copy numbers and transcript levels, respectively, from a single copy of the *egl-237* cellulase-encoding gene in the reduced-genome strains were significantly higher than those in the wild-type strain during stationary phase (Morimoto et al. 2008; Manabe et al. 2011). Taken together, the elevated transcriptional levels of

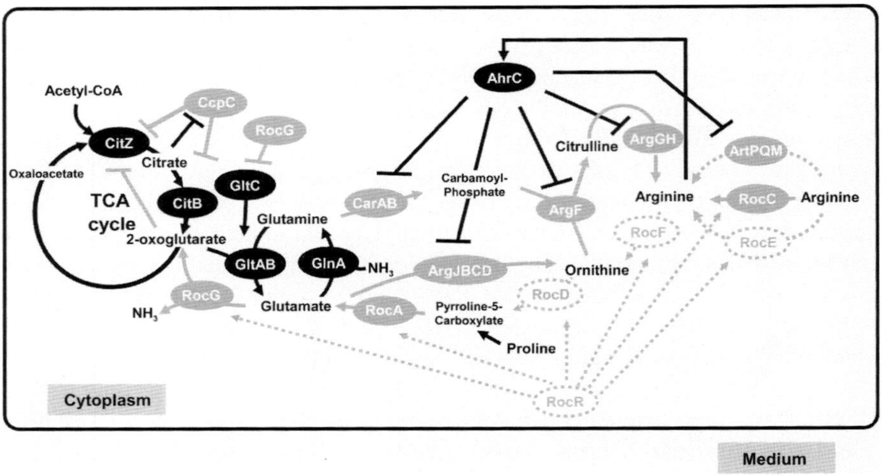

Fig. 1.5 Illustration of the proposed intersection between nitrogen and carbon metabolism in MGB874 during the stationary phase. Proteins are shown as *circles* and their names are indicated. *Open gray circles* indicate the corresponding proteins of genes that were deleted in strain MGB874. *Closed black and gray circles* indicate proteins whose expression or function are activated and repressed, respectively, in MGB874, in comparison to wild-type strain 168. The tiling array data reported previously (Morimoto et al. 2008) were used as a reference. Deletion of the *rocDEF-rocR* region resulted in not only the inactivation of the arginine degradation pathway, but also the repression of the genes related to the arginine synthetic pathway and high-affinity arginine transporter ArtPQM (Makarova et al. 2001; Sekowska et al. 2001) In addition, reduced expression of *rocG* from the absence of transcriptional activator RocR resulted in the activation of the genes encoding aconitase (CitB) and glutamate synthase (GltAB)

egl-237 in reduced-genome strains were possibly the result of the combined effects of increases in plasmid copy number and promoter activity of *egl-237*, although the extent of the contribution remains unknown (Manabe et al. 2011).

1.4 Construction and Characterization of RGF (*R*efined *G*enome *F*actory) Strains

A reduced-genome strain, RGF1334, was constructed by deleting an additional 27 regions in the MGB874 genome (1,334-kbp deletion in total from the wild-type genome). RGF1334 grew more slowly than MGB874, leading to delayed consumption of maltose as a carbon source in the medium. In the late stage of cultivation, RGF1334 continued to utilize maltose and maintained a high level of productivity of recombinant Egl-237 cellulase compared to strains 168 and MGB874. This result suggests that the carbon source utility between for cell growth and for recombinant protein production is changed in RGF1334 compared to wild-type 168 and MGB874. These observations of the properties of RGF1334 raise interesting implications concerning efficient recombinant protein production with less carbon source use in a less-exothermic process.

1.5 Conclusion

We aimed to create a new host suitable for the secretory production of useful proteins using *B. subtilis*. We focused on genome reduction as a new concept for enhancing production of recombinant enzymes in bacteria. To create simplified hosts for recombinant protein production as a new concept, we developed a novel *B. subtilis* strain, MBG874, depleted of 874 kb of the genomic sequence. Actually, the strain exhibited higher protein productivity than the wild-type strain, demonstrating that genome reduction contributed to the creation of bacterial cells with a practical application. The genome-reduced strains should act as new platforms to build in various genetic systems toward the development of *B. subtilis* genome factories for industrial protein production. In addition, analyses of a novel host strain RGF1334, with 1,334 kb of the genomic sequence depleted, raise interesting implications for industrial recombinant protein production with reduced energy and resources. We are currently engaged in future rational deletion studies based on transcriptome data, gene function information, and comparative genomics approaches, with a view toward generating simple, predictable cells containing genes with defined functions as a new platform for the development of bacterial strains for industrial applications. Furthermore, this technology is applicable to many other industrial production host cells, and addition of this technology can be used to substitute a chemistry process.

Acknowledgments The research and development of *Bacillus subtilis* RGF was conducted as part of a subproject, "Development of a Technology for the Creation of a Host Cell" included within the industrial technology project, "Development of Generic Technology for Production Process Starting Productive Function," of the Ministry of Economy, Trade and Industry, entrusted by the New Energy and Industrial Technology Development Organization (NEDO), Japan. We are grateful to Dr. Shu Ishikawa and Dr. Taku Oshima (Nara Institute of Science and Technology) for help in transcriptome analysis, and Mr. Naoki Kondo and Ms. Eri Shimizu (Kao Corporation) for help in metaborome analysis. We also thank Dr. Tatsuro Fujio, Dr. Hideaki Nanamiya (Tokyo University), Dr. Hiroki Yamamoto (Shinshu University), Dr. Masayuki Hashimoto (Shinshu University), Dr. Tatsuya Fukushima (Shinshu University), Dr. Ken Kurokawa (Tokyo Institute of Technology), Dr. Mitsuhiro Itaya (Keio University), Dr. Hironori Niki (National Institute of Genetics), Dr. Hirofumi Yoshikawa (Tokyo University of Agriculture), Dr. Kei Asai (Saitama University), Dr. Hiroki Takahashi (Nara Institute of Science and Technology), Dr. Shigeo Inoue (Kao Corporation), and Dr. Toshiharu Numata (Kao Corporation) for valuable advice in conducting this study.

References

Ara K et al (2007) *Bacillus* minimum genome factory: effective utilization of microbial genome information. Biotechnol Appl Biochem 46:169–178

Bohannon DE et al (1989) Positive regulation of glutamate biosynthesis in *Bacillus subtilis*. J Bacteriol 171:4718–4727

Commichau F et al (2007) A regulatory protein–protein interaction governs glutamate biosynthesis in *Bacillus subtilis*: the glutamate dehydrogenase RocG moonlights in controlling the transcription factor GltC. Mol Microbiol 65:642–654

Fabret C et al (2002) A new mutation delivery system for genome-scale approaches in *Bacillus subtilis*. Mol Microbiol 46:25–36

Hakamada Y et al (2000) Deduced amino acid sequence and possible catalytic residues of a thermostable, alkaline cellulase from an alkaliphilic *Bacillus* strain. Biosci Biotechnol Biochem 64:2281–2289

Kageyama Y et al (2009) Genome reduction in *Bacillus subtilis* and enhanced productivities of recombinant proteins. In: Knudsen WD, Bruns SS (eds) Bacterial DNA, DNA polymerase and DNA helicases. Nova Science, New York

Kakeshita H et al (2010) Enhanced extracellular production of heterologous proteins in *Bacillus subtilis* by deleting the C-terminal region of the SecA secretory machinery. Mol Biotechnol 46:250–257

Kakeshita H et al (2011a) Improvement of heterologous protein secretion by *Bacillus subtilis*. In: Petre M (ed) Advances in applied biotechnology. In Tech (Open access publisher)

Kakeshita H et al (2011b) Propeptide of *Bacillus subtilis* amylase enhances extracellular production of human interferon-α in *Bacillus subtilis*. Appl Microbiol Biotechnol 89:1509–1517

Kakeshita H et al (2011c) Secretion of biologically active human interferon-β by *Bacillus subtilis*. Biotechnol Lett 33:1847–1852

Kobayashi T et al (1995) Purification and properties of an alkaline protease from alkalophilic *Bacillus* sp. KSM-K16. Appl Microbiol Biotechnol 43:473–481

Kobayashi K et al (2003) Essential *Bacillus subtilis* genes. Proc Natl Acad Sci USA 100(8):4678–4683

Kobayashi T et al (2012) Identification and characterization of a novel polysaccharide deacetylase C(PdaC) from *B. subtilis*. J Biol Chem 287:9765–9776

Kodama T et al (2007a) Effect of the *Bacillus subtilis spo0A* mutation on cell wall lytic enzymes and extracellular proteases, and prevention of cell lysis. J Biosci Bioeng 103:13–21

Kodama T et al (2007b) *Bacillus subtilis* AprX involved in degradation a heterologous protein during the late stationary growth phase. J Biosci Bioeng 104:135–143

Kodama T et al (2011) Approaches for improving protein production in multiple protease-deficient *Bacillus subtilis* host strains. In: Petre M (ed) Advances in applied biotechnology. In Tech (Open access publisher)

Kunst F et al (1997) The complete genome sequence of the gram-positive bacterium *Bacillus subtilis*. Nature (Lond) 90:249–256

Liu S et al (2007) The accurate replacement of long genome region more than several hundred kilobases in *Bacillus subtilis*. Genes Genet Syst 82:9

Liu S et al (2008) Introduction of marker-free deletions in *Bacillus subtilis* using the AraR repressor and the ara promoter. Microbiology 154:2562–2570

Makarova KS et al (2001) Conservation of the binding site for the arginine repressor in all bacterial lineages. Genome Biol 2:RESEARCH0013

Manabe K et al (2011) Combined effect of improved cell yield and increased specific productivity enhances recombinant enzyme production in genome-reduced *Bacillus subtilis* strain MGB874. Appl Environ Microbiol 77:8370–8381

Manabe K et al (2012) High external pH enables more efficient secretion of alkaline α-amylase AmyK38 by *Bacillus subtilis*. Microb Cell Fact 11:74

Mizoguchi H et al (2007) *Escherichia coli* minimum genome factory. Biotechnol Appl Biochem 46:157–167

Mizoguchi H et al (2008) Superpositioning of deletions promotes growth of *Escherichia coli* with a reduced genome. DNA Res 15:277–284

Morimoto T et al (2008) Enhanced recombinant protein productivity by genome reduction in *Bacillus subtilis*. DNA Res 4:1–9

Morimoto T et al (2009) A new simple method to introduce marker-free deletions in the *Bacillus subtilis* genome. Genes Genet Syst 84:315–318

Morimoto T et al (2011a) A new simple method to introduce marker-free deletions in the *Bacillus subtilis* genome. Methods Mol Biol 765:345–358

Morimoto T et al (2011b) Simple method for introducing marker-free deletions in the *Bacillus subtilis* genome. In: Williams JA (ed) Strain engineering: methods and protocols. Humana Press, Totowa

Picossi S et al (2007) Molecular mechanism of the regulation of *Bacillus subtilis gltAB* expression by GltC. J Mol Biol 365:1298–1313

Posfai G et al (2006) Emergent properties of reduced-genome *Escherichia coli*. Science 312:1044–1046

Schumann W, Ehrlich SD, Ogasawara N (2000) Functional analysis of bacterial genes: a practice manual. Wiley, Chichester

Sekowska A et al (2001) Extracting biological information from DNA arrays: an unexpected link between arginine and methionine metabolism in *Bacillus subtilis*. Genome Biol 2:RESEARCH0019

Simonen M, Palva I (1993) Protein secretion in *Bacillus* species. Microbiol Rev 57:109–137

Tagami K et al (2012) Expression of a small (p)ppGpp synthetase, YwaC, in the (p)ppGpp(0) mutant of *Bacillus subtilis* triggers YvyD-dependent dimerization of ribosome. Microbiology Open 1:115–134

Westers H et al (2003) Genome engineering reveals large dispensable regions in *Bacillus subtilis*. Mol Biol Evol 20(12):2076–2090

Chapter 2
Minimum Genome Factories in *Schizosaccharomyces pombe*

Hiromichi Kumagai, Mayumi Sasaki, Alimjan Idiris, and Hideki Tohda

Abstract This chapter gives an overview of the "minimum genome factory" (MGF) of the fission yeast *Schizosaccharomyces pombe* (*S. pombe*). The *S. pombe* genome is one of the smallest found in free-living eukaryotes. We engineered a reduction in the number of *S. pombe* genes using a large-scale gene deletion method called the LATOUR method. This method enabled us to identify the minimum gene set required for growth under laboratory conditions. The genome-reduced strain has four deleted regions: 168.4 kb of the left arm of chromosome I; 155.4 kb of the right arm of chromosome I; 211.7 kb of the left arm of chromosome II; and 121.6 kb of the right arm of chromosome II. These changes represent a loss of 223 genes of an estimated 5,100. The 657.3-kb deletion strain was less efficient at taking up glucose and some amino acids from the growth media than the parental strain. This strain also showed increased gene expression of the mating pheromone M-factor precursor and NADP-specific glutamate dehydrogenase. There was also a 2.7-fold increase in the concentration of cellular ATP, whereas levels of heterologously produced proteins, such as the green fluorescent protein and the secreted human growth hormone, increased by 1.7 fold and 1.8 fold, respectively.

Keywords LATOUR method • Minimum genome factory • Protein expression system • *Schizosaccharomyces pombe* (*S. pombe*)

H. Kumagai (✉) • M. Sasaki • A. Idiris
Research Center, Asahi Glass Co., Ltd, 1150 Hazawa-cho, Kanagawa-ku, Yokohama 221-8755, Japan
e-mail: hiromichi-kumagai@agc.com

H. Tohda
Technology Planning Office, Asahi Glass Co., Ltd, Chiyoda-ku, Tokyo 100-8305, Japan

H. Anazawa and S. Shimizu (eds.), *Microbial Production: From Genome Design to Cell Engineering*, DOI 10.1007/978-4-431-54607-8_2, © Springer Japan 2014

2.1 *Schizosaccharomyces pombe*

New pathways for designing microorganisms for industrial-scale production of biological materials have been made possible by significant advances in genome sequencing, bioinformatics, and genetic engineering. The so-called minimum genome factory (MGF) has been created in which unnecessary or detrimental genes are deleted from the microorganism, leaving only the genes necessary for industrial production of the desired molecule. The first examples of MGFs are *Escherichia coli*, *Bacillus subtilis*, and *Schizosaccharomyces pombe* (*S. pombe*) (Ara et al. 2007; Giga-Hama et al. 2007; Fujio 2007; Mizoguchi et al. 2007). This chapter provides an overview of MGFs in *S. pombe*.

S. pombe is a unicellular eukaryote belonging to the Ascomycetes class of fungi. Although *S. pombe* belongs to the same class as the budding yeast *Saccharomyces cerevisiae* (*S. cerevisiae*), it is taxonomically and evolutionarily distant. *S. pombe* reproduces by fission, a process similar to that used by higher eukaryotic cells. Furthermore, *S. pombe* shares many molecular, genetic, and biochemical features with multicellular organisms, making it a particularly useful model for studying the function and regulation of genes from more complex species (Zhao and Lieberman 1995). These attributes make it attractive for industrial fermentation, a process characterized by high cell densities, short fermentation times, and the use of chemically defined media lacking components derived from animal cells. Thus, *S. pombe* is a useful host for heterologous expression of molecules derived particularly from higher organisms.

Figure 2.1 shows the MGF concept. Microorganisms have a variety of genes that are expressed during adaptation to different environmental conditions. These genes are thought to be unnecessary under nutrient-rich growth conditions. Hence, a minimal gene set is required for cellular viability, and identifying this gene set would provide important clues about the evolutionary origins of eukaryotic organisms. In addition, minimal gene sets can be used to construct minimal genome factories for industrial production of biological materials.

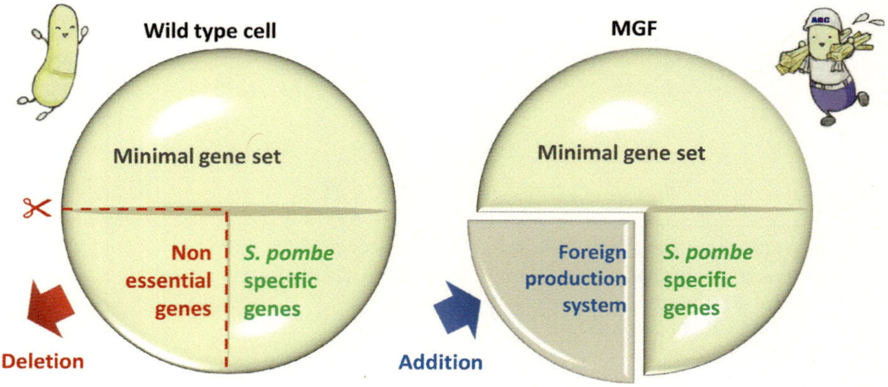

Fig. 2.1 Concept of minimum genome factory

Development of microbial production systems requires computer modeling and simulation of cellular metabolic systems to optimize metabolic networks (Medema et al. 2012). However, understanding the metabolic systems that are predicted to be present in microorganisms can be problematic, as complex intracellular metabolic pathways often interfere with experiments designed to elucidate such systems. Recently, reducing the genome sizes of some microorganisms has been employed as a strategy to simplify their intracellular metabolic pathways while maintaining their growth efficiencies. For successful construction of an MGF, it has been proposed that effective use of intracellular energy can be achieved by eliminating unnecessary genes (Fujio 2007). Comparative genomics supports this hypothesis, and it has been speculated that genome reduction can be a selective process favoring adaptation to low-nutrient environments for effective energy utilization (Moya et al. 2009).

Several genome-reduced microorganisms are reported to have beneficial properties. In *E. coli*, for example, such benefits include improved electroporation efficiency, accurate propagation of recombinant genes and unstable plasmids (Posfai et al. 2006), and an increase in L-threonine production (Mizoguchi et al. 2008; Baba et al. 2006). In *B. subtilis*, increases in production and secretion levels of heterologous enzymes have also been noted following genome reduction (Manabe et al. 2011; Morimoto et al. 2008; Kobayashi et al. 2003). Deletion of part of the *S. cerevisiae* genome, which results in altered carbon metabolism, has also been reported (Murakami et al. 2007).

The *S. pombe* genome sequencing project was completed in 2002 (Wood et al. 2002). The whole genome, which is distributed on three chromosomes, is estimated at 13.8 Mb. In *S. pombe*, a single systematic genome-wide collection of gene deletion mutants has been reported (Kim et al. 2010), and a pilot study reported deleting 100 of its genes (Decottignies et al. 2003). Essential genes in *S. pombe* constitute about 25 % of its total gene content (1,260/5,100), which is a higher percentage than that found in other model organisms because *S. pombe* has one of the smallest gene numbers found among free-living eukaryotes.

Genetic manipulation of *S. pombe* is well established. Hence, we investigated the minimal gene set required by this free-living model eukaryote by deleting as many nonessential genes as possible. From this, we generated *S. pombe* mutants dedicated to heterologous protein production (Sasaki et al. 2013).

2.2 Construction of Genome-Reduced Fission Yeast Strains

We have developed a unique method for chromosomal modification in *S. pombe*. This method, which we have designated the "latency to universal rescue method" (LATOUR method; Fig. 2.2), is an extremely simple method that only requires a negative selectable marker for its application. No foreign sequences remain following chromosomal modification. Using this method, it was easy to delete a chromosomal segment of more than 100 kb containing 33 genes at once (Hirashima et al. 2006). LATOUR is a very useful method for construction of genome-reduced strains and for clean deletion of unnecessary genes. It is also possible to identify genes that are essential in *S. pombe* by this method.

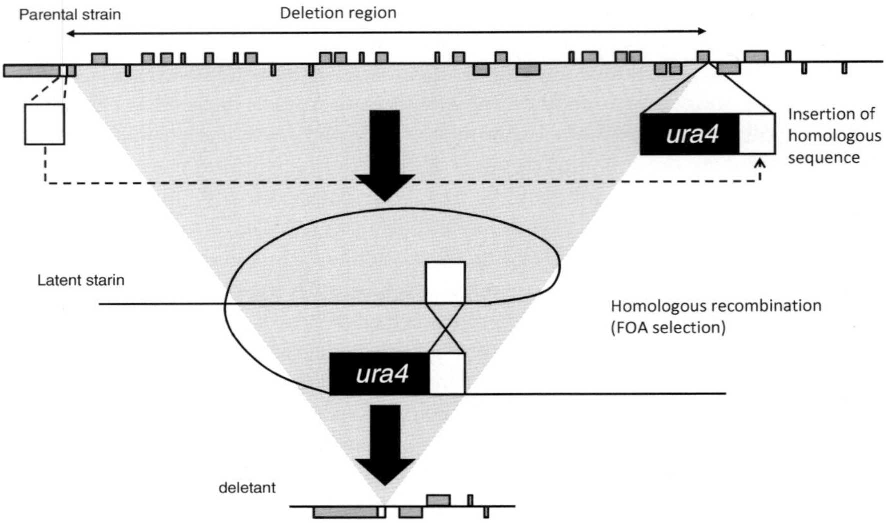

Fig. 2.2 Schematic representation of the deletion by the LATOUR method. The portion including the *ura4* sequence is the introduced modification fragment for homologous recombination. The direct repeats are not contained in the modification fragment, and the *ura4* and target genes are places between direct repeats on the chromosome during the latent stage. The important difference from previous methods is that the target gene to be deleted is retained during the stage in which the modification fragment is introduced

The *S. pombe* genome was reduced by deletion of the terminal regions of chromosomes I and II using the LATOUR method. This method generated a 657.3-kb (5.2 % of the total 12.57-Mb genome size sequenced to date) deletion strain that maintained its growth rate. The genome-reduced strain (called the quadruple-deletion strain) has four deleted genomic regions: 168.4 kb of the left arm of chromosome I; 155.4 kb of the right arm of chromosome I; 211.7 kb of the left arm of chromosome II; and 121.6 kb of the right arm of chromosome II. These deletions correspond to a loss of some 223 genes from the original 5,100 that are estimated to be present in *S. pombe* so far (Sasaki et al. 2013).

Figure 2.3 summarizes the reduced genome size and gene number for this strain. The gene number of the quadruple-deletion strain is currently the smallest among eukaryotic model organisms. The genes that were deleted are summarized in Table 2.1.

We determined that the most-terminal essential genes in the left and right arms of chromosomes I and II were *trs33* (ALT, left arm of chromosome I), *sec16* (ART, right arm of chromosome I), *zas1* (BLT, left arm of chromosome II), and *usp109* (BRT, right arm of chromosome II) (Fig. 2.3). Although it has been reported that the genes SPAC1F8.07c (ALT), SPBC1348.06c (BLT), *alr2* (BLT), and *dea2* (BLT), which are located on the telomeric side of *trs33* and *zas1*, are also essential for growth (Kim et al. 2010), a deletion strain that does not include the genes SPBC1348.06c, *alr2*, or *dea2* exhibited no decrease in growth rate in comparison to

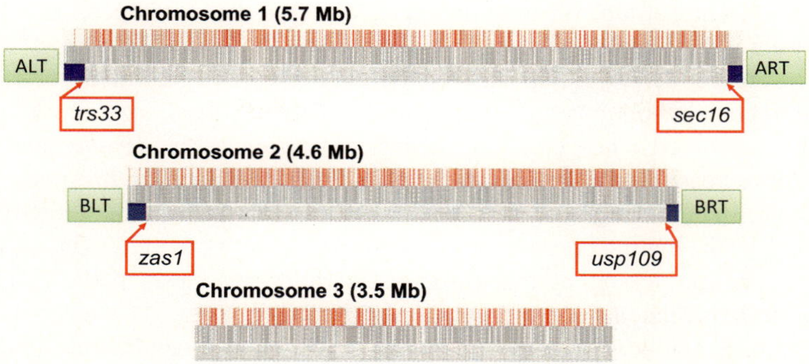

Fig. 2.3 Deletion regions of *Schizosaccharomyces pombe* chromosome. *Red box*, essential gene; *gray*, nonessential; *blue*, deletion region in the quadruple-deletion strain

Table 2.1 Classification of deleted genes in the quadruple-deletion strain

Term	Name	ALT	ART	BLT	BRT	Subtotal
GO:0055085	Transmembrane transport	8	10	20	3	41
GO:0006520	Cellular amino-acid metabolic process	4	5	4	1	14
GO:0006091	Generation of precursor metabolites and energy	3	3	2	0	8
GO:0006486	Protein glycosylation	3	0	0	4	7
GO:0005975	Carbohydrate metabolic process	0	2	2	2	6
GO:0007155	Cell adhesion	2	1	2	1	6
GO:0006355	Regulation of transcription, DNA dependent	2	0	4	0	6
GO:0006310	DNA recombination	2	0	0	1	3
GO:0006766	Vitamin metabolic process	2	0	0	1	3
GO:0006629	Lipid metabolic process	1	0	1	0	2
GO:0007126	Meiosis	2	0	0	0	2
GO:0055086	Nucleobase-containing small molecule metabolic process	0	0	2	0	2
GO:0006351	Transcription, DNA dependent	1	1	0	0	2
GO:0006260	DNA replication	0	0	0	1	1
GO:0070882	Cellular cell wall organization or biogenesis	0	1	0	0	1
GO:0007059	Chromosome segregation	1	0	0	0	1
GO:0051186	Cofactor metabolic process	0	0	1	0	1
GO:0000747	Conjugation with cellular fusion	0	1	0	0	1
GO:0007010	Cytoskeleton organization	0	0	0	1	1
GO:0071941	Nitrogen cycle metabolic process	0	0	0	1	1
GO:0006605	Protein targeting	0	1	0	0	1
GO:0023052	Signaling	0	1	0	0	1
	Others	23	22	16	13	74
	Conserved unknown	5	1	2	3	11
	Sequence orphan	6	1	3	3	13
	Pseudogene	5	0	4	4	13
	Dubious	0	1	0	0	1
	Total deletion gene	70	51	63	39	223

that of the parental strain. Whether this phenotype is a unique property of our laboratory strain is not known.

The quadruple-deletion strain showed a slight decrease in growth rate and smaller cell size in comparison to the parental strain. This reduced cell size may be similar to what is observed when cells are subject to nutrient stress (Fantes and Nurse 1977). An increase in the expression levels of nitrogen starvation-response genes was observed in the quadruple-deletion strain during its logarithmic growth phase. Expression of the mating pheromone M-factor precursor increased in this strain. We postulate that mimicking of the nitrogen starvation response in the quadruple-deletion strain is related to downregulation of TORC1 activity (Matsuo et al. 2007).

The glucose uptake efficiency, microarray analysis, and metabolome analysis in the quadruple-deletion strain were compared against a nonauxotrophic strain. The quadruple-deletion strain showed the following characteristics:

1. Slightly decreased glucose uptake and ethanol production
2. Decreased amino-acid uptake
3. Induction of sexual development genes and genes associated with nonglucose metabolism
4. Increased ATP concentration

2.3 *Schizosaccharomyces pombe* MGF Mutants as Hosts for Recombinant Protein Expression

We compared the production levels of various recombinant proteins using the genome reduced strains and compared them to those of the parental strain (Sasaki et al. 2013). We constructed vectors for heterologous expression of enhanced green fluorescent protein (EGFP), human transferrin, or human growth hormone, and integrated these vectors into the parental and the genome-reduced strains. Next, GFP fluorescence in each of the deletion mutants (per milliliter or cell numbers) was measured. Surprisingly, the expression levels, which were higher in the deletion mutants during all the growth phases, depended on the length of the deletion (i.e., from short to long regions). In the case of EGFP expression, the strain showing the highest productivity was the quadruple-deletion strain. The EGFP production rate in the quadruple-deletion strain increased about 1.7 fold in comparison to that of the parental strain during the logarithmic-phase growth. Based on transcriptome and metabolome analysis, we speculate that the increased EGFP expression in the deletion mutant could be related to an increase in intracellular GTP (which is required for ribosomal activity), thereby leading to increases in translation and EGFP production. Increased intracellular ATP levels could also lead to activation of amino-acid biosynthesis reactions. However, an increase in the concentration of intracellular amino acids was not observed, and some intracellular amino acids in the quadruple-deletion strain decreased. Our results also showed that the expression levels for human growth hormone and human transferrin reached much higher levels in the genome-reduced strains than in the parental strain. Taken together, these results suggest that genome reduction is a valuable tool for construction of heterologous protein production systems.

The LATOUR method is a powerful tool for clean deletion of protease-encoding genes that are a problem during expression and purification of recombinant proteins. As reported previously (Idiris et al. 2006, 2009, 2010), multi-protease gene deletion mutants are useful for producing recombinant proteins that are particularly sensitive to proteases. One such mutant, which we called A8, was constructed by multiply disrupting eight protease genes known to cause protein degradation by using conventional gene deletion technology. Each of the eight target genes was disrupted by substitution with the $ura4^{+}$ selectable marker gene. As a result, eight $ura4^{-}$ sequences were retained in the A8 strain where the eight protease genes were originally located. However, the remaining $ura4^{-}$ genes could prevent the $ura4^{+}$ marker gene being used for further gene deletion or integration experiments. Consequently, we have recently reconstructed the A8 strain using the LATOUR method without retention of any of the *ura4* sequences in the genome.

2.4 Conclusions

We have shown that the MGF approach, in which a parental line has been engineered to contain the minimum number of genes necessary for growth and survival via inactivation of unnecessary or detrimental genes, is effective at creating new lines that can be used to produce a variety of heterologous proteins. The production levels of recombinant EGFP, human transferrin, and human growth hormone in the genome-reduced strains were higher than in the parental strain. These results confirm that the gene deletion technology we have developed is very useful for constructing *S. pombe* mutants for recombinant protein production.

Although much progress has been made in manufacturing recombinant proteins, many issues still need to be solved. The genome-reduced strain is suitable for modeling and simulation of cellular metabolism. However, problems related to system unpredictability are often caused by the complex nature of the intracellular metabolic system, because not all the cellular metabolic intermediates have been completely elucidated. A simplified cellular metabolism, while maintaining satisfactory growth and requisite functions (e.g., high recombinant protein productivity), should lead to successful control of intracellular metabolism in accordance with the metabolic modeling/simulations and thus benefit high-level production of biological materials.

The MGF concept should help advance heterologous protein production systems, whole-cell biocatalysts, and synthetic processes for small molecules. The continued development of MGF technology promises to shed light on issues currently limiting large-scale industrial production of biological materials.

Acknowledgments This study was partly supported by the Ministry of Economy, Trade and Industry (Project for the Development of a Technological Infrastructure for Industrial Bioprocesses on Research and Development of New Industrial Science and Technology Frontiers supported by New Energy and Industrial Technology Development Organization). We are sincerely grateful to the late Yuko Giga-Hama.

References

Ara K, Ozaki K, Nakamura K et al (2007) *Bacillus* minimum genome factory: effective utilization of microbial genome information. Biotechnol Appl Biochem 46(pt 3):169–178

Baba T, Ara T, Hasegawa M et al (2006) Construction of *Escherichia coli* K-12 in-frame, single-gene knockout mutants: the Keio collection. Mol Syst Biol 2:20060008. doi:2010.1038/msb4100050

Decottignies A, Sanchez-Perez I, Nurse P (2003) *Schizosaccharomyces pombe* essential genes: a pilot study. Genome Res 13(3):399–406

Fantes P, Nurse P (1977) Control of cell size at division in fission yeast by a growth-modulated size control over nuclear division. Exp Cell Res 107(2):377–386

Fujio T (2007) Minimum genome factory: innovation in bioprocesses through genome science. Biotechnol Appl Biochem 46(pt 3):145–146

Giga-Hama Y, Tohda H, Takegawa K et al (2007) *Schizosaccharomyces pombe* minimum genome factory. Biotechnol Appl Biochem 46(Pt 3):147–155

Hirashima K, Iwaki T, Takegawa K et al (2006) A simple and effective chromosome modification method for large-scale deletion of genome sequences and identification of essential genes in fission yeast. Nucleic Acids Res 34(2):e11. doi:10.1093/nar/gnj1011

Idiris A, Tohda H, Bi KW et al (2006) Enhanced productivity of protease-sensitive heterologous proteins by disruption of multiple protease genes in the fission yeast *Schizosaccharomyces pombe*. Appl Microbiol Biotechnol 73(2):404–420

Idiris A, Tohda H, Sasaki M et al (2009) Enhanced protein secretion from multiprotease-deficient fission yeast by modification of its vacuolar protein sorting pathway. Appl Microbiol Biotechnol 85(3):667–677

Idiris A, Tohda H, Kumagai H et al (2010) Engineering of protein secretion in yeast: strategies and impact on protein production. Appl Microbiol Biotechnol 86(2):403–417

Kim DU, Hayles J, Kim D et al (2010) Analysis of a genome-wide set of gene deletions in the fission yeast *Schizosaccharomyces pombe*. Nat Biotechnol 28(6):617–623

Kobayashi K, Ehrlich SD, Albertini A et al (2003) Essential *Bacillus subtilis* genes. Proc Natl Acad Sci USA 100(8):4678–4683

Manabe K, Kageyama Y, Morimoto T et al (2011) Combined effect of improved cell yield and increased specific productivity enhances recombinant enzyme production in genome-reduced *Bacillus subtilis* strain MGB874. Appl Environ Microbiol 77(23):8370–8381

Matsuo T, Otsubo Y, Urano J et al (2007) Loss of the TOR kinase Tor2 mimics nitrogen starvation and activates the sexual development pathway in fission yeast. Mol Cell Biol 27(8):3154–3164

Medema MH, van Raaphorst R, Takano E et al (2012) Computational tools for the synthetic design of biochemical pathways. Nat Rev Microbiol 10(3):191–202

Mizoguchi H, Mori H, Fujio T (2007) *Escherichia coli* minimum genome factory. Biotechnol Appl Biochem 46(Pt 3):157–167

Mizoguchi H, Sawano Y, Kato J et al (2008) Superpositioning of deletions promotes growth of *Escherichia coli* with a reduced genome. DNA Res 15(5):277–284

Morimoto T, Kadoya R, Endo K et al (2008) Enhanced recombinant protein productivity by genome reduction in *Bacillus subtilis*. DNA Res 15(2):73–81

Moya A, Gil R, Latorre A et al (2009) Toward minimal bacterial cells: evolution vs. design. FEMS Microbiol Rev 33(1):225–235

Murakami K, Tao E, Ito Y et al (2007) Large scale deletions in the *Saccharomyces cerevisiae* genome create strains with altered regulation of carbon metabolism. Appl Microbiol Biotechnol 75(3):589–597

Posfai G, Plunkett G 3rd, Feher T et al (2006) Emergent properties of reduced-genome *Escherichia coli*. Science 312(5776):1044–1046

Sasaki M, Kumagai H, Takegawa K et al (2013) Characterization of genome-reduced fission yeast strains. Nucleic Acids Res 41(10):5382–5399

Wood V, Gwilliam R, Rajandream MA et al (2002) The genome sequence of *Schizosaccharomyces pombe*. Nature (Lond) 415(6874):871–880

Zhao Y, Lieberman HB (1995) *Schizosaccharomyces pombe*: a model for molecular studies of eukaryotic genes. DNA Cell Biol 14(5):359–371

Chapter 3
The Concept of the *Escherichia coli* Minimum Genome Factory

Hideharu Anazawa

Abstract The research program on the "minimum genome factory (MGF)" was launched in 2001. In this program, several microorganisms have been genetically modified to breed a cell with fewer genes on chromosomes of reduced size and expected less regulation on the gene network to be an ideal platform for a cell-factory. The genome of *Escherichia coli* has been reduced to 3.6 Mbp, by elimination of a total 1.06-Mbp region on various sites. This reduced-genome strain, MGF-01, has been grown to reach a higher level in minimal medium. This strain has the potential to be a general host for recombinant cells with higher productivity by retaining robustness of regulation on the metabolic network and activation of the fundamental biosynthetic pathway.

Using MGF-01 as a host strain, the production of amino acid and ATP are elevated to several fold.

Keywords *Escherichia coli* • Genome size reduction • Harmful unnecessary gene • Minimum genome factory

The term minimum genome factory (MGF) is defined as follows: a mutant host strain having a set of genes required only for fermentation production, its productivity has been greatly improved in fermentative or enzymatic production process compared to its parent strain. For the purpose of further improvement of the material production, there is a completely different concept for mini-genome cells having a gene set for survival. Therefore, as a starting strain for breeding a MGF mutant, a typical microorganism used in the fermentation process is selected.

H. Anazawa (✉)
Japan Bioindustry Association, 2-26-9 Hetchobori, Chuo-ku, Tokyo 104-0032, Japan
e-mail: anazawahdhr@jba.or.jp

H. Anazawa and S. Shimizu (eds.), *Microbial Production: From Genome Design to Cell Engineering*, DOI 10.1007/978-4-431-54607-8_3, © Springer Japan 2014

The word "MGF" is found in a report drafted through discussion between Tatsuro Fujio (Kyowa Hakko Kogyo), Megumi Inoue (Kao), and Hiromichi Kumagai (Asahi Glass) with the Ministry of Economy, Trade and Industry Department of Biological Chemistry (Mizoguchi et al. 2007a). They had discussed organizing a R&D project, "a technology development for production processes with biological function." This project started in April 2001, and the concept was transferred to the next project, "a manufacturing base of advanced performance technology for microbial production," completed in 2011 March.

By these research projects, the concept of MGF has extremely high potential for breeding a hyper-producer microorganism, and they have enough capability for further development. Therefore, we expect that this idea can be expanded to breeding mutant cells, which aims to significantly improve productivity in various organisms such as animal cells and plant cells.

3.1 Introduction

From the beginning of the development and application of recombinant DNA technology, *Escherichia coli* has been used in the fermentative production process for many materials. Commercialization of medical protein drug production systems using *E. coli* is well known, such as granulocyte colony-stimulating factor (G-CSF) or insulin. Many examples are known for mass production by fermentation or an enzymatic conversion process, such as L-lysine or L-aspartic acid from fumaric acid.

One of the reasons for selecting *E. coli* as a starting host strain of MGF breeding is that it always been used in the development and operation of the basic concept and technology of molecular biology. Also, many mutant strains with one gene knocked out and its phenotype have been reported. Knowledge about gene functions and regulation has accumulated. *E. coli* is a microorganism elucidated at the molecular level.

To breed a MGF, there are three key steps: (1) selection of genes for deletion as harmful or unnecessary code, (2) making a mutant strain by removing selected genes from the chromosome, and (3) evaluation of the deletion mutant strains from which harmful or unnecessary genes have been removed. Various types of innovative technology have been required for breeding these novel mutant strains with accurate deletion of target genes from the chromosome on many sites through these steps.

For step (1), it is relatively easy to select candidate genes for deletion as harmful or unnecessary genes from the sequenced whole genome of *E. coli* because many one-gene-knockout mutants have created and much information about gene functions has been reported. In particular, the one-gene-knockout mutant library of *E. coli* that has been created in the Nara Institute of Advanced Science and Technology and Keio University was very useful for evaluating the function of genes that had been classified as functionally unknown (Ito et al. 2005).

For step (2), a technique was established to remove the target gene region of interest. A linear vector for double-crossover insertion with an antibiotic-resistant gene and a negative selection marker, *Bacillus sacB* gene, was flanked with homologous sequences at both ends. Consecutive selection was carried out with two selection markers for creation of a deletion mutant without a selection marker on the target site. This marker-less deletion technology is an important tool for multiple rounds of deletions (Fukiya et al. 2004a, b).

By repeating this operation, the deleted region was expanded. These techniques may be also referred to as advanced methodology of molecular biology techniques that have been developed for *E. coli*.

For step (3), they evaluated the capability of *E. coli* MGF by growth response on minimum media with glucose as a carbon source, not by a specific product indicator.

The intention was to prove the concept of *E. coli* MGF is appropriate for breeding a mutant strain for a hyper-producer for fermentation by deletion of harmful or unnecessary genes from the host strain.

3.2 Selection of the Genes to Be Deleted

The first candidate genes for deletion are the factors that destabilize the genome such as pro-phage, IS, transposons, and clearly harmful genes such as an enzyme degrading a product of fermentation. In addition, also considered for deletion are candidate genes corresponding to environmental change such as temperature, pH, and nutrients for assimilation to a microorganism in a culture. Carbon flow on the biosynthetic pathway, competition, branching, and accumulation of intermediates should be regulated by optimization of corresponding gene expression. The common feature of the organism homeostasis (robustness) acts as a suppressive factor on high-level activation of biosynthetic pathways in hyper-fermentation producer strains. Also, genes of unknown function are candidates for removal from the host cell: 30 % of the ORF in *E. coli* is still unknown as to function.

Many *E. coli* strains including clinical isolates have been sequenced. From these DNA sequence data analyses, conserved genes are considered to be important and necessary genes and nonconserved genes might be unnecessary for *E. coli*. One simple method for comparison of genome-wide conservation is the comparative genome hybridization (CGH) method: 4,071 genes among 4,478 genes deduced from the *E. coli* W3110 genomic sequence have been spotted on a microarray slide and hybridized with two other *E. coli* strains labeled with different fluorescence dyes. From the pattern of refracting fluorescence, it is easy to select conserved regions among these three *E. coli* strains (Fukiya et al. 2004a, b): 2,586 genes have been conserved in all strains, and 1,424 genes were not conserved in 22 strains of *E. coli* examined. These 2,586 genes could be considered as common genes in *E. coli* and should not be deleted from the host genome.

It is also important to evaluate each gene function for *E. coli* MGF.

One-gene-knockout mutant libraries (Baba et al. 2006) and the phenotype analysis system of Bio-Log Co. have been employed (Ito et al. 2005). In the mutant libraries, 4,320 individual genes have been deleted in each cell, and 93 carbon substrates have been examined for their metabolic capabilities with all individual mutants on a plate. By evaluating the metabolic capability for 93 carbon substrates on a plate of each mutant, genes of similar function show similar patterns for the metabolic profile to 93 carbon substrates. Clustering analysis of metabolic profiles show positioning of similar function genes brought closer.

These homologue genes are candidates for deletion. Using this clustering analysis and phenotypic analysis, nine genes of unknown function have similar functions in the vicinity of genes of known function.

One operation of deletion of genes from a genome requires at least 2 weeks and evaluation of the deletion mutant takes longer. Therefore, several deletion regions on the genome at one operation have been designed and the mutants are evaluated by criteria of growth on minimal medium. The best mutant strain is selected for the next mutant design.

3.3 Creation of Genome Reduction in Strain

For *E. coli* gene manipulation, many technical tools have been reported such as the double transposon system of λ phage Red and Cre/loxP (Fukiya et al. 2004a, b).

Using the Red system, Fukiya et al. developed a two-step method for marker-less deletion. An antibiotic-resistant gene and the negative selection marker, *Bacillus sacB* gene, were flanked with homologous sequences at both ends. When this fragment is introduced into the *E. coli* chromosome by electroporation, they could make the deletion by double crossing-over replacement using the antibiotic resistant gene as the selection marker. In the next step, a linear DNA fragment consisting only of homologous sequences directly bound to each other is used for the second double crossing-over replacement. Because the *Bacillus sacB* gene makes *E. coli* sensitive to sucrose, they could select as a marker-less deletion mutant the sucrose-resistant colonies. This marker-less deletion can be applied to multiple rounds of deletions. By repeating this operation, they expanded the deleted region (Mizoguchi et al. 2007a, b).

A vector has been designed for deletion to eliminate as long a region as possible at one operation. However, when several deletion processes have proceeded and been repeated, it becomes difficult to expand a deletion region to be long at one operation.

Twenty-six operations of design and deletion of regions from the *E. coli* genome have accomplished to the MGF-01 mutant strain (Fig. 3.1). This mutant is 3.6 Mbp in genome size, and a 1.03-Mbp region was eliminated, with no effect and no reduced growth on minimal medium (Mizoguchi et al. 2006). The MGF-01 mutant strain shows a unique deletion profile compared to reported deletion mutants in size and region (Posfai et al. 2006). Table 3.1 shows the list of deleted genes in MGF-01. Mobile element and hypothetical genes are deleted in many cases.

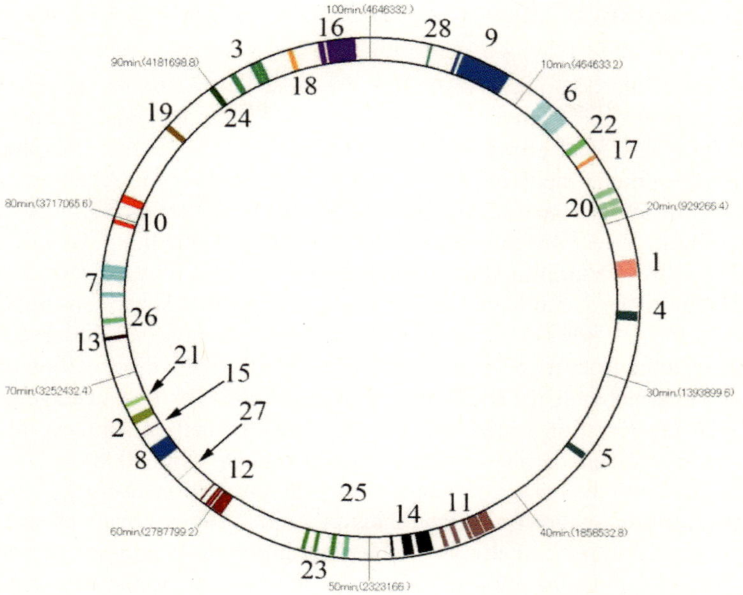

Fig. 3.1 Schematic illustration of operation order of deleted regions in MGF-01 genome indicated by the *numbers*. (From Mizoguchi et al. 2008)

Table 3.1 Distribution of *Escherichia coli* genes categorized by function

Description	Common	MGF-01	Deleted
Amino-acid metabolism	135	130	5
Biosynthesis of cofactors	126	125	1
Cell envelope	189	143	46
Cellular process	102	74	28
Central metabolism	154	114	40
Energy metabolism	364	322	42
Fatty acid metabolism	59	55	4
Nucleotide metabolism	116	103	13
Regulatory functions	99	79	20
Replication	88	78	10
Transport/binding protein	350	253	97
Translation	149	144	5
Transcription	47	42	5
Other categories	213	173	40
Mobile elements	333	83	250
Pseudogenes	69	28	41
Hypotheticals	1,710	1,303	407
Miscellaneous RNA	49	44	5
rRNA	22	22	0
tRNA	86	84	2
Total	4,460	3,399	1,061

(From Mizoguchi et al. 2007a, b)

3.4 Evaluation of the Mutant Strain with Reduced Genome

The genome-reduced strain MGF-01 has grown as well as the parent strain at the exponential phase and continued growing after the parent had entered into the stationary phase. The final growth level has reached 1.5 times higher than the parent strain in M9 minimal medium (Fig. 3.2), an unexpected result. Selection of genes for elimination and design of the deletion region has been carried out by picking out those having no effect on growth of a strain in minimal medium. This result was brought about by performing the delete operation 26 times in many regions and the removed region has accumulated as one strain. Decrease of sugar consumption per cell growth and deletion of carbon dioxide in exhaust gas have been observed. These phenomena come from repression of a futile metabolic pathway of carbon flow and cellular energy metabolism. These features of the strain with a reduced genome are suitable for fermentation because of the high yield of carbon sources; this strain may come to be a general platform as a host (Mizoguchi et al. 2008).

Table 3.2 shows the results of mutation for ATP production (Δ*nlpD-1*) in *E. coli*. As to the effect of mutation on ATP production, MGF-1 shows greater effect than the parent strain: 6 fold per cell and 12 fold per culture broth. In addition, introduction of plasmids for L-threonine biosynthesis to *E. coli* results in productivity in MGF-01 that is more than doubled compared to the parent strain (Hara et al. 2009).

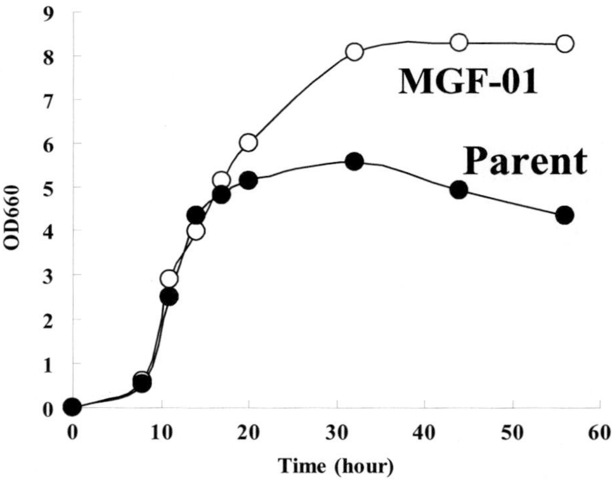

Fig. 3.2 Growth profile of *Escherichia coli* strains in minimum medium (*MM*) measured by OD_{660}. (From Mizoguchi et al. 2007a, b)

Table 3.2 Effect of mutation for ATP synthesis (Δ*nlpD-1*) in MGF-01

Strain	OD	ATP synthetic activity Per OD	ATP synthetic activity Per volume
Parent strain	100	100	100
Parent Δ*nlpD-1*	101	112	116
MGF-01	185	70	130
MGF-01 Δ*nlpD-1*	203	747	1,537

(Hara et al. 2009)

From these results, MGF-01 has potential to increase its productivity by mutations on the chromosome and plasmid in both ways. In other words, the concept of MGF of deleting the genes that are considered unnecessary and harmful in *E. coli* suggests that it is a very useful concept in the breeding of bacteria for fermentation.

A mutant strain (MGF-01), a 1.03-Mbp (22 % of whole genome) reduced genome of *E. coli*, show improved growth about 1.5 fold on minimal medium compared to the parent strain. Furthermore, a mutation for ATP synthesis or introduction plasmid for L-threonine production to MGF-01 showed elevated productivities several times greater than the parent strain.

These results indicate elimination of toxic genes and unnecessary genes is a useful and powerful concept for designing a mutant strain for microbiological production as a breeding host strain.

However, there are some cases showing higher productivity than MGF-01 in a mutant strain on the way to breeding MGF-01 for some other products. In such cases, the deleted region in the next operation step would be restored to the mutant and productivity could be recovered. Thus, there are some effective genes in the region of deletion for the product and this gene becomes a candidate for gene amplification for production.

The precise control of gene expression level is a challenging technology for breeding a fermentation microorganism. Regulation of factors of gene expression control is the next important tool to activate a metabolic pathway at the appropriate time in the growth of microorganisms.

Degradation and competition of intermediate metabolites and excretion of product through a membrane are other interesting research targets for breeding fermentation strains.

A simple genomic structure by deletion of unnecessary and harmful genes leads to clear explanation and reasonable understanding of complex cellular reactions. It is possible to breed a fermentation strain with genome design technology in that the inside condition of the cell does not remain a black box but is measured as a snapshot.

It is still unclear why effects of mutation are increased when a plasmid is introduced for production using MGF-01, but this mutant strain has high potential for the host strain for production. A smaller genome of *E. coli* has been bred by Hirokawa et al. (2013).

This concept has been applied to other organisms, such as *Bacillus subtilis*, as well as fission yeast as described in other chapters, and it might be expanded to various industrial microorganisms, plant cells, and even animal cells. In addition, I considered that when application of a genetically modified microorganism will be used in an open system, the harmful, doubtful, and unnecessary regions of the genome from the host strain can be deleted from the strain to reduce the risk in the environment.

Until now, strengthening activity on biosynthetic pathways by gene amplification was a major breeding methodology by gene recombination techniques. I have considered here a new concept that has been introduced into breeding cells for production by removal of unnecessary genes from the host cell.

References

Baba T et al (2006) Construction of *E. coli* K-12 in-frame single-gene knock out mutants: the Keio collection. Mol Syst Biol 2:20060008

Fukiya S et al (2004a) Extensive diversity in pathogenic *E. coli* and *Shigella* strains revealed by comparative genomic hybridization microarray. J Bacteriol 186(2):3911–3921

Fukiya S et al (2004b) An improved method for deleting large regions of *E.coli* K-12 chromosome using a combination of Cre/loxP and λRed. FEMS Microbiol Lett 234:325–331

Hara KY et al (2009) Systematic genome-wide scanning for genes involved in ATP generation. Metab Eng 11:1–7

Hirokawa Y et al (2013) Genetic manipulation restored the growth fitness or reduced- genome *E. coli*. J Biosci Bioeng 116(1):52–58

Ito M et al (2005) Functional analysis of 1440 *E. coli* genes using the combination of knock-out library and phenotype microarray. Metab Eng 7:318–327

Mizoguchi et al (2006) WO2006/057340

Mizoguchi H, Mori H, Fujio T (2007a) *Escherichia coli* minimum genome factory. Biotechnol Appl Biochem 46(pt 3):157–167

Mizoguchi H et al (2007b) A simple method for multiple modification of *E. coli* K-12 chromosome. Biosci Biotechnol Biochem 71(12):2905–2911

Mizoguchi H et al (2008) Superpositioning of deletions promotes growth of *E. coli* with reduced genome. DNA Res 15:277–284

Posfai G et al (2006) Emergent properties of reduced genome *E. coli*. Science 312:1044–1046

Part II
Whole Genome Manipulation for Genome Design

Chapter 4
Efficient and Accurate Production of De Novo Designed Large-Size Gene Clusters by a Novel *Bacillus subtilis*-Based System

Mitsuhiro Itaya, Shinya Kaneko, and Kenji Tsuge

Abstract Recombinant cells harnessing a set of genes designed to produce useful materials or to respond to environmental changes have been constructed. Thanks to the recent burst of genome sequence information, target genes are not limited to the existing genes; the candidates extend to those from unculturable microbes and de novo designed genes. As synthesis processes become more complex, the number of relevant enzymes can be expected to increase. The increase in the number of genes delivered to industrial microbes has also raised certain technological barriers not encountered previously. Conventional molecular cloning systems rely on *Escherichia coli* strain K-12, which is poorly suited for multiple gene assembly processes, the primary topic of this chapter. DNA cloning in the *E. coli* plasmid requires a plasmid in circular form before transformation. However, reactions to circularize long DNA and to connect multiple DNA fragments are trade-offs in solution and are thus rarely conducted for a single ligation. In contrast, our novel DNA cloning system using *Bacillus subtilis* strain 168 as a final host is offered to solve all plausible stumbling blocks in gene assembly technologies. We call the new system BGM (*Bacillus subtilis-based* gene/genome manipulation). In this chapter, we focus on the development of the BGM system and certain related achievements, highlighting the system's clear conceptual differences from *E. coli* systems.

Keywords BGM vector • Gene assembly • Long-term DNA reservoir • Metabolic engineering • Molecular cloning • OGAB method

---

M. Itaya (✉) • K. Tsuge
Institute for Advanced Biosciences, Keio University, Tsuruoka, Yamagata 997-0017, Japan
e-mail: mita2001@sfc.keio.ac.jp

S. Kaneko
Department of Life Science, Tokyo Institute of Technology,
Nagatsuta, Midori-ku, Yokohama, 226-8501, Japan

H. Anazawa and S. Shimizu (eds.), *Microbial Production: From Genome Design to Cell Engineering*, DOI 10.1007/978-4-431-54607-8_4,

4.1 Introduction

The ability of microbes to produce useful materials is of great interest in the fields of medicine, environmental studies, and business (Peralta-Yahya et al. 2012; Alper and Stephanopoulos 2009; Allen et al. 2013). Microbial fermentation has a central role in the production of pharmaceuticals, enzymes, and chemicals. To meet the demands of industrial production, it is desirable that microbes express the highest efficiency of the terminal metabolites (Zhu et al. 2012). An emerging research field, synthetic biology, is expected to pave the road for new platforms for microbial production. The goals of synthetic biology are to develop the foundational technologies and knowledge that make the engineering of biology safer, more reliable, and more predictable. As synthetic biology progresses from research to business-oriented applications, a clear direction is shown toward the production of useful substances through newly designed synthetic pathways that combine elements of genes such as promoters, terminators, and translational regulation signals (indicated in Fig. 4.1).

Gene circuits that regulate the expression of a full set of genes are also target molecules to be used toward the optimal and stable synthesis of the products in microbes. Computational modeling tools enable better prediction not only for metabolic engineering but also for biological systems before the preparation of gene clusters, and thus these tools are indispensable in synthetic biology (Cambray et al. 2011). The next fundamental question concerns how DNA molecules that include all the predicted information can be produced. There are numerous approaches that start with a number of genes related to synthetic routes, and together these approaches are called metabolic engineering (Ajikumar et al. 2010; Nielsen and

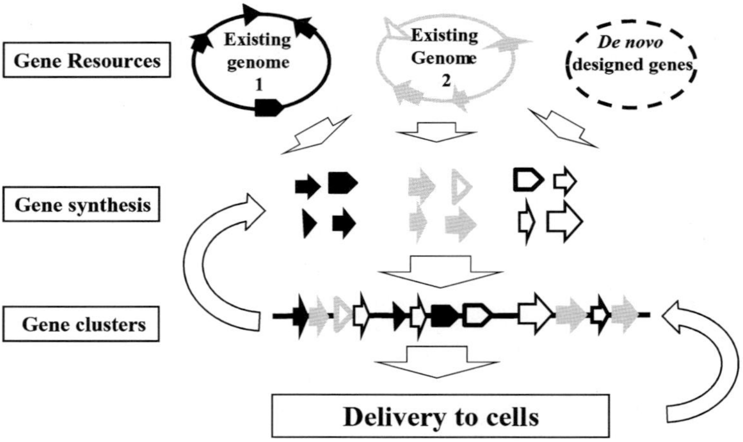

Fig. 4.1 Molecular design from genes to gene clusters. Genes of interest are obtained from various genome sources or are designed de novo. Efficient tools to assemble the genes and effective systems to deliver the genes to cells of interest demand cutting-edge technology. The gene clusters defined in the text are expected to play fundamental roles in metabolic engineering. Rapid and effective reiterative redesigning processes from genes to cells are central issues in the field of synthetic biology

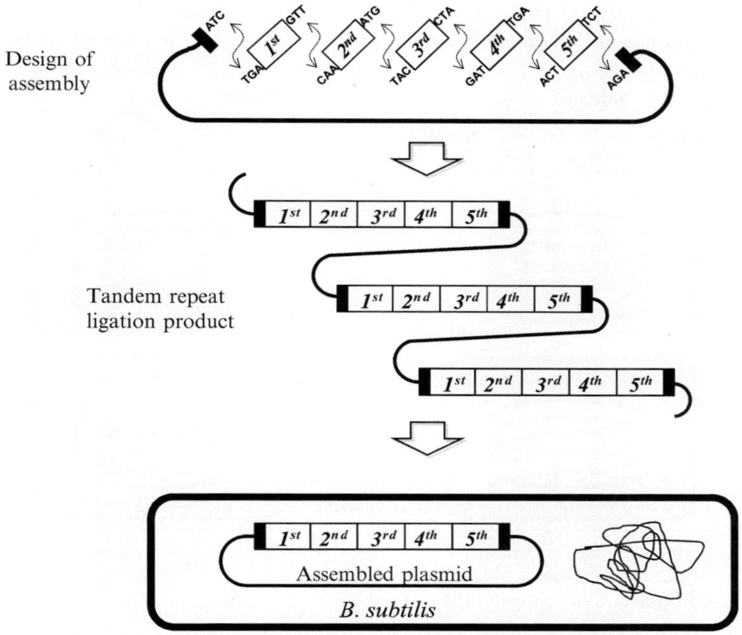

Fig. 4.2 The design of OGAB (ordered gene assembly in *Bacillus subtilis*) blocks. The six OGAB blocks (*top*), five OGAB blocks and one plasmid vector, possess protruding ends designed to give one connected form (*arrows*). These fragments are adjusted to an equimolar ratio to maximize the length of the tandem repeat form ligation products by T4 DNA ligase. The ligation products obtained (*middle*) are added to a competent *B. subtilis* 168 cell to give rise to circular plasmid DNA (*bottom*). The molecular mechanisms of the plasmid establishment from the ligation products are described in Fig. 4.3. Typical catalogue data for the six OGAB blocks are given in the text

Keasling 2011; Park et al. 2012). Gene delivery is a basic and central technology for metabolic engineering, used to convert existing microbes into better producers.

There are two fundamental inquiries about the need for gene assembly and delivery. Do we need to prepare as many delivery plasmids as needed to load all the genes, or do we need to expand the method to assemble all the genes in one DNA molecule? Because a number of plasmids are available only for limited model microbes, a powerful method that can conduct multiple gene assembly in molecular cloning is desired. The assembly of multiple genes into one DNA molecule, hereafter called "gene clusters" (Fig. 4.1), has depended on in vitro micromanipulations followed by cloning techniques using *Escherichia coli*. Conventional *E. coli*-based gene technology that allows the daily handling of several genes at once has contributed greatly to the preparation of small gene circuits composed of several gene blocks (Nandagopal and Elowitz 2011).

However, the greater the number of genes to be connected, the less effectively long gene clusters are obtained. We developed a unique and novel DNA cloning system using *B. subtilis* 168, a gram-positive end spore-forming eubacterium, as a final host to solve all plausible problems associated with the conventional gene

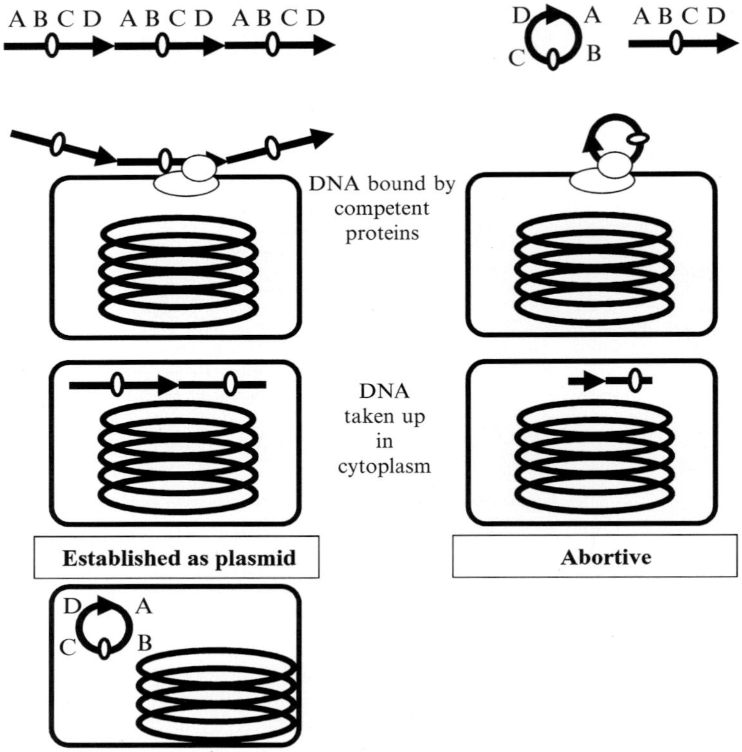

Fig. 4.3 Molecular model used to establish a *B. subtilis* plasmid by transformation. The plasmid unit is illustrated by a replication initiation site (*open circles*), sandwiched by AB and CD sequences. In the transformation of *B. subtilis*, DNA with a tandem repeat of the plasmid unit (*left*) can be processed to form only a circular plasmid. Protein complexes formed in the membrane bind dsDNA non-sequence specifically. ssDNA incorporated in cytoplasm longer than unit length (*left*) only can be repaired to form a circular plasmid. Transformation by DNA with a unit length (*right*) remains too short to be repaired and is abortive. [Find further molecular details are in Itaya (2013).] Tandem repeat ligation products in OGAB method in Fig. 4.2 are equivalent to the structure of *top left*

cluster synthesis methods. Our DNA cloning system is shown in Fig. 4.2. We call it the BGM system, which stands for *Bacillus* gene/genome manipulation (Itaya 2009, 2013; Itaya and Tsuge 2011). The use of *B. subtilis* as the new workhorse in the BGM system enables a large number of gene assemblies at high efficiency, with no errors incorporated during the assembly process because polymerase chain reaction (PCR) amplification steps are omitted. Here we describe the underlying molecular mechanisms and the inherent ability of *B. subtilis* to develop natural competency for the uptake of foreign DNA (Fig. 4.3). Several achievements of our new system are also presented.

4.2 Why Assemble Genes to Form Gene Clusters?

4.2.1 *Gene Delivery on Multiple Plasmids or Gene Cluster in a Single Plasmid*

Metabolic engineering is expected to produce methods to regulate existing pathways and stimulate the production of both primary and secondary metabolites (Park et al. 2012; Keasling 2012). Metabolic pathways are known to be designable if all the relevant genes and intermediate metabolites are well characterized, such as in carotenoid biosynthesis (Nishizaki et al. 2007). This concept is definitely different from the classic cases where targeted genes were limited to those of rate-determining steps. The fundamental goal becomes the construction of an effective metabolic pathway that includes both genes which code enzymes responsible for biochemical reactions and regulatory genes to control the effective expression of those genes. Generally speaking, the more complex the compound structure, the more catalytic reaction steps are needed. More important, the effective delivery of designed metabolic gene clusters into appropriate microbes followed by stabilization is urgently needed (Fig. 4.1).

The location of the delivered gene clusters in the genome or plasmid is another concern because of the different effects of gene expression and regulation as described previously (Itaya and Tsuge 2012). Clustered genes carried by the genome are likely to have advantages in terms of genetic stability, constant copy number, and more choices for location by integration. The current lack of genetic tools for industrial microbes in general means that gene cluster delivery must be performed via plasmid vectors in many cases. Even in plasmids, however, the construction of gene clusters faces difficulties concerning the increased numbers of genes to be assembled and the resulting long DNA. These difficulties are associated with the in vitro ligation before transformation for cloning hosts (detailed in Sect. 4.3) and the physicochemical instability of giant DNA in solution (described in Sect. 4.5).

4.2.2 *How Many Genes Are to Be Assembled?*

The average size of the genes from microbes described in this chapter is approximately 1 kbp. Given a case in which 20 genes are to be assembled into one gene cluster, the DNA size including the vector plasmid would exceed 30 kbp. How many genes are actually expected for a gene cluster? The highest number of genes functionally related in one particular cellular gene network might be components of the ribosome. Suppose an *E. coli* ribosome gene cluster is intended: 53 ribosomal proteins and three ribosomal RNAs are mandatory. Taking into account other accessory and regulatory genes for the ribosome to be functional, the total number of genes for the hypothetical ribosomal gene cluster is more than 60, and the total size would be far above 60 kbp. Specifications of fewer than 50 genes seem sufficient for most gene clusters in the foreseeable future. Examples are shown in Sects. 4.3.3, 4.3.4, 4.3.5. The DNA size is also less than 50 kbp, which is the upper zone for stable in vitro DNA manipulations as described in Sect. 4.5.

4.2.3 **Escherichia coli** *May Not Be an Efficient Giant DNA Assembler*

De novo DNA synthesis is carried out starting with DNA oligomers prepared in vitro and then cloned into an *E. coli* plasmid, followed by sequence confirmation. The current chemically synthesized oligomers of more than 100 bases do not assure sequence fidelity. In addition, amplification by PCR reaction in the connection steps produces inevitable error incorporations and the correct ones should be screened by sequence determination. Small DNA segments (<1,000 bp, or 1 kbp), stable in test tubes, are readily available at relatively low cost from DNA synthesis companies. In contrast, de novo DNA synthesis of several thousand base pairs involves considerable time and expense for the correct clone and by now seems a barely effective technique. The fragility of DNA molecules in solution also increases for DNA larger than 50 kbp. The quality of large DNA decreases primarily because of hydrodynamic shearing under regular laboratory manipulations such as agitation by vortexing and precipitation by the addition of ethanol. Unnoticed nucleases that arise during the biochemical isolation procedure might be a secondary reason for the inability to obtain damage-free DNA of more than 50 kbp (Kaneko et al. 2005). A bacterial artificial chromosome (BAC) vector is frequently used to clone large DNA fragments. Even DNA fragments stably maintained by a BAC vector at about 100–200 kbp are far smaller than the smallest existing genome (Gibson et al. 2008; Itaya and Kaneko 2012).

4.3 *Bacillus subtilis* 168: A Unique Host for Gene Assembly

4.3.1 Ordered Gene Assembly in **B. subtilis** *168*

Since the first recombinant DNA technology (Cohen et al. 1973) performed in an *E. coli* plasmid, the established DNA manipulation method in genetic engineering has been a combined use of circular plasmids and derivatives of *E. coli*. Circular plasmid DNA technology, a standard molecular biology tool, strongly depends on in vitro ligation technology. It is important to recognize that the circularization of a plasmid before delivery to *E. coli* is a central issue and is vital for a multiple DNA assembly protocol. Regarding multiple DNA fragments in vitro, the formation of circular DNA as a final form has rarely been viewed from a thermodynamic and steady-state kinetic perspective. Briefly, two ligation reactions, one intermolecular and the other intramolecular, should occur in one ligation condition. Intermolecular ligations proceed at high DNA concentrations in contrast to intramolecular ligations, which preferentially occur when the DNA concentration is low. As a consequence, circularized DNA possessing multiple DNA fragments connected at their high concentrations have rarely been produced, especially when more than several DNA fragments are used (Tsuge et al. 2003). Because *E. coli* starts replication for the circular double-stranded DNA (dsDNA) form only, the preparation of the final structure before transformation is critical.

Bacillus subtilis 168 is the second most often studied bacterium after *E. coli* at any level (Sonenshein et al. 2002). A striking and unique feature of *B. subtilis* able to develop natural competence was not well recognized as a host for molecular cloning (Itaya 2009, 2013; Itaya and Tsuge 2011; Kaneko and Itaya 2011; Itaya and Kaneko 2012). The natural competence of *B. subtilis* allows plasmid DNA uptake through molecular processes entirely different from those used with *E. coli* (Canosi et al. 1981). In the delivery of plasmid to competent *B. subtilis*, the efficiency differs significantly on plasmid structures prepared by in vitro ligation. As summarized in Figs. 4.2 and 4.3, a competent protein complex formed on the surface of the *B. subtilis* cell binds to exogenous dsDNA and takes up into the cytoplasm only one strand, that is, single-stranded DNA (ssDNA) from a cleavage site, leaving the other strand to be digested to nucleotides (Hahn et al. 2005; Kidane et al. 2009). All dsDNA given to a *B. subtilis* competent cell, regardless of circular or linear forms, must pass through membranes as ssDNA, which are repaired to restore the dsDNA form in the cytoplasm. This simple principle provides two advantages for the synthesis of gene clusters with multiple fragments: one is that there is no need to prepare circular-form DNA in ligation, and the other is that the plasmid can have more than one unit length (Figs. 4.2 and 4.3).

We developed and established a novel unprecedented method to efficiently construct gene assemblies in a manner that is totally different from that used with *E. coli* (Tsuge et al. 2003, 2007; Itaya and Tsuge 2011, 2012) (Fig. 4.2). The method, named OGAB (Tsuge et al. 2003), which stands for "ordered gene assembly in *B. subtilis*," has the advantages already described: that is, the ligation of multiple DNA fragments at high and equivalent molar concentrations yields preferentially linear DNA with tandem repeats of expected unit length (as drawn in Figs. 4.2 and 4.3). All the end sequences of the DNA fragments (named OGAB blocks) to be assembled should be designed to give a single primary structure only. That is, designed protrusions at both ends determine the order and orientation of all the OGAB blocks in a unique fashion, resulting in their tandemly repeated structure by T4 ligase (Fig. 4.2). Such linear DNA of more than single-unit length is a prerequisite and a good substrate for the plasmid to be established in *B. subtilis*, namely, for an OGAB-mediated plasmid.

Any protrusion sequence other than a nonpalindromic sequence can be used for the end of an OGAB block. We preferentially use 3′-protrusion ends for this purpose, as they are provided by many type IIS restriction endonucleases. They are, for example, *Alw*NI (5′-CAGNNN/CTG-3′), *Bsa*XI [5′-/$(N)_9$AC$(N)_5$CTCC$(N)_{10}$/-3′, 5′-/$(N)_7$GGAG$(N)_5$GT$(N)_{12}$/-3′], *Bgl*I (5′-GCCNNNN/NGGC-3′), *Dra*III (5′-CACNNN/GTG-3′), *Pfl*MI (5′-CCANNNN/NTGG-3′), and *Sfi*I (5′-GGCCNNNN/NGGCC-3′). Their cut sites, indicated as "/", include variable Ns (N = A, T, G, or C) that allow many arbitrary choices. OGAB blocks and a carryon plasmid vector possessing designed protrusion ends are connected in tandem repeat form using T4 DNA ligase. Tandem repeat DNAs with various numbers of units are formed in vitro given all OGAB blocks at equal molar concentrations, precisely adjusted. Because the adjustment of the molar concentration is critical to form a good substrate (Itaya and Tsuge 2011), SYBR Green I and a fluorescent plate reader are necessary if the number of OGAB blocks is greater than seven. Semiquantitative measurements of the DNA amount fluoresced by ethidium bromide are still sufficient on small numbers of OGAB blocks (fewer than six; Tsuge and Itaya unpublished data). Ligation by T4 ligase takes a surprisingly short time, as little as 5 min, in the presence of 10 % polyethylene glycol and 150 mM NaCl. The ligation product added directly to a competent *B. subtilis* cell is processed by *B. subtilis*, yielding a high number of transformants selected by a vector plasmid-associated antibiotic resistance marker.

4.3.2 Advantages of Simple Ligation and No Errors

The advantages of a gene cluster structure including all the relevant genes make the delivery protocol simpler, as argued in Sect. 4.2. The research flow of synthetic biology is represented by reiterative cycles composed of design, construction, and testing. In the design step, systems biology should play certain roles based on knowledge about enzymes and gene expression profiles. In the construction step, it is critical to produce gene clusters as real DNA molecules with an exact designed structure as fast as possible, and to produce as many as possible. In the last (testing) step, the effective delivery of gene clusters for biological testing depends largely on the cells of interest (see Fig. 4.1).

Typically, in an OGAB protocol using six OGAB blocks (Fig. 4.2), 10^4 transformants were obtained per microgram of DNA in total, 98 % of which carried the correct gene cluster as reported (Tsuge et al. 2003). The remarkably high yield allows us to obtain the designed gene cluster by screening just a few transformants; this was the case for up to 13 OGAB blocks (Tsuge and Itaya unpublished results). More importantly, the resultant gene clusters had acquired no mutations, as confirmed in our previous studies (Tsuge et al. 2003, 2007; Nishizaki et al. 2007; Hiroe et al. 2012). The exceptional advantage of the OGAB method stems from its exclusion of the PCR-mediated amplification steps throughout the OGAB assembly; essentially no mutations are acquired during the transformation of *B. subtilis* (Tsuge and Itaya unpublished observations). Aside from arguments about the detailed molecular mechanisms of competent development specific to *B. subtilis* (Hahn et al. 2005; Kidane et al. 2009), the extremely high fidelity of gene clusters produced by OGAB assembly will reduce the practical costs and energy involved.

A limitation of the OGAB method is that some difficulty may be encountered when adjusting the working concentrations of the OGAB blocks. Inevitable deviations in the DNA concentration obtained by experimental measurement cause the adjustment of large numbers of OGAB blocks to become tricky and, as a consequence, a subtle misadjustment in concentration can have an adverse effect.

Taking this limitation into account, we considered whether the OGAB method could be performed by an apparatus. A discussion of this direction in our research is presented in Sect. 4.4.

Achievements with the OGAB method are described in the next three sections. Gene clusters for the biosynthesis of the antifungal pesticide plipastatin were tested in a *B. subtilis* strain from which all seven relevant genes were deleted from its genome, as described in Sect. 4.3.3 (Tsuge et al. 2007; Itaya and Tsuge 2012). Similarly, clustered genes were tested for carotenoid pigment synthesis (Nishizaki et al. 2007) (Sect. 4.3.4) and for polyhydroxyalkanoate (PHA) production (Sect. 4.3.5) (Hiroe et al. 2012). These experiments were all conducted to elucidate the underlying rules of gene expression brought about by operon structure regulated by a single promoter. Our results demonstrated that genes positioned closer to the promoter were more highly expressed than genes positioned away from the promoter. In addition, the gene expression levels were positively correlated with the protein expression levels (Tsuge and Itaya unpublished results). The ubiquity of the relationship between gene expression, protein expression, and promoter activity is under investigation.

4.3.3 Plasmid-Borne Nonribosomal Peptide Synthetase Cassette

Bacillus subtilis and close relatives are known as producers of cyclic lipopeptides. Plipastatin (Fig. 4.4a) produced by *B. subtilis* strain 168 is a potent antifungal pesticide used for rice disease. Its peptide portion, composed of ten amino acids with both L and D configurations, is synthesized in a nonribosomal manner, processed by huge template enzyme complexes encoded in the *ppsABCDE* operon. Because the nonribosomal peptide synthetases (NRPSs) are composed of a tandem array of similar enzyme units, the *ppsABCDE* operon is large (38.6 kb) (Fig. 4.4b) (Tsuge et al. 2001). Several repetitive DNA sequences contribute the enzymes with high molecular weights; for instance, three internal 1.5-kb regions are almost identical DNA sequences (identity >97 %). A small regulator gene, *degQ* (0.6 kb), is known to activate *ppsABCDE* (Tsuge et al. 2001). In addition, the *sfp* gene (1.0 kbp) encoding 4′-phosphopantetheinyl transferase is required for the conversion of nascent template synthetase to its active form (Tsuge et al. 2001). These two small genes and one large operon are located in different genome loci, equally separated from each other. We used these three DNAs to examine the effectiveness of the OGAB method on size difference of OGAB blocks and to estimate the influence possibly caused by highly repetitive sequences in NRPSs. For the *degQ* and *sfp* genes, *Sfi*I recognition sequences were added at both ends during their preparation by PCR amplification. Those *Sfi*I sites were designed to generate protrusion sequences as OGAB blocks. However, as the *ppsABCDE* operon is too large and possesses too many repetitive sequences to conduct PCR-mediated preparation, the operon was prepared in a *B. subtilis* plasmid by the *Bacillus* recombinational transfer (BReT) method (Tsuge et al. 2007). The BReT method is a unique and original manipulation method specific for the BGM system (Tsuge and Itaya 2001) through which *ppsABCDE* operon DNA (38.6 kbp)

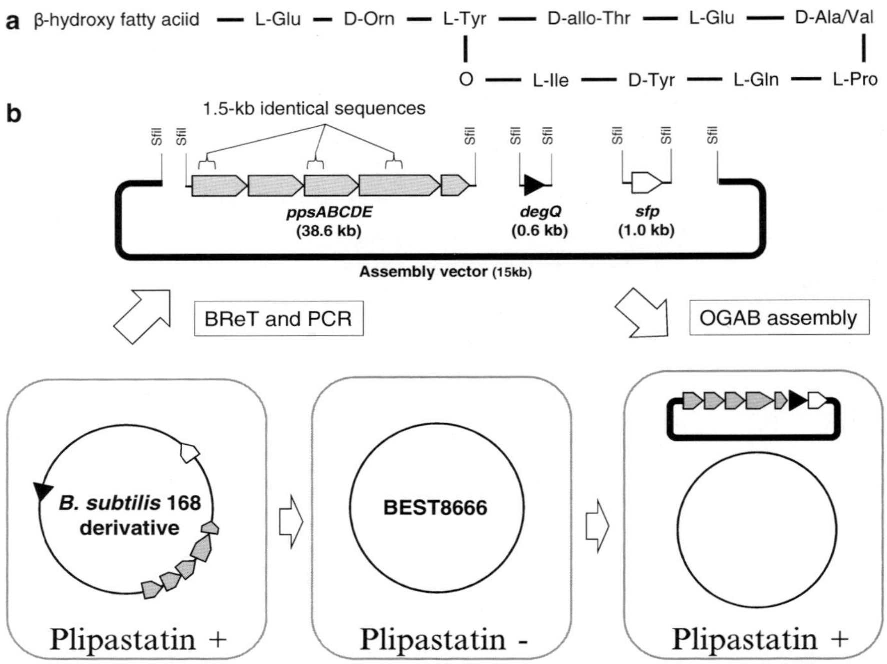

Fig. 4.4 (**a**) Structure of plipastatin. The peptide portion is synthesized by nonribosomal peptide synthetases (NRPSs). (**b**) Assembly of plipastatin production genes by the OGAB method. At least three genes are considered requisite for plipastatin production. *ppsABCDE* is the largest, encoding a 38.6-kb-long NRPS operon. Because of the nature of NRPS genes, there are several repetitive sequences in the operon. The location of one of the most obvious repeat sequence units is indicated by 1.5-kb identical sequences. The other accessory genes *sfp* (1.0 kb) and *degQ* (0.6 kb) are one or two orders of magnitude smaller than *ppsABCDE*. It should be emphasized that the size differences of OGAB blocks is marginal, as demonstrated elsewhere (Tsuge et al. 2007; Itaya and Tsuge 2012). These genes are obtained by BReT or PCR using *B. subtilis* strain 168 by adding *Sfi*I recognition sequences to generate protrusions for OGAB assembly. We assembled these three OGAB blocks using BEST8666, a plipastatin nonproducer, based on the deletion of all relevant genes. The resulting strains restored plipastatin production by the assembled genes in the plasmid

was obtained as the last OGAB block (Tsuge et al. 2007; Itaya and Tsuge 2012). Last, a large amount of DNA was prepared in vitro by digestion at the two *Sfi*I sequences, which were also designed and added to its ends. The designed gene cluster with three DNA blocks was successfully obtained in *B. subtilis* strain BEST8666, a non-plipastatin producer that lacks *ppsABCDE*, *degQ*, and *sfp* to avoid sequence duplication (Fig. 4.4b). The yield was greater than 80 %; that is, more than 80 % of the transformants selected by the OGAB vector-linked antibiotic. The strain BEST8666 carrying the plipastatin synthesis gene cluster in plasmid turned out to be a plipastatin producer as predicted, indicating that the assembled genes are fully functional (Tsuge et al. 2007). The outcome not only demonstrated the feasibility of the OGAB method for NRPS genes but also showed a negligible rate of error incorporation during OGAB manipulations with a series of blocks that varied by two orders of magnitude in size.

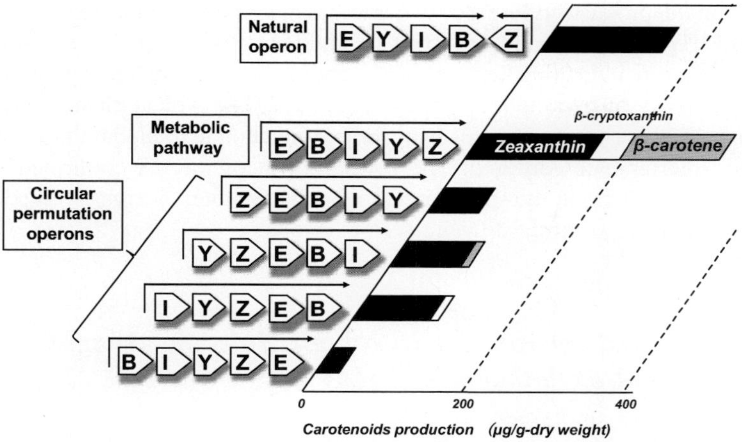

Fig. 4.5 Carotenoid production by newly designed operons. The metabolic pathway of zeaxanthin biosynthesis uses five enzymes: *CrtE*, *CrtB*, *CrtI*, *CrtY*, and *CrtZ*, in this order (Misawa et al. 1990). *Left:* Natural operon[*crtE*(E)-*crtY*(Y)-*crtI*(I)-*crtB*(B) and *crtZ*(Z)] (*top*) composed of two transcriptional units. A metabolic pathway operon (*middle*) was constructed to possess five genes aligned in their order of appearance in the zeaxanthin metabolic pathway in a polycistronic manner. Four polycistronic operons were constructed (*bottom*) by changing the order of the five genes with circular permutations. *Right*: The production levels of zeaxanthin and intermediates β-cryptoxanthin and β-carotene are indicated by *closed*, *open*, and *gray-colored boxes*, respectively

4.3.4 Plasmid-Borne Operons for Carotenoid Synthesis

Carotenoids are one of the best known pigments, coloring from red to yellow; they are found in bacteria, fungi, algae, and higher plants (Nishizaki et al. 2007). Carotenoids have potential vitamin A activity and have attracted great attention for their prominent antioxidant activity in human eyes. They are produced by a broad range of organisms, including bacteria. The carotenoid synthesis genes we used and describe here were cloned from a marine bacterium, *Erwinia uredovora*, by an *E. coli* cloning system as clones rendering the host colony red in color (Misawa et al. 1990).

The natural carotenoid synthesis gene cluster is composed of six genes responsible for zeaxanthin production in *E. coli*. They are separately included in two transcriptional units: one includes four genes (*crtE*, *ertY*, *crtI*, and *crtB*) in a polycistronic operon-like manner, and the other possesses *crtZ*. To explore functional gene clusters, we separately cloned the five genes required for zeaxanthin synthesis and used them as OGAB blocks to reconstruct a single operon form as designed. The fundamental design form was these five genes aligned in a polycistronic manner under a single Pr promoter borrowed from *E. coli* lambda phage DNA (Fig. 4.5). Typically, five genes were aligned exactly according to their catalytic order in the metabolic pathway of zeaxanthin biosynthesis made by our OGAB method. Four additional operons were constructed similarly where the order of the genes slides in a circular permutation manner (Nishizaki et al. 2007). These five operons were delivered in

E. coli, and the production levels of two intermediates and the final zeaxanthin were measured (Fig. 4.5). The amount of these second metabolites in *E. coli* varied from operon to operon, indicating that gene order could be a factor in the regulation of the apparent activity of particular metabolic pathways. The typical cluster, aligned in metabolic pathway order, showed the highest production in total, but the amounts of the other intermediates differed. The measurement of mRNA confirmed that its abundance, maximum at the spot closest to the Pr promoter, decreased steadily and was lowest at the spot furthest from the Pr promoter.

4.3.5 Designed Operons for Ultrahigh Molecular Weight Polyhydroxylalkanoate (PHA)

The biodegradable plastic PHA is a promising alternative to petroleum plastics. A homopolymer of (*R*)-3-hydroxybutyrate [P(3HB)], the most common type of PHA, is produced by the bacterium *Ralstonia eutropha*. An operon for P(3HB) synthesis was cloned from the bacterium and is widely used to confer *E. coli* P(3HB) production (Agus et al. 2006). The natural operon is composed of three genes aligned in the order of one polymerase gene (*phaC*) and two monomer-supplying genes (*phaA* and *phaB*). The PHA product from the original strain is brittle and rigid because of low molecular weight. Taking into account the carotenoid synthesis operons and our unpublished results, we attempted to formulate operons to produce ultrahigh molecular weight (UHMW)-P(3HB) by altering the gene order. We constructed six operons covering all permutations for the three genes shown in Fig. 4.6 by the OGAB method (Hiroe et al. 2012): these were made in *B. subtilis* and delivered to *E. coli*, a host suited for PHA production. The molecular weights of the products clearly differed depending on the gene order in the operons (Fig. 4.6). The abundance of mRNA and protein expressed from operons measured in *E. coli* decreased at the spot distant from the promoter. This observation accounted for how the gene order affected the product molecular weight distributions. Because of the negative correlation between PhaC activity and total production amount, the best operon for UHMW-P(3HB) in terms of a balance between molecular weight and production was *phaB-phaC-phaA* (Hiroe et al. 2012).

4.4 Possible Semiautomatic Apparatus for the OGAB Method

As mentioned in Sect. 4.3.2, it might be possible to automate the OGAB method. To test this notion, we divided the steps shown in Fig. 4.2 into smaller steps and reexamined the process in relationship to automation. We assumed that the initial OGAB blocks of less than 1 kbp are all sequence verified and supplied by an *E. coli* plasmid commercially provided. The OGAB blocks are designed to be cut off from the plasmid by the appropriate restriction enzymes listed in Sect. 4.3.1 to give end sequences for OGAB assembly.

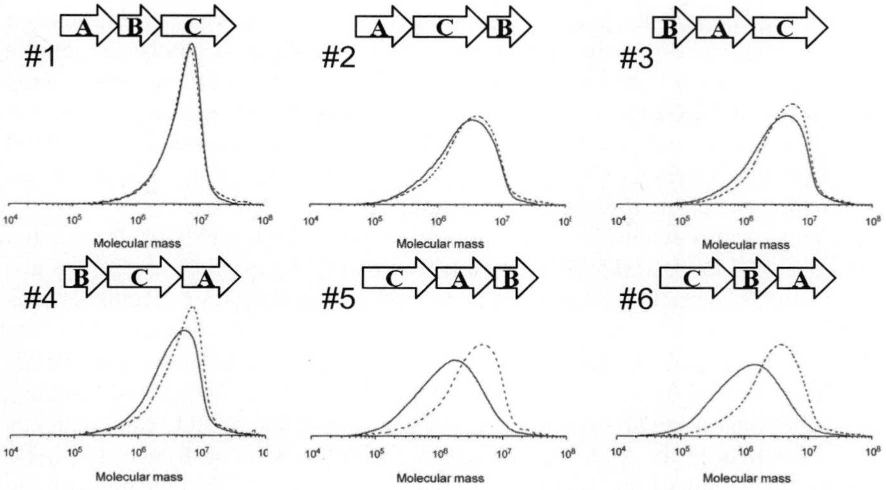

Fig. 4.6 Molecular weight distributions of P(3HB) synthesized by recombinant *E. coli* harboring the plasmids. All six operons of the three genes, *phaA*, *phaB* and *phaC*, shown by *open arrows* marked *A*, *B*, and *C*, were constructed in the *B. subtilis–E. coli* shuttle vector pGETS109. *E. coli* carrying these plasmids (1–6) produced PHA with the molecular weight measured at two cultivation times, 12 h (*dashed lines*) and 72 h (*solid lines*)

The steps that may be suitable for automation are the purification of DNA fragments from plasmid by gel electrophoresis, the measurement of DNA concentrations, the remixing with equal molar ratios, the ligation, the addition of ligation product to competent *B. subtilis*, and the selection of transformants by appropriate antibiotics. Screening for colonies could be connected to another existing automation apparatus.

In light of the effective OGAB block concentrations described in Sect. 4.3 and the fragile nature of the physicochemical properties of long gene clusters in solution as described in Sect. 4.2.2, care should be taken to assemble at most approximately 12 blocks not exceeding 50 kbp in length. Programs that allow the design of any given sequence for appropriate OGAB blocks are under construction and will be installed in prototype instruments (Tsuge and Itaya unpublished data).

4.5 Rapid Preparation of Large High-Quality OGAB Blocks

4.5.1 Isolation of a Giant Plasmid from B. subtilis

In particular, OGAB blocks larger than 10 kbp may provide an alternative choice; that is, the first gene clusters made by using the OGAB method are readily used as OGAB blocks for the second assembly, in a sequential manner. Because OGAB blocks require only precisely designed end sequences, all the initial OGAB blocks in use should be well designed for the reiterative OGAB assembly. For this purpose, the quality in addition to the quantity of the OGAB blocks is crucial to yield the desired giant gene clusters with high accuracy. We made it a rule to prepare large circular plasmids, thought to be equivalent to the first gene clusters, by an ultracentrifugation technique from *E. coli* and *B. subtilis* studies (Kaneko et al. 2005). Although ultracentrifugation supplies ultrapure DNA, the number of samples is limited to the number of the slots of the centrifuge rotors.

A rapid and reproducible technique for obtaining high yields of damage-free OGAB blocks from *B. subtilis* is essential for a seamless assembly process to give larger gene clusters. To the best of our knowledge, no isolation kit specific for plasmid from *B. subtilis* is commercially available at this time, and thus we tested several kits suited for *E. coli* plasmids for a variety of large-plasmid DNA from *B. subtilis*. In a pilot case study, we obtained a promising result using an 80-kbp plasmid able to shuttle between *B. subtilis* and *E. coli*. Among the available plasmid extraction kits, we used primarily the Midi Plus Ultrapure Plasmid Extraction System (Viogene-BioTek, New Taipei City, Taiwan) to purify DNA from *E. coli* in the quickest and simplest manner that could be applied to large DNA extraction from *B. subtilis*. The result stimulated the rapid isolation of the first OGAB gene clusters made in *B. subtilis*. Figure 4.7a shows the separation of eight *Sph*I fragments ranging from 2.4 to 25 kbp of the plasmid by pulsed-field gel electrophoresis.

4.5.2 Rapid Preparation of Large DNA Fragments of OGAB Quality

We examined the isolation of the *Sph*I fragments from an agarose gel plug by two conventional methods to determine their quality as OGAB blocks. One method was to elute the DNA fragments from agarose plugs, and the other was to use DNA extraction kits and agarose gels. An agarose plug containing a DNA fragment was transferred in a small dialysis membrane bag (50 Å pore size; filtrate materials >14,000 molecular weight) in the presence of 0.5× TBE buffer. The bag was soaked in the same buffer and subjected to electrophoresis for 17 h overnight to elute all DNA fragments from the agarose plug. The DNA in solution from the dialysis bag was concentrated by ethanol precipitation. Figure 4.7b shows the results of the recovery of the six fragments that were larger than 3.7 kbp. The other isolation was carried out using commercially available kits and the same agarose plugs. All the kits examined, that is, the Wizard SV Gel and PCR Clean-Up System (Promega, Madison, WI, USA), the QIAquick Gel Extraction Kit (Qiagen, Valencia, CA, USA), and the centrifugal Ultrafree-DA kit (EMD Millipore, Billerica, MA, USA) certify effective isolation of DNA fragments less than 10 kbp, and no responsibilities for much larger DNA. The QIAEX II Gel Extraction Kit (Qiagen) gave satisfactory results under our technical conditions for large-sized DNA (>3 kbp) (Fig. 4.7c). The use of the appropriate kit facilitated the purification of OGAB-grade DNA more quickly and conveniently than the dialysis tube method, up to 50 kbp.

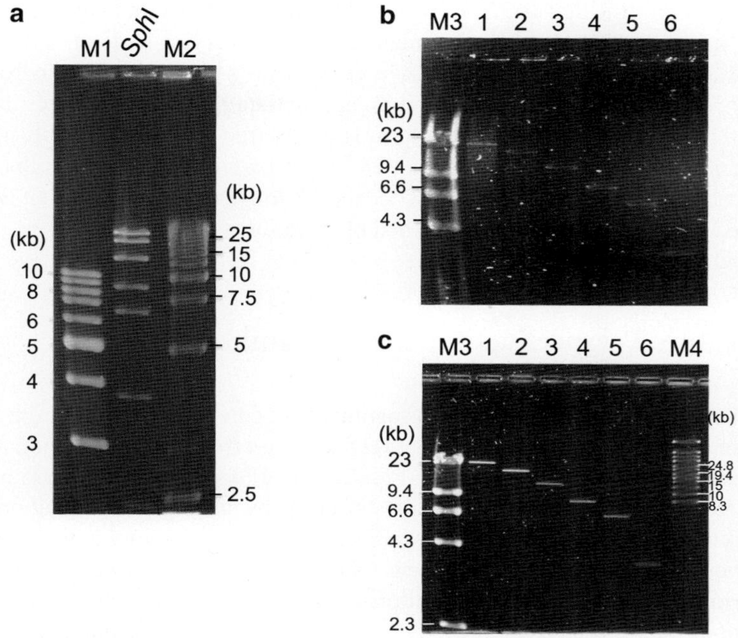

Fig. 4.7 (**a**) A plasmid derived from pGEST101*spc* (Kuroki et al. 2007) was purified using the Midi Plus Ultrapure Plasmid Extraction System (Viogene-BioTek), digested by *Sph*I, and resolved by CHEF. Lanes *M1* and *M2* are size markers. Running conditions: 3 Vcm^{-1}, 12-s pulse time, 20-h running time at 14 °C. (**b**) Isolation of *Sfi*I fragments using a small dialysis membrane bag. *Lane M3* is the size marker. Running conditions: 3 Vcm^{-1}, 6-s pulse time, 20-h running time at 14 °C. (**c**) Isolation of each fragment using the QIAEX II Gel Extraction Kit (Qiagen). *Lanes M3* and *M4* are size markers. Running conditions were the same as in (**b**)

4.6 Long-Term Storage of Valuable DNA Resources in the BGM System

An important aspect of the BGM system is the long-term storage of assembled gene clusters. Small DNA such as OGAB blocks are readily synthesized on demand and do not need to be stored. However, the gene clusters obtained by effective OGAB methods are invaluable because they are not readily reproduced. Not only the complete blocks but also their intermediate blocks are valuable as DNA resources for future use. Naked DNA, in particular large-sized naked DNA, is fragile in solution and susceptible to contaminating nucleases, shearing forces during repeated freezing and thawing, and other types of damage. Storing DNA in host cells is a promising technique. *E. coli* transformant cells should be preserved at –80 °C in the presence of glycerol or dimethyl sulfoxide (DMSO), or at room temperature after being lyophilized. Both these preservation techniques require certain facilities, and space, and in frozen stock in particular, the cost of long-term storage is thus high.

In sharp contrast, *B. subtilis* is capable of forming spores that survive for long periods in many hazardous environments, including drying out on a benchtop. Spores start to germinate instantly when exposed to a nutrition-rich environment and will form colonies the next day on an appropriate plate. BGM spores are thus expected to function as an ideal long-term and cost-free reservoir of DNA, needing no special equipment or facility. The stability of certain DNA for months inside spores was reported (Kaneko et al. 2005), and far longer term preservation is being examined (Kaneko and Itaya unpublished observations).

4.7 Future Perspectives for Gene Assembly

We are in the middle of a burst of information about genome sequences from a wide variety of natural cells (Youssef et al. 2011) and even from human individuals with unique characteristics (Zong et al. 2012). Current life science studies are moving rapidly from examining biology systems perturbed by small genetic variations such as knockin or knockout mutations toward more synthetic biology approaches featuring the use of de novo designed pieces (Nandagopal and Elowitz 2011). Aided by bioinformatics analyses, both conventional and new approaches are committed to the same mission: to unveil global cellular networks of gene products, RNA, protein, and metabolites that support life activities. Along with these academic pursuits, approaches based on synthetic biology have been stimulating the production of a vast range of chemicals by cells, mostly by microbes. Particularly, if target materials are more complex, not only because of structure but also because of inaccessibility through chemical synthesis, they may be appropriate targets for production by microbes. Our BGM system provides a novel and powerful pipeline for DNA synthesis. Given any genes, regardless of whether they exist in nature or are designed, the BGM system can synthesize entire gene clusters that are testable in cells, with significant structural flexibility.

We hope that the progress reported in this chapter will help the field of metabolic engineering be less dependent on *E. coli* and stimulate a movement toward the use of safe and sound BGM systems where *B. subtilis* 168 plays an essential and central role (Itaya 2009, 2013). Nevertheless, we greatly appreciate the utility of *E. coli*, without which no initial OGAB blocks could be prepared.

The tremendous diversity of the plant and animal kingdoms allows a great variety of enzymes to exist, and they may be synthesized. In the search for additional useful genes and gene clusters, meta-genomic approaches are expected to unveil new resources for synthetic gene clusters. The single-cell genomics approach to determine whole sequences from DNA is still insufficient for the completion of a whole genome (Youssef et al. 2011; Zong et al. 2012), and the impact of complete genome sequences from viable but nonculturable (VNC) microbes for gene clusters or genome design is invaluable (Itaya 2013).

Strategies and methodologies to modify individual enzyme activities are beyond the scope of this chapter, but it should be addressed here that any genes are targeted by the scenario in this chapter.

References

Agus J, Kahar P, Abe H, Doi Y, Tsuge T (2006) Molecular weight characterization of poly[(*R*)-3-hydroxybutyrate] synthesized by genetically engineered strains of *Escherichia coli*. Polym Degrad Stab 91:1138–1146

Ajikumar PK, Xiao WH, Tyo KE et al (2010) Isoprenoid pathway optimization for taxol precursor overproduction in *Escherichia coli*. Science 330:70–74

Allen HK, Levine UY, Looft T et al (2013) Treatment, promotion, commotion: antibiotic alternatives in food-producing animals. Trends Microbiol 21:114–119

Alper H, Stephanopoulos G (2009) Engineering for biofuels: exploiting innate microbial capacity or importing biosynthetic potential? Nat Rev Microbiol 7:715–723

Cambray G, Mutalik VK, Arkin AP (2011) Toward rational design of bacterial genomes. Curr Opin Microbiol 14:624–630

Canosi U, Iglesias A, Trautner TA (1981) Plasmid transformation in *Bacillus subtilis*: effects of insertion of *Bacillus subtilis* DNA into plasmid pC194. Mol Gen Genet 181:434–440

Cohen S, Chang A, Boyer H et al (1973) Construction of biologically functional bacterial plasmids in vitro. Proc Natl Acad Sci USA 70:3240–3244

Gibson DG, Benders GA, Andrews-Pfannkoch C et al (2008) Complete chemical synthesis, assembly, and cloning of a *Mycoplasma genitalium* genome. Science 319:1215–1220

Hahn J, Maier B, Haijema BJ, Sheetz M, Dubnau D (2005) Transformation proteins and DNA uptake localize to the cell poles in *Bacillus subtilis*. Cell 15:59–71

Hiroe A, Tsuge K, Nomura CT et al (2012) Rearrangement of gene order in the phaCAB operon leads to effective production of ultrahigh-molecular-weight poly[(*R*)-3-hydroxybutyrate] in genetically engineered *Escherichia coli*. Appl Environ Microbiol 78:3177–3184

Itaya M (2009) Recombinant genomes: novel resources for systems biology and synthetic biology. In: Pengcheng F, Latterich F, Panke S (eds) Systems biology and synthetic biology. Wiley, Hoboken. doi:10.1002/9780470437988.ch5

Itaya M (2013) Tools for genome synthesis. In: Zhao H (ed) Synthetic biology tools and applications. Academic Press/Elsevier, San Diego, pp 225–242

Itaya M, Kaneko S (2012) Integration of stable extra-cellular DNA released from *Escherichia coli* into the *Bacillus subtilis* genome vector by culture mix method. Nucleic Acids Res 38:2551–2557

Itaya M, Tsuge K (2011) Construction and manipulation of giant DNA by a genome vector. Methods Enzymol 498:427–447

Itaya M, Tsuge K (2012) Towards bacterial genome reconstruction: comparison of expression profiles of genes from different cellular DNA vehicles, plasmids and genomes. In: Sadaie Y, Matsumoto K (eds) *Escherichia coli* and *Bacillus subtilis*: the frontiers of molecular microbiology revisited. Research Signpost, Kerala, pp 297–317. ISBN 978-81-308-0492-7

Kaneko S, Itaya M (2011) Production of multi-purpose BAC clones in the novel *Bacillus subtilis* based host systems. In: Chatterjee P (ed) Bacterial artificial chromosomes. InTech Open, Croatia, pp 119–136

Kaneko S, Akioka M, Tsuge K et al (2005) DNA shuttling between plasmid vectors and a genome vector: systematic conversion and preservation of DNA libraries using the *Bacillus subtilis* genome (BGM) vector. J Mol Biol 349:1036–1044

Keasling J (2012) Synthetic biology and the development of tools for metabolic engineering. Metab Eng 14:189–195

Kidane D, Carrasco B, Manfredi C et al (2009) Evidence for different pathways during horizontal gene transfer in competent *Bacillus subtilis* cells. PLoS Genet 5(9):e1000630. doi:10.1371/journal.pgen.1000630

Kuroki A, Ohtani N, Tsuge K et al (2007) Conjugational transfer system to shuttle giant DNA cloned by the *Bacillus subtilis* genome (BGM) vector. Gene (Amst) 399:72–80

Misawa N, Nakagawa K, Kobayashi S et al (1990) Elucidation of the *Erwinia uredovora* carotenoid biosynthetic pathway by functional analysis of gene products expressed in *Escherichia coli*. J Bacteriol 172:6704–6712

Nandagopal N, Elowitz MB (2011) Synthetic biology: integrated gene circuits. Science 333:1244–1248

Nielsen J, Keasling JD (2011) Synergies between synthetic biology and metabolic engineering. Nat Biotechnol 29:693–695

Nishizaki T, Tsuge K, Itaya M et al (2007) Metabolic engineering of carotenoid biosynthesis in *Escherichia coli* by ordered gene assembly in *Bacillus subtilis* (OGAB). Appl Environ Microbiol 73:1355–1361

Park SJ, Lee TW, Lim SC et al (2012) Biosynthesis of polyhydroxyalkanoates containing 2-hydroxybutyrate from unrelated carbon source by metabolically engineered *Escherichia coli*. Appl Microbiol Biotechnol 93:273–283

Peralta-Yahya PP, Zhang F, Cardayre SB et al (2012) Microbial engineering for the production of advanced biofuels. Nature (Lond) 488:320–328

Sonenshein AL, Hoch JA, Losick R (eds) (2002) *Bacillus subtilis* and its closest relatives: from genes to cells. ASM Press, Washington, DC

Tsuge K, Itaya M (2001) Recombinational transfer of 100-kb genomic DNA to plasmid in *Bacillus subtilis* 168. J Bacteriol 183:5453–5456

Tsuge K, Akiyama T, Shoda M (2001) Cloning, sequencing, and characterization of the iturin A operon. J Bacteriol 183:6265–6273

Tsuge K, Matsui K, Itaya M (2003) One step assembly of multiple DNA fragments with designed order and orientation in *Bacillus subtilis* plasmid. Nucleic Acids Res 31(21):133

Tsuge K, Matsui K, Itaya M (2007) Production of the non-ribosomal peptide plipastatin in *Bacillus subtilis* regulated by 3 relevant gene blocks assembled in a single movable DNA segment. J Biotechnol 129:592–603

Youssef NH, Blainey PC, Quake SR et al (2011) Partial genome assembly for a candidate division OP11 single cell from an anoxic spring (Zodletone Spring, Oklahoma). Appl Environ Microbiol 77:7804–7814

Zhu L, Zhu Y, Zhang Y et al (2012) Engineering the robustness of industrial microbes through synthetic biology. Trends Microbiol 20:94–101

Zong C, Lu S, Chapman AR et al (2012) Genome-wide detection of single-nucleotide and copy-number variations of a single human cell. Science 338:1622–1626

Chapter 5
Development and Application of Novel Genome Engineering Technologies in *Saccharomyces cerevisiae*

Yu Sasano, Minetaka Sugiyama, and Satoshi Harashima

Abstract Various genome engineering technologies have been developed in the yeast *Saccharomyces cerevisiae*. One such key technology is PCR-mediated chromosome-splitting technology, designated PCS. The aim of PCS is to cut a chromosome at any chosen site into two smaller pieces and to make these "newly generated chromosomes" behave as functional chromosomes. PCS splits a chromosome very efficiently (more than 70 %) and allows repeated splitting because the built-in Cre/*lox*P site-specific recombination system facilitates the use of marker recycling. Subsequently, PCR-mediated chromosome deletion (PCD) technology was developed as a derivative technology of PCS. PCD facilitates the deletion of any chromosomal region, irrespective of an internal or terminal location. Genome reorganization (GReO) technology was also developed on the basis of PCS. In GReO, a huge variety of genomic constitutions are generated from a master strain harboring a few dozen mini-chromosomes constructed by PCS when it undergoes combinatorial mini-chromosome loss during mitosis. PCD and GReO technology have been exploited in the breeding of yeast. Several strains with a large-scale deletion ranging from 400 to 500 kb were constructed by PCD. Some of these strains produced twofold more ethanol and glycerol than the parental strain. The gene expression profiles revealed that the physiological adjustment from fermentative to oxidative metabolism, including stimulation of mitochondrial function, does not occur in these strains. GReO technology has been successfully used to create ethanol-tolerant strains showing a more than tenfold-higher specific growth rate in the presence of 11 % ethanol as compared with the parental strain. Thus, these genome engineering technologies provide a new tool for breeding novel yeasts useful for industrial applications.

Y. Sasano • M. Sugiyama • S. Harashima (✉)
Department of Biotechnology, Graduate School of Engineering, Osaka University, Suita, Osaka, Japan
e-mail: harashima@bio.eng.osaka-u.ac.jp

H. Anazawa and S. Shimizu (eds.), *Microbial Production: From Genome Design to Cell Engineering*, DOI 10.1007/978-4-431-54607-8_5, © Springer Japan 2014

Keywords Bioethanol • Biotechnological application • Chromosome-splitting technology • Ethanol tolerance • Genome engineering • Genome reorganization technology • Glycerol • *Saccharomyces cerevisiae*

5.1 Traditional and New Breeding Technologies

The ultimate goal of biotechnology is to establish a superior system for the production of a desired biomaterial in the highest yield. One of the fundamental issues for this purpose is to breed the best strain for the planned bioproduction. To create such strains of microorganisms, biotechnologists have employed four breeding technologies, namely, mutagenesis, conjugation, cell fusion (cytoplasmic fusion), and recombinant DNA technology. Among these breeding technologies, recombinant DNA technology made important contributions in the late twentieth century, especially so far as proteins, the primary product of genes, are concerned. It is becoming increasingly apparent, however, that the bioproduction of various chemicals is desired for sustainable society and that the most useful cellular traits needed for this bioproduction are controlled by multiple genes; as a result, new breeding technologies that facilitate the simultaneous manipulation of a large number of genes are urgently needed.

Although a few such new breeding technologies, including "global transcription machinery engineering" and "artificial transcription factor engineering," have emerged, we considered that genome engineering technology, which originally was intended to manipulate chromosomes and genomes on a large scale, might also provide opportunities for breeding novel strains of microorganisms that could not be created by conventional breeding technologies. On the basis of this idea, we have developed several genome engineering technologies for an industrially important microorganism, *Saccharomyces cerevisiae* (Fig. 5.1). This chapter focuses on some of these newly developed genome engineering technologies and their application to breeding.

5.2 Large-Scale Genome Manipulation in *S. cerevisiae*

Since the invention of recombinant DNA technology, numerous applications of this technology have been developed to manipulate individual genes, including site-directed mutagenesis, gene disruption, allele substitution, and overexpression of genes (often collectively called genetic engineering). With these various approaches, gene function has been intensively studied. Further development of recombinant DNA technology has also made it possible to manipulate intact chromosomes (chromosome engineering) and genomes (genome engineering) through the construction of small deletions, translocations, inversions, and fragmentation of chromosomes. However, the development of methods for manipulating chromosomes and genomes on a large scale lags far behind analogous developments for the manipulation of individual genes. This situation motivated us to devise the following genome engineering approach, which we call chromosome-splitting technology.

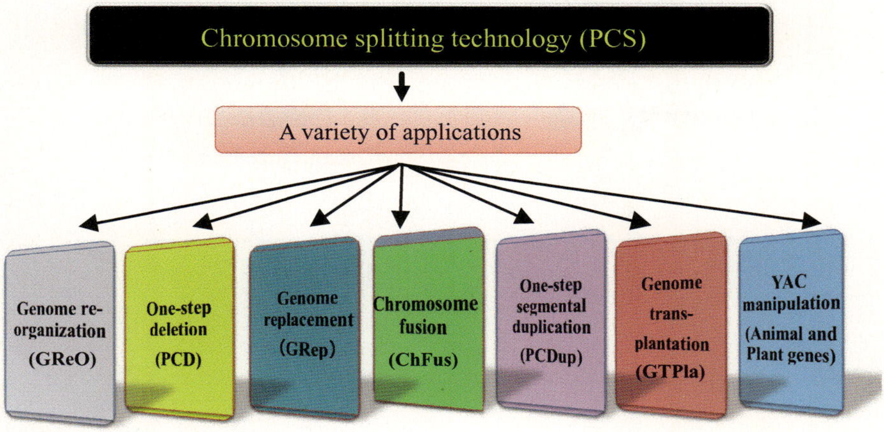

Fig. 5.1 Polymerase chain reaction (PCR)-mediated chromosome-splitting (PCS) technology and its various applications to genome engineering. *GReO*, genome reorganization technology; *PCD*, PCR-mediated chromosomal deletion technology; *GRep*, genome replacement technology; *ChFus*, chromosomal fusion technology; *PCDup*, PCR-mediated chromosomal duplication technology; GTPla genome transplantation technology; *YAC manipulation*, manipulation for animal and plant chromosome cloned on YAC

5.2.1 PCR-Mediated Chromosome-Splitting Technology

The aim of chromosome-splitting technology is to cut a chromosome, at any desired site, into two smaller pieces and to make these pieces behave as functional chromosomes. For their replication and faithful transmission to daughter cells during mitotic growth and meiosis, chromosomes must have three elements: namely, a centromere (*CEN*), telomere (*TEL*), and autonomously replicating sequence (ARS). Therefore, if chromosomes are newly generated, these elements must be provided by some means. Chromosome-splitting technology can satisfy these requirements (Sugiyama et al. 2005). This technology exploits homologous recombination and uses two DNA fragments, herein called fragment I and fragment II (Fig. 5.2a). Each pair of fragments (i.e., fragment I and fragment II) has a centromere and a selection marker for transformants, respectively, flanked by a target sequence and six copies of the *Tetrahymena* telomere sequence, 5′-CCCCAA-3′, which is known to function, when present as multiple copies, as a telomere seed in *S. cerevisiae* (Murray et al. 1988). These two DNA fragments are prepared by PCR and then introduced into a yeast cell by a single transformation. If homologous recombination takes place between each of the two introduced fragments and their respective targets simultaneously, chromosome splitting occurs in a manner such that both split chromosomes harbor the three essential elements. We called this technology PCS, which stands for PCR-mediated chromosome splitting (Sugiyama et al. 2005). This method had been further improved by adding an ARS sequence to the fragments so that they can be applied to small chromosomal regions that do not harbor an intrinsic ARS element (Yamagishi et al. 2008)

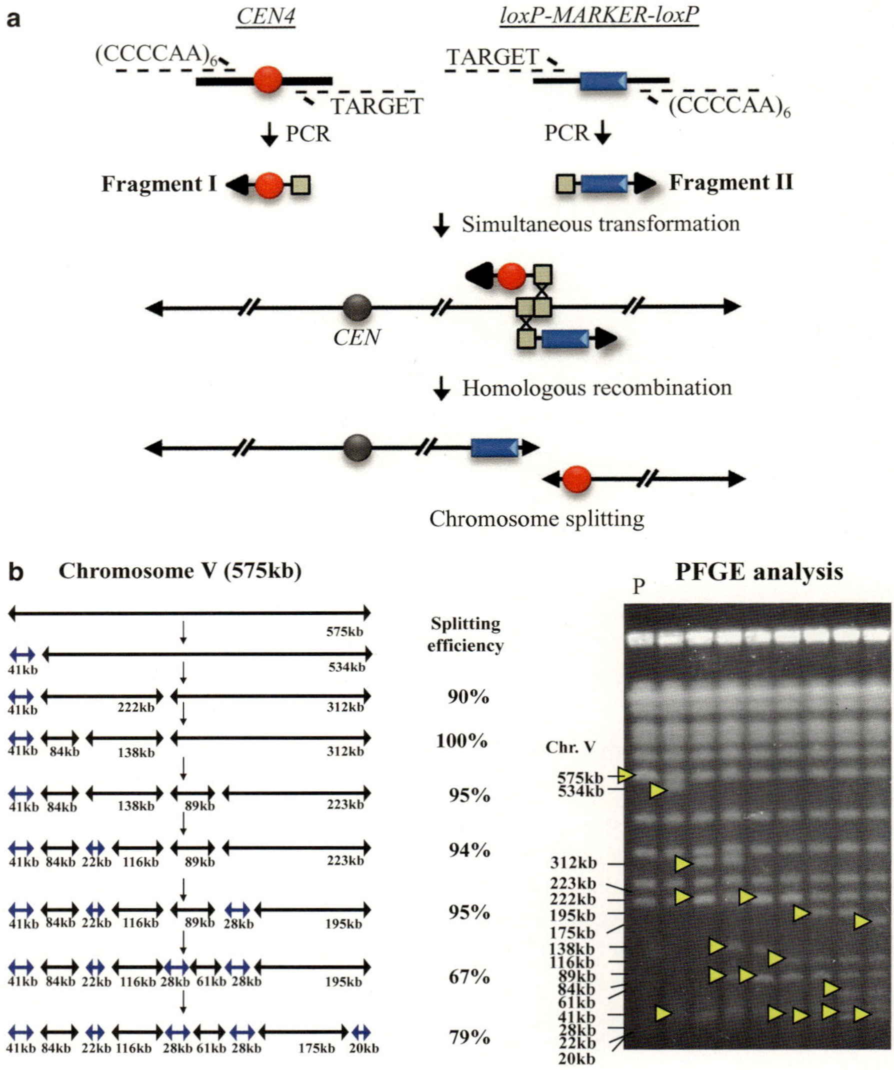

Fig. 5.2 Genome manipulation by PCS technology. **(a)** Chromosome splitting by the PCS method. The centromere (*CEN4*) and a selection marker (*loxP-MARKER-loxP*) are amplified to prepare fragments I and II by PCR using two primers: one with a sequence homologous to the target site (TARGET), and the other with six copies of the 5′-CCCCAA-3′ repeat sequence [$(CCCCAA)_6$]. To prepare fragments I and II with 400 bp of a long homology extension, a two-step PCR with overlap extension is performed. First, two target fragments of 400 bp with an additional 30-bp sequence are amplified by PCR. The resulting PCR products are then mixed with either a *loxP-MARKER-loxP* or *CEN4* construct sandwiched between the same 30-bp sequence and the six copies of the 5′-CCCCAA-3′ repeat sequence. These are prepared by PCR and pooled in advance as stock solutions (Sugiyama et al. 2005). Second, overlap-extension PCR is performed to generate fragments I and II with long homology extension. Simultaneous integration of the two fragments results in splitting of a chromosome at the target site into two monocentric chromosomes. **(b)** Sequential splitting of chromosome V by PCS. The chromosomes were resolved by pulsed-field gel electrophoresis (PFGE). The nine new chromosomes were derived from chromosome V (575 kb) by sequential splitting. *Arrowheads* in the PFGE analysis indicate chromosomes newly generated by sequential splitting. P, parental strain

To make PCS more efficient and to enable multiple splitting of a *S. cerevisiae* chromosome, two widely available techniques have been further incorporated into the PCS method. Because only a limited number of selection markers for transformation are available, the Cre/*loxP* site-specific recombination system (Güldener et al. 1996) was introduced to allow repeated splitting of the chromosome. The Cre/*loxP* system provides an excellent tool to eliminate an integrated marker from the chromosome and enables its reuse for multiple splitting. The other system is overlap-extension PCR for enhancing the frequency of homologous recombination (Dillon and Rosen 1990). Because a longer target sequence increases the frequency of homologous recombination at the target site, fragments I and II with ~400 bp of homologous sequence of the target region are prepared by two-step PCR: that is, a first PCR to amplify the target sequence, followed by a second overlap-extension PCR to fuse the first PCR product with a fragment consisting of the centromere and telomere seed or marker and telomere seed. These fragments are then used for transformation (see Fig. 5.2a legend for details).

Figure 5.2b shows an example of multiple splitting of chromosome V by PCS technology. The PCS method was applied eight successive times to split chromosome V into nine mini-chromosomes. High splitting frequencies (more than 67 %) were obtained in all splitting experiments. During the course of these splitting experiments, the marker was removed from strains with split chromosomes by expressing Cre recombinase, and the same marker was then used for successive transformations. Because chromosome-splitting technology is simple and efficient, permits repeated manipulation of a *S. cerevisiae* genome on a large scale, and has a wide range of applications such as deletion, fusion, transplantation, segmental duplication of a particular region of a chromosome, and genome reorganization (GReO) (described below) (Fig. 5.1), the PCS method presents a new way for breeding and analysis of genome function in *S. cerevisiae.*

5.2.2 PCR-Mediated Chromosome Deletion Technology

A new method allowing one-step direct deletion of an internal chromosomal region carrying many nonessential genes was developed by applying the chromosome-splitting technology in haploid *S. cerevisiae* (Sugiyama et al. 2008). The resulting method, designated PCR-mediated chromosomal deletion (PCD), consists of PCR preparation of fragments I and II and a single transformation per deletion event (Fig. 5.3a). To delete an internal region of chromosome, fragments I and II are simultaneously introduced into a yeast cell. When chromosome splitting occurs simultaneously at two distantly positioned sites on the same chromosome, the internal region between the target sites loses its centromere and telomeres, thereby disappearing from yeast cells during mitotic growth. Transformation of fragment II into a yeast cell can also induce a terminal deletion of a chromosome (Fig. 5.3b).

The performance of the PCD method was evaluated. Four internal and two terminal regions ranging from 16 to 38 kb in length and containing between 10 and 19 nonessential genes were subjected to deletion in haploid cells (Sugiyama et al.

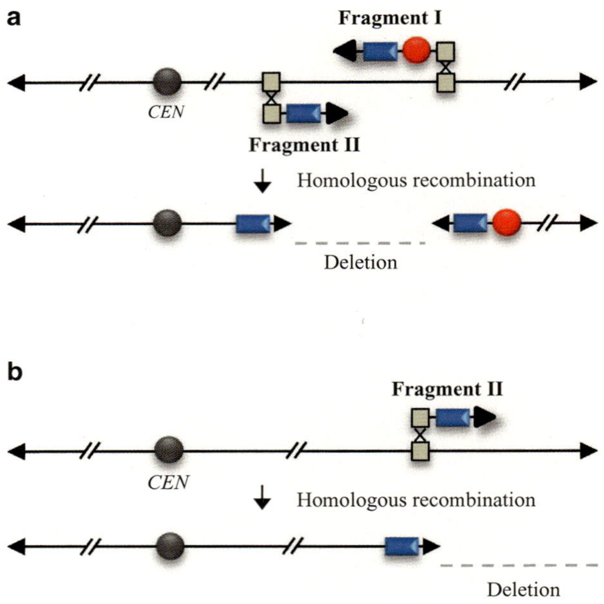

Fig. 5.3 Genome manipulation by PCR-mediated chromosome deletion (PCD) technology. **(a)** Internal deletion of a chromosome by the PCD method. The pair of fragments I and II consist of, respectively, *loxP-MARKER#1-loxP-CEN4* and *loxP-MARKER#2-loxP* flanked by 400 bp of target sequence and six copies of a 5′-CCCCAA-3′ repeat sequence. These two fragments are generated by two-step PCR with overlap extension as described in Fig. 5.2. Simultaneous integration of he two fragments results in deletion of a given chromosomal region coupled with chromosome splitting. **(b)** Terminal deletion of a chromosome by the PCD method. Transformation of fragment II results in deletion of a terminal region of chromosome

2008). The results showed that, of the six targeted regions, five were successfully eliminated at high efficiency. Unexpectedly, one internal region was not deleted although that region harbors only nonessential genes. It was subsequently found that the region contains a potentially synthetic lethal combination of genes essential for yeast growth, although none of the genes in the region is in itself essential. Because strains undergoing chromosomal and genomic alterations, especially the type of chromosomal deletion described herein, can be exploited as important resources for undiscovered phenotypes, the PCD method provides a powerful tool both for dissecting the function of the *S. cerevisiae* genome and for breeding yeast.

5.2.3 Genome Reorganization Technology

A haploid *S. cerevisiae* cell has 16 chromosomes ranging in size from 200 to 2,000 kb. When a chromosome is smaller than 50 kb (mini-chromosome), it is lost at relatively high frequency during mitotic cell divisions (Surosky et al.

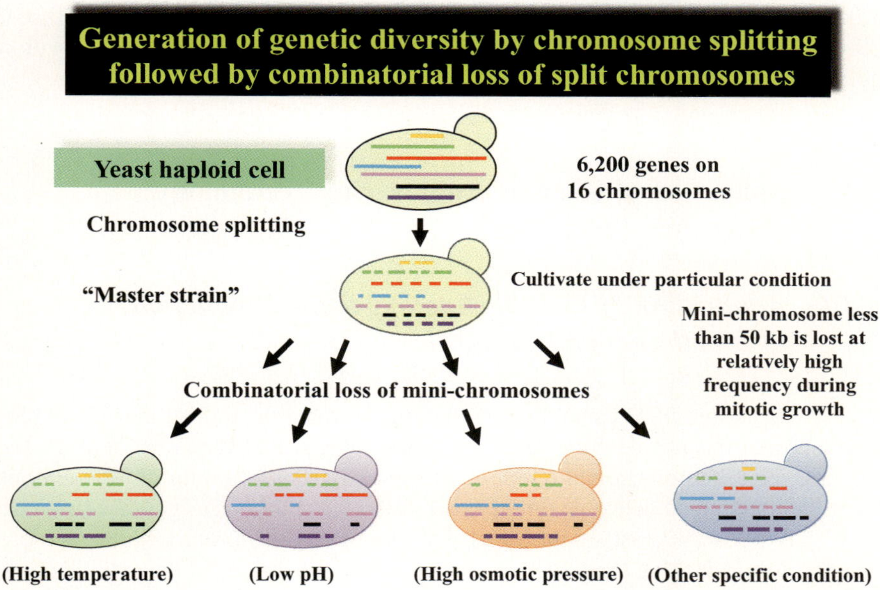

Fig. 5.4 Concept of GReO technology. A master strain with many mini-chromosomes is constructed by repeated chromosome splitting. Such mini-chromosomes can be lost at relatively high frequencies during cultivation if they are dispensable under particular conditions such as high temperature, low PH, or high osmotic pressure. On this basis, a cell population with vast genetic diversity is generated by the combinatorial loss of mini-chromosomes. Consequently, this cell population can be screened for cells showing desired phenotypes

1986). On the basis of this size-dependent chromosome loss, we predicted that yeast cells containing a particular set of mini-chromosomes would be able to generate novel genomic constitutions through combinatorial loss of the mini-chromosomes (Fig 5.4). By culturing yeast strains harboring mini-chromosomes under different conditions, this GReO approach could be used to generate a yeast cell population with a vast number of different subsets of mini-chromosomes adapted to the particular conditions employed. We designated this approach "GReO" technology.

To validate this idea, chromosomes IV and XI were split into 16 chromosomes by the PCS method (Ueda et al. 2012). Among the 16 chromosomes, seven were mini-chromosomes (less than 50 kb) and harbored no essential genes. The resulting strain was cultivated in nutrient-rich medium and analyzed for combinatorial loss of mini-chromosomes. During cultivation, cells with various combinations of mini-chromosomes arose, indicating that genomic reorganization could be achieved by splitting chromosomes to generate mini-chromosomes followed by combinatorial loss. This result indicates that yeast strains with countless combinations of the genome can be generated by GReO technology even if the strain subjected to this approach harbors only a few dozen mini-chromosomes (e.g., 1,000, 1,000,000, and

1,000,000,000 combinations will be generated by combinatorial loss of 10, 20, and 30 mini-chromosomes, respectively). Consequently, yeast cell populations with a great diversity of phenotypes can be generated.

5.3 Application of Genome Engineering Technology to Microbial Production

5.3.1 Application of PCD Technology to the Production of Bioethanol and Glycerol

We have exploited PCD technology in the improvement of strains for efficient bioethanol production by deleting yeast genes on a large scale (Murakami et al. 2007). We targeted 18 chromosomal regions that were adjacent to termini. These regions were predicted to be deletable on the basis of the *Saccharomyces* genome database (SGD) (http://www.yeastgenome.org/). By a repeated deletion procedure, we constructed strains lacking these regions in various combinations. In the strain lacking all 18 regions, approximately 5 % (~532 kb) of the genome was lost. The deletion strains did not show any significant growth defect on nutrient agar plate at 30 °C as compared with the parental strain. In addition, some of the strains showed almost the same stress tolerance as the parental strain under various conditions including high temperature, heat shock, acid, alkaline, ethanol, high osmolarity, and UV irradiation. Furthermore, one of the strains produced 1.8-fold more ethanol and another produced 2-fold more glycerol than the parental strain. The gene expression profiles revealed that physiological adjustment from fermentative to oxidative metabolism, including stimulation of mitochondrial function, does not occur in these strains. This alteration probably accounted for the accumulation of glucose metabolites in these strains. Thus, the PCD method can be a useful tool for breeding superior strains for microbial production by generating a novel cell with sets of genes exhibiting appropriately altered regulation for enhanced metabolite production.

5.3.2 Application of GReO Technology to Breeding Stress-Resistant Yeast

To validate the idea that GReO technology can be applied to breeding yeast strains that are useful for microbial production, we focused on creating an ethanol-tolerant yeast (ETY) because ethanol tolerance is a highly desirable trait for bioethanol production. Using the PCS method, we constructed a *S. cerevisiae* strain (SH6310) containing 31 chromosomes including 12 artificial mini-chromosomes (Park et al. 2012). Strain SH6310 was continuously cultivated in nutrient-rich medium containing 6–10 % ethanol to induce the random loss of mini-chromosomes. Ultimately, we obtained an ETY series of strains showing a specific growth rate more than tenfold higher in the presence of 8 % ethanol as compared with the parental strain. These strains also exhibited increased tolerance to 10 % ethanol. Investigation of the constitution of mini-chromosomes by PCR and Southern hybridization analysis suggested that loss of two particular mini-chromosomes (25 and 35 kb) derived from chromosome VIII and chromosome XIV, respectively, was responsible for the enhanced ethanol tolerance. These strains are promising for efficient bioethanol production because they might retain fermentation ability even in the presence of a high concentration of ethanol. Thus, GReO technology also provides a new tool for breeding novel yeasts useful for industrial applications.

5.4 Future Perspective

Herein, we have described some newly developed genome engineering technologies, namely PCS, PCD, and GReO, and their applications to microbial production. Other genome engineering techniques, which were not described owing to space limitations, include PCR-mediated chromosomal duplication (PCDup), genome replacement (GRep), and genome transplantation (GTpla) technology (Sugiyama et al. 2009) (Fig. 5.1). The aim of PCDup is to create segmental duplication of a desired chromosomal region, whereas that of GRep is to replace a particular chromosome region of one strain with the corresponding region from another (Sugiyama et al. 2006). GTpla is a rather straightforward application of PCS, PCD, and PCDup because desired chromosomal regions are prepared by PCS, PCD or PCDup and subsequently isolated by pulsed-field gel electrophoresis (PFGE) and transferred into other strains.

Because animal and plant chromosomes cloned on a yeast artificial chromosome (YAC) can be manipulated in a similar way and transplanted back into their respective host cells, the genome engineering technologies that we have described might be useful to analyze the function of particular regions of animal and plant chromosomes (Kim et al. 2005; Kim et al. 2006). Thus, advances in genome engineering technologies are expected to accelerate the breeding of novel strains for biotechnological purposes and also to reveal the functions of currently uncharacterized chromosomal regions in *S. cerevisiae* and possibly in other organisms as well.

References

Dillon PJ, Rosen CA (1990) A rapid method for the construction of synthetic genes using the polymerase chain reaction. Biotechniques 9:298–299

Güldener U, Heck S, Fiedler T, Beinhauer J, Hegemann JH (1996) A new efficient gene disruption cassette for repeated use in budding yeast. Nucleic Acids Res 24:2519–2524

Kim YH, Sugiyama M, Kaneko Y, Fukui K, Kobayashi A, Harashima S (2005) A versatile and general splitting technology for generating targeted YAC subclones. Appl Microbiol Biotechnol 69:65–70

Kim YH, Sugiyama M, Kaneko Y, Fukui K, Kobayashi A, Harashima S (2006) A polymerase chain reaction-mediated yeast artificial chromosome-splitting technology for generating targeted yeast artificial chromosomes subclones. Methods Mol Biol 349:103–115

Murakami K, Tao E, Ito Y, Sugiyama M, Kaneko Y, Harashima S, Sumiya T, Nakamura A, Nishizawa M (2007) Large scale deletions in the *Saccharomyces cerevisiae* genome create strains with altered regulation of carbon metabolism. Appl Microbiol Biotechnol 75:589–597

Murray AW, Claus TB, Szostak JW (1988) Characterization of two telomeric DNA processing reactions in *Saccharomyces cerevisiae*. Mol Cell Biol 8:4642–4650

Park AH, Sugiyama M, Harashima S, Kim YH (2012) Creation of an ethanol-tolerant yeast strain by genome reconstruction based on chromosome splitting technology. J Microbiol Biotechnol 22:184–189

Sugiyama M, Ikushima S, Nakazawa T, Kaneko Y, Harashima S (2005) PCR-mediated repeated chromosome splitting in *Saccharomyces cerevisiae*. Biotechniques 38:909–914

Sugiyama M, Yamamoto E, Mukai Y, Kaneko Y, Nishizawa Y, Harashima S (2006) Chromosome-shuffling technique for selected chromosomal segments in *Saccharomyces cerevisiae*. Appl Microbiol Biotechnol 72:947–952

Sugiyama M, Nakazawa T, Murakami K, Sumiya T, Nakamura A, Kaneko Y, Nishizawa M, Harashima S (2008) PCR-mediated one-step deletion of targeted chromosomal regions in haploid *Saccharomyces cerevisiae*. Appl Microbiol Biotechnol 80:545–553

Sugiyama M, Yamagishi K, Kim YH, Kaneko Y, Nishizawa M, Harashima S (2009) Advances in molecular methods to alter chromosomes and genome in the yeast *Saccharomyces cerevisiae*. Appl Microbiol Biotechnol 84:1045–1052

Surosky RT, Newlon CS, Tye BK (1986) The mitotic stability of deletion of chromosome III in yeast. Proc Natl Acad Sci USA 83:414–418

Ueda Y, Ikushima S, Sugiyama M, Matoba R, Kaneko Y, Matsubara K, Harashima S (2012) Large-scale genome reorganization in *Saccharomyces cerevisiae* through combinatorial loss of mini-chromosomes. J Biosci Bioeng 113:675–682

Yamagishi K, Sugiyama M, Kaneko Y, Nishizawa M, Harashima S (2008) Construction and characterization of single-gene chromosomes in *Saccharomyces cerevisiae*. J Biosci Bioeng 106:563–567

Chapter 6
Genome Design of Actinomycetes for Secondary Metabolism

Kiyoko T. Miyamoto and Haruo Ikeda

Abstract An industrial microorganism, *Streptomyces avermitilis*, which is the producer of the anthelmintic agent avermectin, has been constructed as a versatile model host for heterologous expression of genes encoding secondary metabolite biosynthesis. Several kinds of the entire biosynthetic gene clusters for secondary metabolites were successively cloned and introduced into a versatile model host, *S. avermitilis*. Most *S. avermitilis* transformants carrying the entire gene cluster produced metabolite(s) caused by the expression of the introduced biosynthetic gene clusters. Productivity of the metabolite(s) in some transformants of the versatile host was higher than that of the original producers.

Keywords Genome • Heterologous expression • Secondary metabolites • *Streptomyces*

6.1 Introduction

Microbial production represented by amino-acid fermentation, nucleotide fermentation, and industrial production of bioactive compounds has made large progress in Japan. The large bacterial phylum *Actinobacteria* is a diverse group of high G+C mol%, gram-positive bacteria. Notably, because the order *Actinomycetales* in the phylum undergoes much more complex morphological differentiation than the other bacteria, the process and regulatory mechanism of morphological development

K.T. Miyamoto • H. Ikeda (✉)
Kitasato Institute for Life Sciences, Kitasato University,
1-15-1 Kitasato, Minami-ku, Sagamihara, Kanagawa 252-0373, Japan
e-mail: ikeda@ls.kitasato-u.ac.jp

H. Anazawa and S. Shimizu (eds.), *Microbial Production: From Genome Design to Cell Engineering*, DOI 10.1007/978-4-431-54607-8_6, © Springer Japan 2014

have attracted many researchers and been intensively studied. The genus *Streptomyces*, a member of the order *Actinomycetales*, is characterized by not only its morphogenesis but also its ability to produce a great variety of secondary metabolites. Some *Streptomyces* strains are industrially important producers of bioactive compounds including antibiotics.

6.2 *Actinomycetales*, a Producer of Secondary Metabolites

The discovery of the anti-tuberculosis agent, streptomycin, from the culture broth of *Streptomyces griseus* by S.A. Waksman in 1944 represented an important milestone in natural product chemotherapy (Waksman 1953) and heralded the beginning of targeting the genus *Streptomyces* and related *Actinomycetales* microorganisms for antibiotics and biologically active substances. Although the variety of secondary metabolites produced by *Actinomycetales* has been known through screening study for 70 years after the discovery of streptomycin, the complete nucleotide sequence of *Actinomycetales* genome has proved the potential of the secondary metabolic proficiency of these microbes. In particular, the genomic analysis of industrial microorganisms will be of great use for understanding complex biosynthetic pathways of industrially important secondary metabolites and for elucidating mechanisms that enable efficient production of the metabolites at the level of industrial production. Until recently, most investigations of secondary metabolite production were limited in mutational analysis and transcriptional analysis of biosynthetic genes for certain secondary metabolites in its original producer. However, the construction of versatile hosts for heterologous expression of genes encoding secondary metabolite biosynthesis would open up a new strategy to elucidate cryptic metabolites encoded by a silent biosynthetic pathway.

Streptomyces avermitilis was discovered in 1977 by Kitasato Institute and produces an anthelmintic macrocyclic lactone, avermectin, which was found by collaborative work between Kitasato Institute and Merck Sharp and Dohme Research Laboratories before being applied to commercial use (Burg et al. 1979). We published the draft genome sequence of *S avermitilis* in 2001, which allowed the first genome-level glimpse into the extensive secondary metabolic dexterity of streptomycetes (Omura et al. 2001). In 2003, we reported the complete sequence of the *S. avermitilis* genome as a first case of an industrial microorganism for antibiotic production (Ikeda et al. 2003). One of the striking features of *Actinomycetales* is their ability to produce a wide variety of secondary metabolites, and a further important feature is having a metabolic system that enables production of secondary metabolites at the industrial level. In other words, *Actinomycetales* possesses a primary metabolism providing an adequate source of the energy and precursors required for secondary metabolism, and the supply from primary metabolism in *Actinomycetales* could be efficiently synchronized with the secondary metabolic pathway.

6.3 Properties of *Actinomycetales* Genomes

At the end of the past century, the genome sequence of *Mycobacterium tuberculosis*, the causative agent of tuberculosis, was first completed among members of the phylum *Actinobacteria* (Cole et al. 1998). Thereafter, at the beginning of this century, complete genome sequences of two *Streptomyces* species were published. To date, 11 *Streptomyces* genomes, *S. avermitilis* MA 4680 (Omura et al. 2001; Ikeda et al. 2003), *S. bingchenggensis* BCW-13 (Wang et al. 2010), *S. cattleya* NRRL 80574 (Barbe et al. 2011), *S. cattleya* DSM 46488, *S. coelicolor* A3(2) (Bentley et al. 2002), *S. flavogriseus* ATCC 33331, *S. griseus* IFO 13350 (Ohnishi et al. 2008), *S. hygroscopicus* subsp. *jinggangensis* 5008, *S. scabiei* 87.22 (Bignell et al. 2010), *S. violaceusniger Tu* 4113, and *Streptomyces* sp. SirexAA, have been completed and deposited in the NCBI database. Taken together with the draft genome sequences in the other public database [*S. albus* J1074, *S. clavuligerus* ATCC 27064 (Medema et al. 2010), *S. ghanaensis* ATCC 14672, *S. griseoflavus* Tu4000, *S. hygroscopicus* ATCC 53653, *S. lividans* TK24, *S. pristinaespiralis* ATCC 25486, *S. roseosporus* NRRL 11379, *S. roseosporus* NRRL 15998, *S. sviceus* ATCC 29083, *S. venezuelae* ATCC 10712 (Pullan et al. 2011), *S. viridochromogenes* DSM 40736, *Streptomyces* sp. AA4, *Streptomyces* sp. C, *Streptomyces* sp. E14, *Streptomyces* sp. Mg1, *Streptomyces* sp. SPB74, and *Streptomyces* sp. SPB78), genomic information of 29 *Streptomyces* species is available. Furthermore, more than 170 *Actinobacteria* genomes have been deposited in the NCBI database to date.

In contrast to most other eubacterial genomes, the chromosomes of *Streptomyces* microorganisms form linear structures that are predominant genetic elements in eukaryotes. Both ends, containing unique terminal-inverted repeats (TIRs), bind basic terminal proteins. Among members of the phylum *Actinobacteria*, all *Streptomyces* strains, *Kitasatospora setae* (Ichikawa et al. 2010), and a few *Rhodococcus* microorganisms have linear chromosomes (McLeod et al. 2006). Another principal feature of *Streptomyces* chromosomes is its genetic instability, resulting in loss of ability of morphological differentiation and production of secondary metabolites. Although complete genome sequences of *Streptomyces* microorganisms have not elucidated a major determinant of the genetic instability, one of the causes of the instability is likely to be the structure of both ends of the linear chromosomes. Both ends of chromosomes of *Streptomyces* and *Kitasatospora* microorganisms consist of TIRs containing the most distal 180-bp sequence with potential to form unique and complex secondary structures (Hung et al. 1998). The sizes of the TIRs range from 49 bp of *S. avermitilis* MA 4,680 to 132,910 bp of *S. griseus* IFO 13350. The longer TIRs may have a higher frequency of DNA rearrangements involving gene duplication, elimination, and acquisition between both TIRs. In fact, *S. griseus* IFO 13350 and *K. setae* KM-6054, containing more than 100-kb TIRs at both ends, showed a high degree of genetic instability. In contrast, before the availability of the genome sequence, *S. avermitilis* had been known to show relatively stable phenotypical properties including morphological differentiation and production of secondary metabolites such as avermectins and oligomycin.

Table 6.1 Genetic instability of genome-sequenced *Streptomycetaceae* microorganisms containing linear chromosomes under high-temperature growth

Microorganism	Terminal-inverted repeats (TIR) (bp)	Appearance of *bld* mutants[a]	
		28 °C (%)	37 °C (%)
Streptomyces avermitilis MA-4680	49	1/2,675 (0.03)	4/2,567 (0.15)
S. coelicolor A3(2)	21,653	9/2,056 (0.43)	197/2,781 (7.15)
S. griseus IFO 13350	132,910	26/2,138 (1.2)	893/2,013 (44.4)
Kitasatospora setae KM-6054	127,148	54/2,064 (2.6)	908/2,367 (38.4)

[a]About 200 spores of each microorganism were inoculated onto an inorganic salts-starch agar plate After each plate was incubated at 28 °C or 37 °C for 6 days, bald (*bld*) mutants were counted

Whole genome sequences of *S. avermitilis* revealed that the TIRs of this strain are a mere 49 bp in size. Mutants affecting morphological differentiation sometimes arise under stressful conditions such as high temperature and hyperosmolarity. An examination of the appearance of bald mutants, which fail to form aerial hyphae on the solid medium in high-temperature incubation, demonstrated that *S. avermitilis* generated bald mutants with a quite low frequency compared with other genome-sequenced *Streptomycetaceae*, *S. coelicolor* A3(2), *S. griseus* IFO 13350, and *K. setae* KM-6054 (Table 6.1). These findings indicated that genetic instability could be closely related to the length of the TIRs on both ends of chromosome.

6.4 Construction of Large-Deletion *S. avermitilis* for the Heterologous Expression of Biosynthetic Gene Clusters for Exogenous Secondary Metabolites

In the heterologous expression of exogenous gene clusters for secondary metabolism, a competition between exogenous and endogenous biosynthetic pathways for their common precursors and energy could occur. This possibility was confirmed in the synthesis of oligomycin and avermectin, both of which are major endogenous polyketide compounds in *S. avermitilis*. Because the biosynthetic pathways of these two compounds share the same precursors, malonyl-CoA and methylmalonyl-CoA, the genetically blocked mutant in avermectin biosynthesis resulted in loss of avermectin production and overproduction of oligomycin (Tanaka et al. 2009). Therefore, the deletion of biosynthetic gene clusters of endogenous major secondary metabolites is preferred for the synthesis of exogenous secondary metabolites by heterologous expression of the biosynthetic gene cluster. In general, the *Streptomyces* chromosomes contain a centrally located replication origin (*oriC*), an internal core region of 6–6.5 Mb in which essential genes are located, and 1-Mb subtelomeric regions at both ends of the chromosome (Ikeda et al. 2003; Ohnishi et al. 2008). The subtelomeric regions contain strain-specific genes as well as genes encoding secondary metabolite biosynthesis, whereas no essential genes are located in these two regions. In contrast to other *Streptomyces* genomes, the chromosome of

S. avermitilis was asymmetrical in structure: the *oriC* on the chromosome is shifted 780 kb away from the center toward the right end and subtelomeric regions are of two different sizes, 2 Mb at the left end and 0.5 Mb at the right end, respectively (Ikeda et al. 2003). The left 2-Mb subtelomeric region contains many biosynthetic gene clusters for secondary metabolites including the endogenous major metabolites avermectin and filipin. From these observations, we expected that the deletion of the large left-subtelomeric region of *S. avermitilis* would not affect either growth or the primary metabolism that is essential for the supply of precursors for secondary metabolism and would confer loss of production in many secondary metabolites including endogenous major metabolites. Therefore, we aimed to construct the large-deletion mutant of *S. avermitilis* (Komatsu et al. 2008, 2010).

To delete more than the 1.4-Mb segment from the left subtelomeric region of the *S. avermitilis* genome, we used two complementary strategies. The first approach used general homologous recombination involving two homologous segments containing *sav6* and *sav1205*, respectively (Fig. 6.1a). The desired mutants should have a deletion of 1,487,159 bp spanning from 7,733 bp to 1,494,899 bp. This large-deletion event turned out to have taken place at extremely low frequency, and almost all the progeny that were generated by this homologous recombination strategy contained irregular deletions. Only two correct deletion mutants, designated SUKA2, were isolated from three trial experiments.

The second approach involved site-specific recombination systems. When the site-specific recombination using the Flp/*FRT* system of a 2-μm plasmid of *Saccharomyces cerevisiae* (Broach and Hichs 1980) was examined, we found that the expression of *flp*-encoding site-specific recombinase in *S. avermitilis* gave a significant negative effect on growth and most progeny able to grow were deficient in the *flp* gene. However, successful site-specific recombination was conducted using the Cre/*loxP* system of bacteriophage P1 (Strenberg et al. 1981). The *cre* gene encoding site-specific recombinase was expressed under the control of two distinct promoters: a constitutive promoter of *ermE*, an erythromycin resistance gene in *Saccharopolyspora erythraea* NRRL 2338, and an inducible promoter of *xylA* encoding a xylose isomerase involved in xylose metabolism in *S. avermitilis*. Expression of the *cre* gene under the control of either promoter yielded recombination between two *loxP* sites with no effect on growth of *S. avermitilis*. For construction of the large-deletion mutant, two *loxP* sites were first introduced in the same orientation into the *S. avermitilis* wild-type strain at 79,454 nt and 1,595,564 nt, respectively, by stepwise homologous recombination events. The desired deletion mutants were efficiently generated after induction of *cre* expression, and all 24 of the resultant progeny that were tested harbored the identical 1,516,020-bp deletion. These desirable large-deletion mutants were designated as SUKA3 (Fig. 6.1b). The two prototype mutants, SUKA2 and SUKA3, were further modified by stepwise deletion of specific regions of the genome such as gene clusters for endogenous secondary metabolites located in the core region of the chromosome, and consequently a series of large-deletion mutants were constructed (Fig. 6.2). All the large-deletion derivatives were able to grow on minimum medium composed of glucose and inorganic salts, suggesting that the regions deleted from *S. avermitilis*

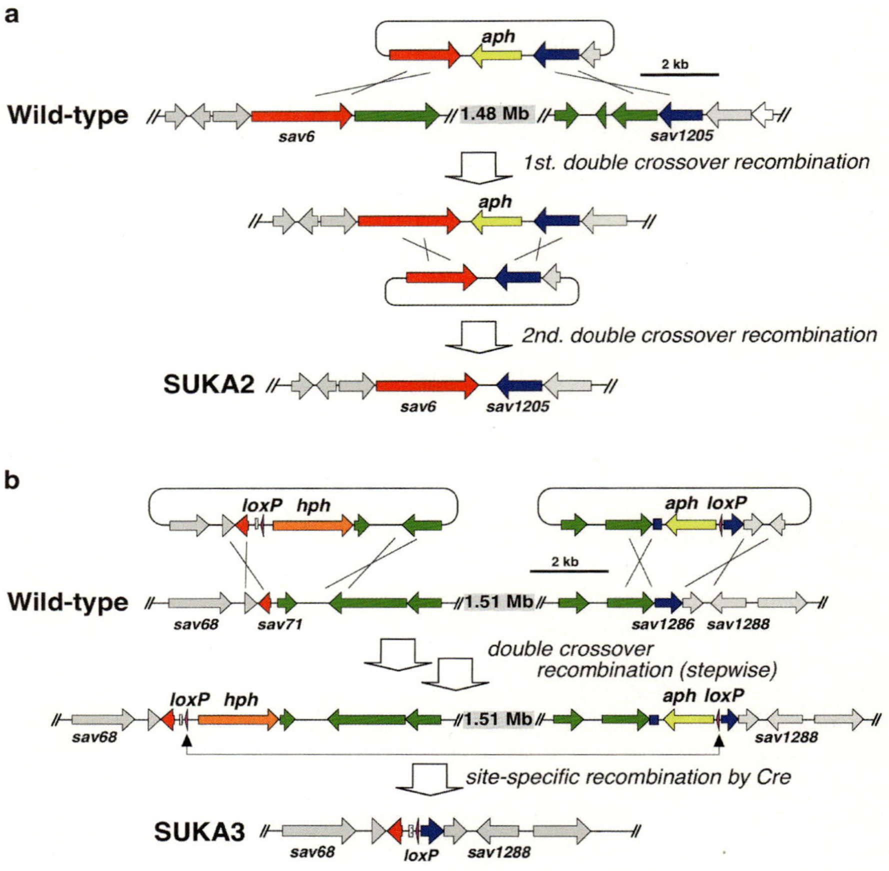

Fig. 6.1 Strategy for construction of large-deletion mutants, *Streptomyces avermitilis* SUKA2 and SUKA3, using homologous recombination (**a**) and site-specific recombination using Cre-*loxP* (**b**)

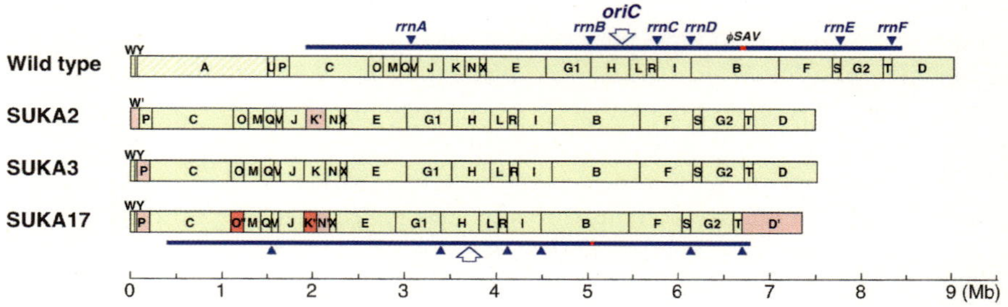

Fig. 6.2 *Ase*I physical maps of wild-type *S. avermitilis* and its large-deletion mutants. *Shaded boxes* on the physical maps indicate the introduction of deletion(s). *Thick blue bars* correspond to the central core region of the *Streptomyces* genome. *Open arrows* and *filled triangles* indicate the replication origin and 16S-23S-5S rRNA operon (*rrn*), respectively. *Red bars* indicate the locus of putative prophage ϕSAV

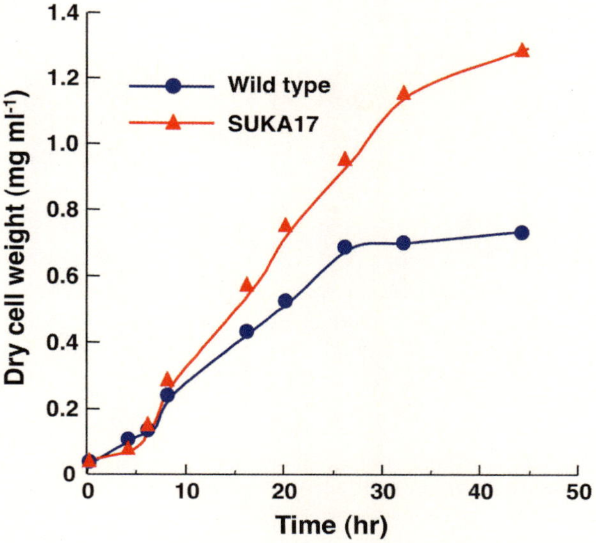

Fig. 6.3 Growth of wild-type *S. avermitilis* and its large-deletion mutant SUKA17 in liquid culture. The microorganisms were grown at 28 °C in synthetic medium containing glucose and casamino acids. The culture broth was sampled, and dry cell weight was measured

chromosome included no essential genes. Intriguingly, a large-deletion mutant, SUKA17, grew faster on a sporulation medium and showed enhanced growth rate and biomass in liquid culture compared with wild-type *S. avermitilis* (Fig. 6.3), even though the chromosome size of SUKA17 is 7,352,064 bp, corresponding to 81.46 % of the wild-type chromosome (9,025,608 bp).

6.5 Heterologous Expression of Exogenous Gene Clusters for Secondary Metabolites

Introduction of DNA into *S. avermitilis* occurs at high frequency, and well-established techniques using phage-derived integrase are available for site-specific integration of the large DNA fragment into the *S. avermitilis* chromosome. Until now, many exogenous biosynthetic gene clusters have been introduced into *S. avermitilis* large-deletion mutants, and efficient production of the corresponding foreign metabolites has been successfully conducted in the *S. avermitilis* transformants. Here we describe three examples: streptomycin, an anti-tuberculosis agent from *S. griseus* IFO 13350, the β-lactam antibiotic cephamycin C from *S. clavuligerus* ATCC 27064, and the broad-spectrum antibiotic chloramphenicol from *S. venezuelae* ATCC 10712. Fragments of about 40 kb containing the entire set of genes involving biosynthesis of streptomycin, cephamycin C, and chloramphenicol were inserted into the integration vector, respectively, and the desired recombinant plasmids were introduced into both wild-type *S. avermitilis* and the large-deletion mutant SUKA17. The resultant transformants produced streptomycin, cephamycin C (Komatsu et al. 2010), and chloramphenicol (Komatsu et al. 2013), respectively

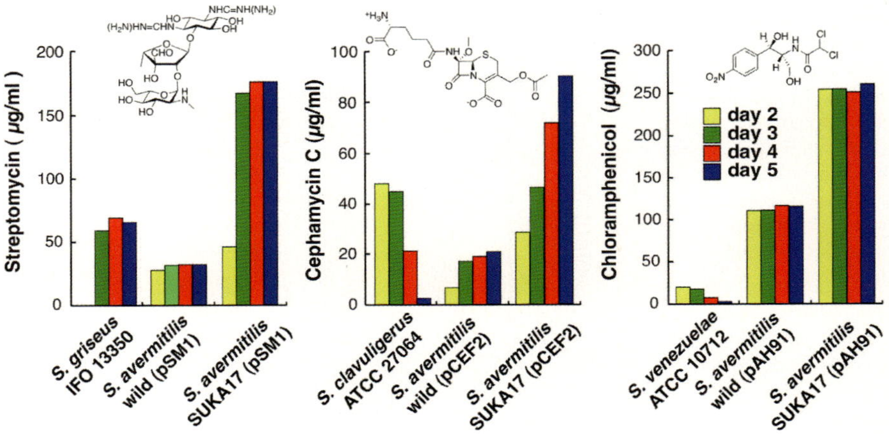

Fig. 6.4 Production profiles of streptomycin, cephamycin C, and chloramphenicol in original producers *S. griseus* IFO 13350, *S. clavuligerus* ATCC 27064, and *S. venezuelae* ATCC 10712, respectively, and in transformants of wild-type *S. avermitilis* and its large-deletion mutant SUKA17 carrying each biosynthetic gene cluster: pSM1, pCEF2 (Komatsu et al. 2010), and pAH91 (Gomez-Escribano and Bibb 2011), respectively

(Fig. 6.4). In all three cases, the productivity of exogenous metabolite in the *S. avermitilis* SUKA17 transformants was higher than that of the wild-type transformant. Because SUKA17 lacks the biosynthetic gene clusters for the principal endogenous secondary metabolites of *S. avermitilis*, the natural precursors and biochemical energy of the host are apparently efficiently used in the biosynthesis of exogenous metabolites. Interestingly, the productivity of streptomycin, cephamycin C, and chloramphenicol in the SUKA17 transformants was higher than those of original producing microorganisms, *S. griseus* IFO 13350, *S. clavuligerus* ATCC 27064, and *S. venezuelae* ATCC 10712, respectively. Recently some *S. coelicolor* A3(2) strains lacking biosynthetic gene clusters for major endogenous secondary metabolites were constructed and a cosmid containing the biosynthetic gene cluster for chloramphenicol was introduced into the *S. coelicolor* A3(2) mutants. The resultant transformants were successful to produce chloramphenicol; however, the production level of chloramphenicol in the *S. coelicolor* transformants were 16 % of that in *S. avermitilis* SUKA17 carrying the biosynthetic gene cluster for chloramphenicol (Gomez-Escribano and Bibb 2011).

More than ten exogenous biosynthetic gene clusters for secondary metabolites have been introduced into the genetically engineered *S. avermitilis*, and almost all the *S. avermitilis* transformants produced the corresponding exogenous metabolites. A few transformants were unable to produce metabolites, but their production was sometimes restored by the expression of biosynthetic genes using an alternative promoter or the expression of a regulatory gene in the gene cluster that controls the expression of biosynthetic genes in the cluster using an alternative promoter. An understanding of the higher-level regulatory system that controls the expression of pathway-specific regulatory genes for individual biosynthetic gene clusters should be essential to control and improve the secondary metabolite production (Kitani et al. 2011), because only a few cases have been studied to date. Although we confirmed the feasibility of using large-deletion derivatives of *S. avermitilis* as a host for heterologous expression of genes encoding secondary metabolite biosynthesis, the *S. avermitilis* derivatives might not be suitable for all cases. Therefore, we are expecting to construct genome-engineered *Streptomyces* microorganisms other than *S. avermitilis*.

References

Barbe V et al (2011) Complete genome sequence of *Streptomyces cattleya* NRRL 8057, a producer of antibiotics and fluorometabolites. J Bacteriol 193:5055–5056

Bentley SD et al (2002) Complete genome sequence of model of actinomycete *Streptomyces coelicolor* A3(2). Nature (Lond) 417:141–147

Bignell DR et al (2010) *Streptomyces scabies* 87–22 contains a coronafacic acid-like biosynthetic cluster that contributes to plant-microbe interactions. Mol Plant Microbe Interact 23:161–175

Broach JR, Hichs JB (1980) Replication and recombination functions associated with the yeast plasmid, 2μ circle. Cell 21:501–508

Burg RW et al (1979) Avermectins, new family of potent anthelminthic agents: producing organism and fermentation. Antimicrob Agents Chemother 15:361–367

Cole ST et al (1998) Deciphering the biology of *Mycobacterium tuberculosis* from the complete genome sequence. Nature (Lond) 393:537–544

Gomez-Escribano JB, Bibb MJ (2011) Engineering *Streptomyces coelicolor* for heterologous expression of secondary metabolite gene clusters. Microb Biotechnol 4:207–215

Hung CH et al (1998) The telomeres of *Streptomyces* chromosomes contain conserved palindromic sequences with potential to form complex secondary structures. Mol Microbiol 28:905–916

Ichikawa N et al (2010) Genome sequence of *Kitasatospora setae* NBRC 14216^T: an evolutionary snapshot of the family *Streptomycetaceae*. DNA Res 17:393–406

Ikeda H et al (2003) Complete genome sequence and comparative analysis of the industrial microorganism *Streptomyces avermitilis*. Nat Biotechnol 21:526–531

Kitani S et al (2011) Avenolide, a *Streptomyces* hormone controlling antibiotic production in *Streptomyces avermitilis*. Proc Natl Acad Sci USA 108:16410–16415

Komatsu M et al (2008) Identification and functional analysis of genes controlling biosynthesis of 2-methylisoborneol. Proc Natl Acad Sci USA 105:7422–7427

Komatsu M et al (2010) Genome-minimized *Streptomyces* host for the heterologous expression of secondary metabolism. Proc Natl Acad Sci USA 107:2646–2651

Komatsu M et al (2013) Engineered *Streptomyces avermitilis* host for heterologous expression of biosynthetic gene cluster for secondary metabolites. ACS Synth Biol 2:384–396

McLeod MP et al (2006) The complete genome of *Rhodococcus* sp. RHA1 provides insights into a catabolic powerhouse. Proc Natl Acad Sci USA 103:15582–15587

Medema MH et al (2010) The sequence of a 1.8-Mb bacterial linear plasmid reveals a rich evolutionary reservoir of secondary metabolic pathways. Genome Biol Evol 2:212–224

Ohnishi Y et al (2008) Genome sequence of the streptomycin-producing microorganism *Streptomyces griseus* IFO 13350. J Bacteriol 190:4050–4060

Omura S et al (2001) Genome sequence of an industrial microorganism *Streptomyces avermitilis*: deducing the ability of producing secondary metabolites. Proc Natl Acad Sci USA 98:12215–12220

Pullan S et al (2011) Genome-wide analysis of the role of GlnR in *Streptomyces venezuelae* provides new insights into global nitrogen regulation in actinomycetes. BMC Genomics 12:175

Strenberg N et al (1981) Bacteriophage P1 site-specific recombination. II. Recombination between loxP and the bacterial chromosome. J Mol Biol 150:487–507

Tanaka Y et al (2009) Antibiotic overproduction by *rpsL* and *rsmG* mutants of various actinomycetes. Appl Environ Microbiol 75:4919–4922

Waksman SA (1953) Streptomycin: background, isolation, properties, and utilization. Science 118:259–266

Wang XJ et al (2010) Genome sequence of the milbemycin-producing bacterium *Streptomyces bingchenggensis*. J Bacteriol 192:4526–4527

Part III
Application of Omics Information and Construction of Mutant Libraries

Chapter 7
Application Methodology of Whole Omics Information

Myco Umemura and Masayuki Machida

Abstract Sequencing genomes and successive functional genomic analyses are powerful means to explore cellular function, including metabolism. Transcriptome analysis has allowed deciphering the mechanism to bring out the potential of degrading raw materials by traditional solid-state cultivation (SSC). Secondary metabolites biosynthesis (SMB) is also an attractive target for the study of filamentous fungi. A significantly larger number of SMB genes than expected has been found by sequencing their genomes to date. In spite of their value, weak and unstable expression and limited conditions for expression and high diversity even between closely related species have made the analysis of SMB genes difficult. Several bioinformatics tools to identify clustered SMB genes have been developed depending on the existence of so-called core genes known to be responsible for synthesizing the basic structure of polyketides and non-ribosomal peptide. Comparative genomics of filamentous fungi indicated enrichment of SMB genes on non-syntenic blocks (NSBs) and suggested large number of SMB gene clusters without the core genes remaining unaddressed. Combining computational analyses of omics information using next-generation sequencer (NGS), DNA microarray, LC/MS, and so on will open the possibility of exploring SMB genes possessing novel biosynthesis functions.

Keywords Comparative genomics • Filamentous fungi • Next-generation sequencing • Non-syntenic blocks • Secondary metabolism

---

M. Umemura • M. Machida (✉)
National Institute of Advanced Industrial Science and Technology (AIST),
Central 6, Higashi 1-1, Tsukuba, Ibaraki 305-8566, Japan
e-mail: m.machida@aist.go.jp

H. Anazawa and S. Shimizu (eds.), *Microbial Production: From Genome Design to Cell Engineering*, DOI 10.1007/978-4-431-54607-8_7, © Springer Japan 2014

7.1 Genome Sequencing of Filamentous Fungi

7.1.1 Sequencing Strategy

Since the sequencing of the *Saccharomyces cerevisiae* genome (Goffeau et al. 1996; Goffeau et al. 1997) was completed in 1996, microorganisms not regarded as a model organism emerged as the next targets for sequencing. At the time when this discussion took place, success of a whole genome shotgun approach for the *Haemophilus influenzae* bacterium (Roach et al. 1995) encouraged researchers to quickly sequence the whole genome of microorganisms including those used in industry to keep and further raise their competitiveness. Although a conventional approach such as BAC-by-BAC sequencing required a long time for completion even before sequencing, the novel shotgun approach could generate large sequence information directly in a short period. The resulted sequences would be a draft or even worse; the information, however, should be of great benefit to industry at least to explore useful genes and to find candidate genes responsible to a particular phenotype by a simple comparative genomic approach. Once a draft sequence was obtained, modeling, functional analysis, and annotation of genes on the genome could be quickly performed by using computational technologies that had advanced dramatically in the past 10 years.

7.1.2 Sequencing Targets

In the eukaryotic research field, fungal genomes were thought to be the next industrially promising targets, same as the human genome. Their genome sizes are generally two to three times larger than that of yeast, and GC content and other characteristics of ribosomal DNA, telomere, centromere, and so on seemed similar to those of yeast. The first minimum tiled cosmid library of *Aspergillus nidulans* was generated in 1997 by Prade et al. (1997). After extensive discussion regarding the first target to be sequenced for filamentous fungi in conferences and workshops, mostly held in the United States, *A. nidulans* and *Neurospora crassa* were selected as candidates.

Sequencing of the *A. nidulans* genome was launched by the cosmid-by-cosmid sequencing approach by Keller et al. at Texas A&M University in 1998. However, in August 1998, just after the announcement of *A. nidulans* sequencing initiation by the academic sector, it was announced that the *A. nidulans* genome sequencing had been completed by a whole genome shotgun approach by Cereon Genomics LLC. Although the sequence from Cereon Genomics was not made public at that time, the sequencing effort of *A. nidulans* at Texas A&M University was suspended. Completion of sequencing the *Aspergillus fumigatus* genome was also announce by a company, Elitra Pharmaceuticals, in 2001.

Several years later, sequencing the *A. nidulans* genome was reinitiated by Broad Institute, to which the draft sequence composed of 3.5X Sanger reads were donated from Cereon Genomics. As a result, three *Aspergillus* genomes, from *Aspergillus oryzae*, *A. fumigatus*, and *A. nidulans*, were sequenced almost in parallel in 2003. This accomplishment did not simply mean acquisition of three filamentous fungal genomes but also allowed the first detailed comparative analysis of eukaryotic genomes of the microorganisms that were known to produce various secondary metabolites. Sequencing the three *Aspergillus* genomes was completed in 2005 by the collaboration of a large number of researchers worldwide (Galagan et al. 2005; Machida et al. 2005; Nierman et al. 2005).

7.2 Expressed Sequence Tag (EST) Analysis of *Aspergillus oryzae*

Before sequencing the *A. oryzae* genome, more than 16,000 ESTs prepared from various cultivation conditions including rich complete medium, starved condition, and solid-state cultivations (SSC) were accumulated (Akao et al. 2007), the last condition of which was widely used in the Japanese fermentation industry. By using the cDNA clones that had been sequenced, the DNA microarray harboring approximately 2,000 cDNA probes was designed and prepared. Although only the genes with relatively high expression strength were used for the preparation by this approach, the cDNA microarray allowed the analysis of key features of primary metabolism of *A. oryzae* in SSC. Interestingly, it was found that the cell were starved in SSC in spite of a huge amount of raw materials surrounding the *A. oryzae* mycelia. This finding was interpreted that the cells could not effectively take up nutrition, probably because of low water activity.

7.3 Comparative Genomics

Comparative analysis of the three *Aspergillus* genomes showed clear differences in genome size among the three species; that is, the *A. oryzae* genome was approximately 25 % larger than the other two genomes. Detailed analysis of the genomes revealed that *A. oryzae* had extra genetic elements in a mosaic manner throughout its genome, which were the so-called non-syntenic blocks (NSBs) (Fig. 7.1). Interestingly, functional analysis of the genes revealed localization of secondary metabolism and secretory hydrolase genes highly enriched on NSBs. Furthermore, transcriptome analyses by EST sequencing and with DNA microarray harboring oligonucleotides probes designed from all the approximately 12,000 genes indicated that transcriptional expression of the NSB genes was generally weaker than that of the genes on syntenic blocks (SBs) (Machida et al. 2005) (Fig. 7.1).

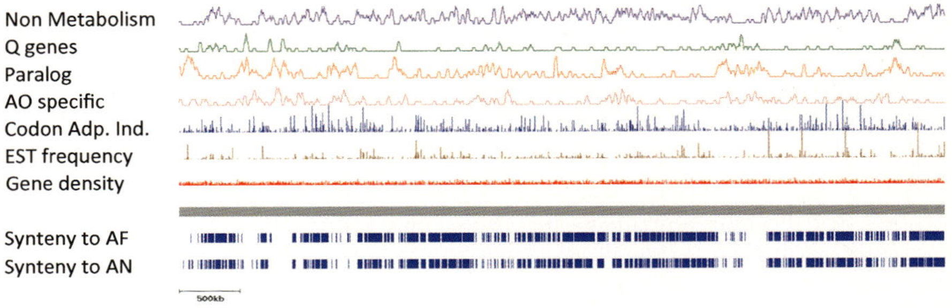

Fig. 7.1 Functional distribution of genes on a chromosome. Gene densities for non-metabolism, Q genes (secondary metabolism), Paralog, AO (*Aspergillus oryzae*) specific (genes), value of codon adaptation index, expressed sequence tag (EST) frequency, gene density (for all genes), synteny to AF [*blue*, synteny block (SB); *white*, non-syntenic block (NSB)], and synteny to AN (same as synteny to AF) on the *A. oryzae* chromosome I are plotted

Comparison of transcriptional expression profiles indicated a clear difference in the manner of regulation between NSB and SB genes (Tamano et al. 2008). Transcription of the SB genes, which played a basic role in cellular function, was found to be induced in average at 42 °C as compared to 30 °C, presumably because of heat shock response. In contrast, NSB genes were induced on average when the expression in SSC was compared to that in liquid medium. Most highly induced NSB genes in SSC were those encoding secretory hydrolases and transporters, which were involved in degradation and successive uptake of raw materials in fermentation. Interestingly, the *A. oryzae* genome showed expansion of those genes that were highly accumulated on NSBs. This finding strongly suggested that SSC was an ideal method to apply the potential of *A. oryzae* to efficiently utilize raw materials. This style of fermentation is thought to have originated 3,000–2,000 years ago in China and was imported into Japan at that period (Murakami 1980; Gomi and Abe 2007). It was surprising that such a rational method for highly efficient fermentation had been established without even the knowledge of microbiology. Further, this example clearly indicates that genomics is a powerful tool to decipher the secrets hidden in traditional excellent technologies that have been optimized over their long history.

7.4 Exploring Genes for Active Degradation of Biodegradable Plastics

The first practical application of the *A. oryzae* DNA microarray was to identify the genes responsible for decomposing biodegradable plastic. In contrast to ordinary plastics, biodegradable plastics can be easily degraded in nature after use. However, the amount of plastics, once widely consumed, would be too excessive for nature to degrade in a limited area. Abe et al. found that *A. oryzae* could grow by degrading the plastic. They analyzed the transcriptome with the cDNA microarray harboring 2,000 cDNAs as probes. *A. oryzae* was grown in media containing pelleted/emulsified polybutylene succinate co-adipate (PBSA), 1,4-butanediol, and succinate, and the genes induced specifically in the presence of PBS but not in the latter two compounds were screened. Consequently, they identified two genes strongly induced by the addition of PBSA, encoding cutinase (CutL1) and hydrophobin (RolA). The former protein degrades PBSA by its esterase activity, which was confirmed by the degradation of PBSA membrane with purified CutL1 in vitro. The latter protein, RolA, did not have enzymatic activity for degradation itself but was found to enhance the CutL1 activity. RolA, known as a detergent protein responsible for the water-repellent surface of aerial hyphae, interestingly, was found to recruit CutL1 to the surface of solid PBSA, enhancing the enzymatic activity. This RolA function in terms of PBSA degradation was revealed by using the quartz crystal microbalance (QCM) method, FITC-labeled RolA, and so on (Maeda et al. 2005; Takahashi et al. 2005). CutL1 might have been detected by using conventional biochemical methods by measuring the enzymatic activity of PBSA degradation; however, detection of RolA would be far more difficult than the detection of enzymatic activity by the conventional biochemical approach. This example of analysis clearly indicated the potential of genomics opening a novel approach to explore genes responsible for biological function of interest.

7.5 Identification of Kojic Acid Biosynthesis Genes

Kojic acid (5-hydroxy-2-(hydroxymethyl)-4-pyrone, KA) was first isolated from a koji culture, SSC of steamed rice inoculated with *A. oryzae* by Saito more than 100 years ago in Japan (Saito 1907). It was used as an antibiotic in the 1950s and is now being used in cosmetics as a skin-whitening reagent. In spite of its industrial value, neither its metabolic pathway nor the genes responsible for its biosynthesis had been known for longer than 100 years. The extensive analysis of condition for KA production revealed that KA was produced 4–7 days after cultivation without addition of nitrate ion in the medium.

Three transcriptome analyses, each of which was composed of a pair of conditions with remarkable difference in KA productivity, using a DNA microarray composed of oligonucleotide probes for all the genes predicted from the *A. oryzae* genome, showed candidates for genes responsible for KA production. However, no genes that were commonly induced in all the three experiments were detected. Therefore, the candidate genes were disrupted one by one followed by the measurement of KA productivity. Of the 20 candidates observed for disruption, 2 genes, AO090113000136 and AO090113000138, showed significant loss of KA productivity when disrupted. Not an annotated gene with significant similarity to AO090113000136 in the database, but only a motif known as FAD-dependent oxidoreductases (Pfam ID: PF01266) was found. Significantly decreased but still remaining KA productivity by the disruption of AO090113000138 having

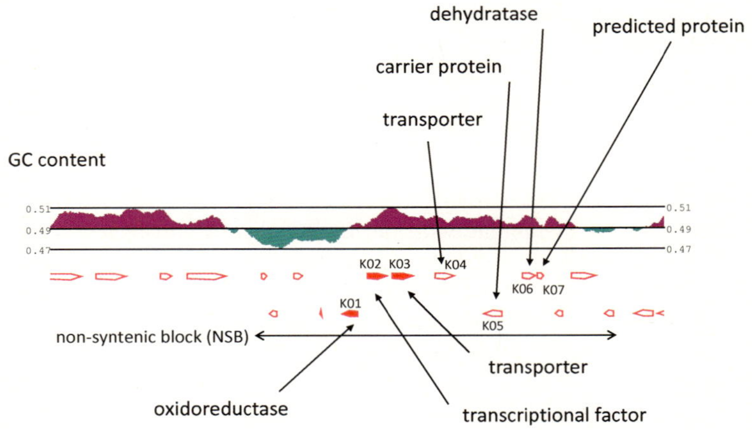

Fig. 7.2 Gene cluster responsible for kojic acid (KA) production. Three genes responsible for KA production and surrounding genes are shown together with GC content and the position of the non-syntenic blocks (NSB)

homology to a known transporter gene was thought to be caused by the loss of secretion of KA synthesized inside the cell. Between the two genes, a gene with fungal transcriptional factor motif, AO090113000137, was found (Fig. 7.2). Although AO090113000137 was not listed as a candidate from the transcriptome experiments, disruption and overexpression of the gene led to complete loss and increase of KA productivity (Marui et al. 2011). Consequently, the three genes were concluded to form a cluster of genes responsible to KA biosynthesis. It should be noted that the cluster was located on NSB similarly to the other known secondary metabolism gene clusters.

7.6 Expansion and High Diversity of Fungal Secondary Metabolism Genes

Extensive efforts of sequencing genomes of various filamentous fungi revealed presence of significantly larger number of secondary metabolite biosynthesis (SMB) genes existing than those expected from the knowledge before the sequencing. Cultivation conditions specific and often different from those ordinarily used in laboratory have made the analysis of SMB genes more difficult than those involved in primary metabolism or basic cellular functions. Weak and unstable expression of SMB genes (weak and unstable productivity of corresponding metabolites as well) may often increase the difficulty of the analysis.

Genomics have shed light on the SMB gene analysis by uncovering SMB genes hidden in the genome. Although only a few SMB gene clusters with known products have been assigned for each species to date in general, 17–35 polyketide synthase (PKS), 14–24 nonribosomal peptide synthetase (NRPS), and 2–10 dimethylallyl tryptophan synthase (DMAT) genes, which play crucial roles on SMB, have been found from sequenced *Aspergillus* genomes (Rank et al. 2010). This finding strongly suggests that a filamentous fungus has potential to produce far more variation of compounds than those already identified. To address the function of the predicted SMB genes including the number of genes indispensable to complete their biosynthesis, various experimental methods must be taken into consideration. Activation of silent gene clusters by overexpressing transcription factors associated in the clusters or other pleiotropic regulators and by cocultivation of microorganisms have been successfully applied to analyze the gene clusters for aspyridones A/B and terrquinone A (Brakhage and Schroeckh 2011).

7.7 Toward Highly Efficient Analysis and Utilization of Fungal SMB Genes

7.7.1 Next-Generation Sequencing and De Novo Assembling

Next-generation sequencers (NGSs) first introduced in the market in 2005 had great impact on the quick and cost-effective analysis of genomes. Sequencing throughput of the recent NGSs has reached approximately 200–600 gigabases within about 2 weeks. The amount of the sequence read is enough to determine 10–50 genomes from a filamentous fungus without a reference sequence in general. As an example, SOLiD 5500xl sequencer has been successfully applied to de novo sequencing of a fungal genome (Umemura et al. 2013). Effective optimization of parameters generated de novo assembly showing more than 1.7 and more than 3.4 Mb for maximum values of N50 and scaffold, respectively. The resulting approximately 300 scaffolds more than 500 bases in length covered more than 98.5 % of the reference genome of *Aspergillus oryzae* sequenced by the conventional Sanger method. The base call accuracy was 99.97 % for CDSs (including introns), and 99.7 % for the entire assembly of the reference. The fact that even SMB gene clusters longer than 70 kbp, including PKS/NRPS genes, were accurately reconstructed indicates that this approach certainly suits genome-wide analysis of SMB gene clusters. Current throughput of this platform is to determine 12 fungal genomes in a single run taking only approximately 2 weeks by a single NGS.

7.7.2 Prediction of SMB Gene Clusters

Once the genome is sequenced by the NGS, including successive de novo assembling, genes on the genome are predicted and annotated automatically. To predict SMB gene clusters, several recently developed algorithms, CLUSEAN (Weber et al. 2009), SMURF (Khaldi et al. 2010), antiSMASH (Medema et al. 2011), and the method recently developed by Andersen et al. (Andersen et al. 2013), may be used. Any currently published methods predict SMB gene clusters absolutely require the presence of a so-called "core gene" or a "backbone enzyme" gene in the cluster such as those encoding PKS, NRPS, or DMAT; this means that those currently available methods are not applicable to discovery of the SMB gene clusters that have novel biosynthetic functions independent of either of the typical enzymes. Genome sequencing and comparative analyses of filamentous fungi demonstrated the existence of approximately 2,000–4,000 genes uniquely existing in each *Aspergillus* species with the genome size of 30–37 Mb. KOG analysis assigned functional information only to approximately one fourth (~500–1,000) of the NSB genes mostly related to metabolism, especially to SMB, secretory hydrolases, and transporters. No functional information was given for the remaining three fourths (~1,500–3,000) of all the NSB genes. Those functionally unknown genes are possibly composed of a large number of novel SMB genes for which no functional information may have been obtained because of difficulty in analysis. Considering that only approximately 50 SMB gene clusters have been predicted by the currently developed method to date, as just mentioned, at least several times more SMB gene clusters remain unaddressed because of the lack of effective methods for their analysis. Based on this observation together with our previous effort to discover the KA biosynthesis gene cluster without genes of any known specific function, we are currently focusing on the prediction of SMB gene clusters without functional knowledge, which is the so-called motif-free de novo prediction method. Comparative genomic analysis using species harboring and lacking productivity of the material of interest, which is a well-known general approach, is thought to be one of the most potential and practical methods to predict novel SMB gene clusters at present when no functional information is available.

7.7.3 Combination of Omics Information

Comparative analysis of extrolites profiles obtained from mass spectrometry following liquid chromatography/mass spectroscopy (LC/MS) separation is a powerful method to identify the product using the disruptant or overexpression strain of the genes included in a SMB gene cluster candidate (Fig. 7.3). However, possible incompatibility of the LC/MS method to the target and unstable and/or low productivity of secondary metabolites in general makes the analysis far more difficult than expected. Further, to determine the detailed chemical structure of a novel compound, labor-intensive and time-consuming experiments, mid- to large-scale preparation, and successive purification of the target compounds are still required for the analysis using high-resolution nuclear magnetic resonance (NMR). Nevertheless, owing to significant advances in analytical technology, the compositional formula

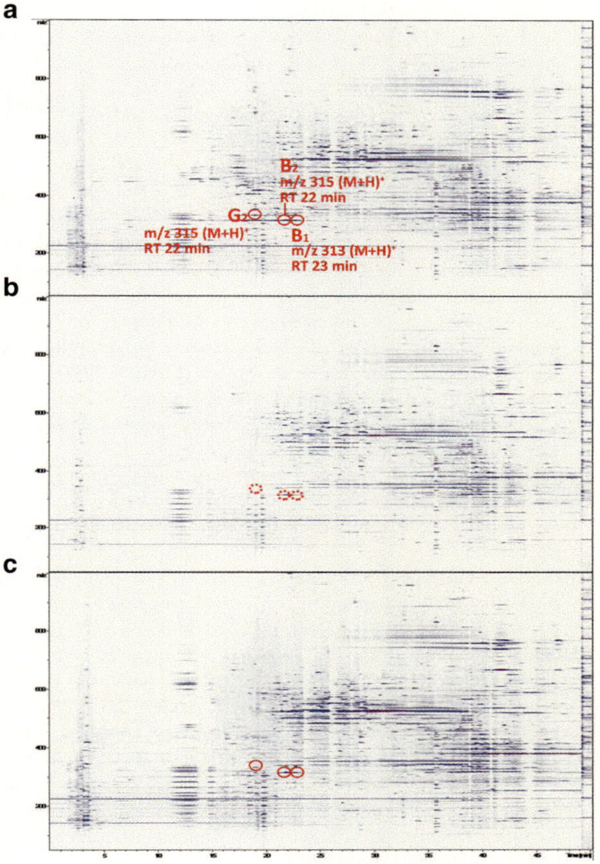

Fig. 7.3 Extrolites profiling by LC/MS. Extrolites from the *Aspergillus flavus pyrG* revertant, disruptant of the aflatoxin gene cluster (**Δ**AFLA_139200-440), and disruptant of hybrid PKS/NRPS gene (**Δ**AFLA_139490) in the CPA gene cluster were extracted by ethyl acetate followed by LC/MS profiling equipped with ODS column. Compounds that specifically disappeared from the aflatoxin disruptant are represented by *red circles*

of the whole and the fragmented compounds can be obtained quickly using high-resolution MS (HR-MS). Extremely high throughput sequencing and extrolites profiling using HR-MS in combination with successive flexible computational analyses as just described should open the way to dramatically accelerate the analysis of fungal SMB genes.

References

Akao T, Sano M, Yamada O, Akeno T, Fujii K, Goto K, Ohashi-Kunihiro S, Takase K, Yasukawa-Watanabe M, Yamaguchi K, Kurihara Y, Maruyama J, Juvvadi PR, Tanaka A, Hata Y, Koyama Y, Yamaguchi S, Kitamoto N, Gomi K, Abe K, Takeuchi M, Kobayashi T, Horiuchi H, Kitamoto K, Kashiwagi Y, Machida M, Akita O (2007) Analysis of expressed sequence tags from the fungus *Aspergillus oryzae* cultured under different conditions. DNA Res 14(2):47–57

Andersen MR, Nielsen JB, Klitgaard A, Petersen LM, Zachariasen M, Hansen TJ, Blicher LH, Gotfredsen CH, Larsen TO, Nielsen KF, Mortensen UH (2013) Accurate prediction of secondary metabolite gene clusters in filamentous fungi. Proc Natl Acad Sci USA 110(1):E99–E107. doi:10.1073/pnas.1205532110 1205532110 [pii]

Brakhage AA, Schroeckh V (2011) Fungal secondary metabolites: strategies to activate silent gene clusters. Fungal Genet Biol 48(1):15–22. doi:10.1016/j.fgb.2010.04.004 S1087-1845(10)00068-X [pii]

Galagan JE, Calvo SE, Cuomo C, Ma LJ, Wortman JR, Batzoglou S, Lee SI, Basturkmen M, Spevak CC, Clutterbuck J, Kapitonov V, Jurka J, Scazzocchio C, Farman M, Butler J, Purcell S, Harris S, Braus GH, Draht O, Busch S, D'Enfert C, Bouchier C, Goldman GH, Bell-Pedersen D, Griffiths-Jones S, Doonan JH, Yu J, Vienken K, Pain A, Freitag M, Selker EU, Archer DB, Penalva MA, Oakley BR, Momany M, Tanaka T, Kumagai T, Asai K, Machida M, Nierman WC, Denning DW, Caddick M, Hynes M, Paoletti M, Fischer R, Miller B, Dyer P, Sachs MS, Osmani SA, Birren BW (2005) Sequencing of *Aspergillus nidulans* and comparative analysis with *A. fumigatus* and *A. oryzae*. Nature (Lond) 438(7071):1105–1115

Goffeau A, Barrell BG, Bussey H, Davis RW, Dujon B, Feldmann H, Galibert F, Hoheisel JD, Jacq C, Johnston M, Louis EJ, Mewes HW, Murakami Y, Philippsen P, Tettelin H, Oliver SG (1996) Life with 6000 genes. Science 274(5287):546–567

Goffeau A et al (1997) The yeast genome directory. Nature (Lond) 387:5–105

Gomi K, Abe K (2007) Food products fermented by *Aspergillus oryzae*. In: The aspergilli: genomics, medical applications, biotechnology, and research methods (mycology). CRC, Boca Raton, pp 427–437

Khaldi N, Seifuddin FT, Turner G, Haft D, Nierman WC, Wolfe KH, Fedorova ND (2010) SMURF: genomic mapping of fungal secondary metabolite clusters. Fungal Genet Biol 47(9):736–741. doi:10.1016/j.fgb.2010.06.003 S1087-1845(10)00105-2 [pii]

Machida M, Asai K, Sano M, Tanaka T, Kumagai T, Terai G, Kusumoto K, Arima T, Akita O, Kashiwagi Y, Abe K, Gomi K, Horiuchi H, Kitamoto K, Kobayashi T, Takeuchi M, Denning DW, Galagan JE, Nierman WC, Yu J, Archer DB, Bennett JW, Bhatnagar D, Cleveland TE, Fedorova ND, Gotoh O, Horikawa H, Hosoyama A, Ichinomiya M, Igarashi R, Iwashita K, Juvvadi PR, Kato M, Kato Y, Kin T, Kokubun A, Maeda H, Maeyama N, Maruyama J, Nagasaki H, Nakajima T, Oda K, Okada K, Paulsen I, Sakamoto K, Sawano T, Takahashi M, Takase K, Terabayashi Y, Wortman JR, Yamada O, Yamagata Y, Anazawa H, Hata Y, Koide Y, Komori T, Koyama Y, Minetoki T, Suharnan S, Tanaka A, Isono K, Kuhara S, Ogasawara N, Kikuchi H (2005) Genome sequencing and analysis of *Aspergillus oryzae*. Nature (Lond) 438(7071):1157–1161

Maeda H, Yamagata Y, Abe K, Hasegawa F, Machida M, Ishioka R, Gomi K, Nakajima T (2005) Purification and characterization of a biodegradable plastic-degrading enzyme from *Aspergillus oryzae*. Appl Microbiol Biotechnol 67(6):778–788

Marui J, Yamane N, Ohashi-Kunihiro S, Ando T, Terabayashi Y, Sano M, Ohashi S, Ohshima E, Tachibana K, Higa Y, Nishimura M, Koike H, Machida M (2011) Kojic acid biosynthesis in *Aspergillus oryzae* is regulated by a Zn(II)(2)Cys(6) transcriptional activator and induced by kojic acid at the transcriptional level. J Biosci Bioeng 112(1):40–43. doi:S1389-1723(11)00114-9 [pii] 10.1016/j.jbiosc.2011.03.010

Medema MH, Blin K, Cimermancic P, de Jager V, Zakrzewski P, Fischbach MA, Weber T, Takano E, Breitling R (2011) antiSMASH: rapid identification, annotation and analysis of secondary metabolite biosynthesis gene clusters in bacterial and fungal genome sequences. Nucleic Acids Res 39(Web Server issue):W339–W346. doi:10.1093/nar/gkr466 gkr466 [pii]

Murakami H (1980) Koji-gaku (in Japanese). Brewing Society of Japan, Tokyo

Nierman WC, Pain A, Anderson MJ, Wortman JR, Kim HS, Arroyo J, Berriman M, Abe K, Archer DB, Bermejo C, Bennett J, Bowyer P, Chen D, Collins M, Coulsen R, Davies R, Dyer PS, Farman M, Fedorova N, Feldblyum TV, Fischer R, Fosker N, Fraser A, Garcia JL, Garcia MJ, Goble A, Goldman GH, Gomi K, Griffith-Jones S, Gwilliam R, Haas B, Haas H, Harris D, Horiuchi H, Huang J, Humphray S, Jimenez J, Keller N, Khouri H, Kitamoto K, Kobayashi T, Konzack S, Kulkarni R, Kumagai T, Lafton A, Latge JP, Li W, Lord A, Lu C, Majoros WH,

May GS, Miller BL, Mohamoud Y, Molina M, Monod M, Mouyna I, Mulligan S, Murphy L, O'Neil S, Paulsen I, Penalva MA, Pertea M, Price C, Pritchard BL, Quail MA, Rabbinowitsch E, Rawlins N, Rajandream MA, Reichard U, Renauld H, Robson GD, de Rodriguez CS, Rodriguez-Pena JM, Ronning CM, Rutter S, Salzberg SL, Sanchez M, Sanchez-Ferrero JC, Saunders D, Seeger K, Squares R, Squares S, Takeuchi M, Tekaia F, Turner G, Vazquez de Aldana CR, Weidman J, White O, Woodward J, Yu JH, Fraser C, Galagan JE, Asai K, Machida M, Hall N, Barrell B, Denning DW (2005) Genomic sequence of the pathogenic and allergenic filamentous fungus *Aspergillus fumigatus*. Nature (Lond) 438(7071):1151–1156

Prade RA, Griffith J, Kochut K, Arnold J, Timberlake WE (1997) In vitro reconstruction of the *Aspergillus* (= Emericella) *nidulans* genome. Proc Natl Acad Sci USA 94(26):14564–14569

Rank C, Larsen TO, Frisvad JC (2010) Functional systems biology of *Aspergillus*. In: Gomi K, Machida M (eds) *Aspergillus* molecular biology and genomics. Caister Academic, Hethersett, pp 173–198

Roach JC, Boysen C, Wang K, Hood L (1995) Pairwise end sequencing: a unified approach to genomic mapping and sequencing. Genomics 26(2):345–353. doi:0888-7543(95)80219-C [pii]

Saito K (1907) Über die Säurebinding von *Aspergillus oryzae*. Bot Mag (Tokyo) 21:7–11

Takahashi T, Maeda H, Yoneda S, Ohtaki S, Yamagata Y, Hasegawa F, Gomi K, Nakajima T, Abe K (2005) The fungal hydrophobin RolA recruits polyesterase and laterally moves on hydrophobic surfaces. Mol Microbiol 57(6):1780–1796

Tamano K, Sano M, Yamane N, Terabayashi Y, Toda T, Sunagawa M, Koike H, Hatamoto O, Umitsuki G, Takahashi T, Koyama Y, Asai R, Abe K, Machida M (2008) Transcriptional regulation of genes on the non-syntenic blocks of *Aspergillus oryzae* and its functional relationship to solid-state cultivation. Fungal Genet Biol 45(2):139–151

Umemura M, Koyama Y, Takeda I, Hagiwara H, Ikegami T, Koike H, Machida M (2013) Fine De Novo Sequencing of a Fungal Genome Using only SOLiD Short Read Data: Verification on *Aspergillus oryzae* RIB40 PloS One 8(5):e63673

Weber T, Rausch C, Lopez P, Hoof I, Gaykova V, Huson DH, Wohlleben W (2009) CLUSEAN: a computer-based framework for the automated analysis of bacterial secondary metabolite biosynthetic gene clusters. J Biotechnol 140(1–2):13–17

Chapter 8
Application of Genomics in Molecular Breeding of the *koji* Molds *Aspergillus oryzae* and *Aspergillus sojae*

Tadashi Takahashi

Abstract Recent results obtained from comparative genome analysis of *koji* molds, functional analysis of transcription factors and their application in breeding, and chromosomal engineering of the *koji* molds by deleting unessential chromosomal regions are presented. Comparative genome analysis of *koji* molds demonstrated genomic differences between *Aspergillus oryzae* and *Aspergillus sojae*. The number of putative protease genes, α-amylase genes, and glutaminase genes in *A. oryzae* was different from that in *A. sojae*. These differences may be responsible for different properties of the species because these genes play important roles in fermentation processes. Analysis of putative transcription factor genes of *A. oryzae* was conducted by disruption and overexpression of the genes. Several new genes implicated in conidiospore formation were discovered. We generated strains with minimized chromosomes using efficient gene-targeting techniques and large chromosomal deletion methods. As a result, a strain with a minimized chromosome 7 was constructed. Moreover, the strain with a minimized chromosome 7 was successfully fused with a strain with a reduced chromosome 8.

Keywords *Aspergillus oryzae* • *Aspergillus sojae* • Comparative genomics • Gene targeting • Minimal genome • Transcription factor

---

T. Takahashi (✉)
Research and Development Division, Kikkoman Corporation,
399 Noda, Noda City 2780037, Japan
e-mail: ttakahashi@mail.kikkoman.co.jp

H. Anazawa and S. Shimizu (eds.), *Microbial Production: From Genome Design to Cell Engineering*, DOI 10.1007/978-4-431-54607-8_8, © Springer Japan 2014

8.1 Introduction

The *koji* molds *Aspergillus oryzae* and *Aspergillus sojae* are extensively used in the production of traditional Japanese fermented food. The *koji* molds have recently been used for the production of enzymes in the biotechnology industry because of their ability to secrete large amounts of proteins as well as their safety as host strains because of their long-standing use in traditional food production. Until recent years, functional analysis of the *koji* mold genomes was delayed compared with that of other microorganisms despite their industrial importance for the following reasons: (1) the *koji* molds retain a multinuclear state throughout their life cycle, (2) a sexual cycle has never been observed in the molds, and (3) the frequency of homologous recombination (HR) is extremely low. The whole-genome sequencing of the *koji* mold *A. oryzae* was completed and made available at the end of 2005 (Machida et al. 2005). As a result, application of postgenomic technologies such as transcriptomics and proteomics has become possible for this organism. However, the many unknown genes found in a genome make it difficult to develop such technologies for practical use. The disruption of genes related to nonhomologous end-joining (NHEJ), *ku70*, *ku80*, and *ligD*, allowing a high frequency of HR, has been reported for a *Neurospor*a strain (Ninomiya et al. 2004; Ishibashi et al. 2006). Takahashi et al. (2006) have developed a high-frequency gene-targeting system based on the *ku* disruption strains of *A. oryzae* and *A. sojae*. The rate of research and development involving the genes of the *koji* molds has improved to a great extent as a consequence of these studies. Further, next-generation sequencing technologies have affected the strategies for improvements in the breeding of the *koji* mold strains. The *koji* mold strains used in industrial fermentation include not only the strains isolated from various locations or selected through long periods of food production but also the strains isolated by classical mutagenesis. Much time and effort is necessary to create *koji* mold strains with superior phenotypes using classical mutagenesis. It is difficult to determine which gene mutation results in which particular phenotypic change. Moreover, generation of a strain with multiple mutations that are beneficial for its improvement is almost impossible. Cheaper and faster genome analysis using a next-generation sequencer will enable a more feasible detection of mutation points in the strains and will improve the breeding techniques for the *koji* mold strains. Here we present the recent results obtained from comparative analysis of the genome, analysis of transcription factors, and construction of the minimum genome of the *koji* molds based on fundamental research on the development of strains required for soy sauce production.

8.2 Comparative Genome Analysis of the *Koji* Molds

Genome analyses of the three strains of *Aspergillus* (*A. oryzae*, *A. nidulans*, and *A. fumigatus*) were performed in 2005 (Machida et al. 2005; Galagan et al. 2005; Nierman et al. 2005). The results have shown that the genome size of *A. oryzae* is approximately 37 Mb and is approximately 30 % larger than that of *Aspergillus*

Table 8.1 Comparison of the genome characteristics of *Aspergillus sojae* and *Aspergillus oryzae*

	A. oryzae RIB40	*A. sojae* NBRC4239
Size of assembly (Mb)	37.6	39.5[a]
GC content (%)	48.2	48.1[a]
Number of tRNA genes	270	275[a]
Number of ORFs	12,074	13,033[a]
Average ORF size (bp)	449.8	455.9[a]

[a]Sato et al. (2011)

nidulans and *Aspergillus fumigatus*. The predicted gene number in *A. oryzae* is approximately 12,000. In particular, the number of putative protease genes, important for soy sauce production, is 134 in *A. oryzae*, which is approximately 50 % more than in the other two *Aspergillus* strains. Moreover, there are three copies of the α-amylase genes that are required for starch degradation in the genome of *A. oryzae*. In general, there are two types of *koji* molds: *A. oryzae*, used extensively for the production of sake, miso, and soy sauce, and *A. sojae*, used mainly for the production of soy sauce. *A. sojae* has high protease activity and low amylase activity compared with *A. oryzae*. Because of the low content of organic acids, the pH of the degradation product of *A. sojae* is higher than that of the degradation product of *A. oryzae*. These differences are important in soy sauce production. Comparison of the genomic sequences of the different *koji* mold strains is useful not only for finding the genes causing such differences but also for acquiring the information necessary for breeding. We performed genome sequencing of *A. sojae* using a next-generation sequencer (Sato et al. 2011). We found that the genome size of *A. sojae* is approximately 39 Mb and the number of predicted genes is approximately 13,000, which is greater than that of *A. oryzae* (Table 8.1). Although the total number of predicted protease genes is almost the same in both strains, some predicted protease genes are specifically present in *A. sojae*. The predicted α-amylase gene found in *A. sojae* exists only as one copy, which may explain the low starch degradation rate of *A. sojae*. Analysis of the effects of such a difference in gene composition of the strains on the quality of fermented products is under consideration.

Identification of glutaminase genes was one of the results of genome analysis and is useful for the breeding of the *koji* molds. Glutaminases are enzymes that convert L-glutamine to L-glutamate and are important in soy sauce production. Accordingly, a *koji* mold strain with high activity of both proteases and glutaminases would be ideal for soy sauce making. However, such a strain is difficult to breed. Furthermore, strains with salt-tolerant glutaminase that would retain its activity in *moromi*, a mixture of salty water and *koji*, are favored for soy sauce production. However, glutaminases of the *koji* molds are not salt tolerant: therefore, it would be necessary to modify properties of those enzymes. The results of enzymatic studies predicted that the *koji* molds have multiple glutaminases. Genome analysis showed that *A. oryzae* has 12 glutaminase genes categorized into four types, and *A. sojae* has 10 glutaminase genes of four types (Table 8.2). Ito et al. (2011, 2012) cloned the *gahA* gene of *A. sojae* that has homology to the glutaminase gene of *Cryptococcus* yeast and

Table 8.2 Number of glutaminase genes in *A. sojae* and *A. oryzae*

	A. oryzae RIB40	*A. sojae* NBRC4239
Type1 (Gah type)	4[a]	3
Type2 (Ggt type)	4	3
Type3 (Gta type)	3[b]	3
Type4 (Gls type)	1[c]	1

[a]Ito et al. (2012)
[b]Koibuchi et al. (2000)
[c]Masuo et al. (2005)

revealed that the GahA enzyme is a new type of glutaminase with a peptidoglutaminase activity. The *koji* mold strains bred by improvement of the glutaminase genes are expected to be utilized for improved soy sauce production and for the development of soy sauce-derived products of high quality.

8.3 Functional Analysis of Transcription Factors and Their Application in Breeding

Transcription factors are proteins that bind to the promoter or enhancer regions of DNA and promote or repress DNA transcription in response to environmental changes. The *koji* mold *A. oryzae* is predicted to have approximately 12,000 genes. However, the function of many of the genes remains unknown, causing difficulty in breeding of the mold. To examine the function of the genes, phenotypic changes in the strain were analyzed following gene disruption or overexpression. However, it would be difficult to conduct such analysis for all the genes of *A. oryzae*. Moreover, in many cases, phenotypic changes are not observed after a single gene disruption. In contrast, disruption of transcription factors is expected to show clear phenotypic changes because transcription factors generally regulate multiple genes at the same time. In addition, most transcription factors regulate the expression of gene clusters that encode enzymes associated with the synthesis or degradation of a particular component. For example, to degrade cellulose, many enzymes other than cellulases are necessary. Therefore, to breed a strain that degrades cellulose with higher efficiency, many genes need to be cloned and expressed at high levels. If a transcription factor that regulates the expression of all genes related to cellulose degradation is found, multiple genes associated with cellulose degradation will be effectively expressed by stimulating the expression of the gene of the transcription factor in the strain.

Transcription factors were searched in the genome of *A. oryzae* using the transcription factor motifs registered in a DNA database. As a result, 600 candidate transcription factor genes were found. We tried to construct disruption or overexpression strains for all these genes.

First, we analyzed transcription factors related to the regulation of conidiospore formation because it was relatively easy to determine phenotypic changes in the strain mutants for these genes. The conidiospore of the *koji* molds is called "*tane-koji*" and is used as a starter in the production of traditional Japanese fermented food products such as soy sauce, sake, and miso. Understanding the mechanisms of regulation of conidiospore formation by the *koji* molds will enable the production of improved quality of "*tane-koji*" and isolation of superior conidiospore formation mutants. However, most of these mechanisms are still unknown. Thus, we analyzed the regulation of conidiospore formation in the *koji* molds and found that the main pathway in *A. oryzae* was similar to that in the model fungus *A. nidulans*. We also found that all the known transcription factors related to conidiospore formation were conserved between these strains (Ogawa et al. 2010). Moreover, we found several new transcription factors that affect conidiospore formation of *A. oryzae*. Overproducers of conidiospores were screened using a library of strains with disrupted or overexpressed transcription factor genes: two genes, *pceA* and *sclR*, were found (Jin et al. 2011). DNA microarray analysis of these conidiospore-overproducing strains revealed an increased expression level of *brlA*, a key transcription factor gene for the regulation of conidiospore formation. However, overexpression of *brlA* showed abnormal phenotypes (Yamada et al. 1999), indicating that appropriate expression of this transcription factor gene is important for the regulation of conidiospore formation.

8.4 Chromosomal Engineering of *koji* Mold Through Deletion of Unessential Chromosomal Regions

As described previously, the genomic sequence of *A. oryzae* was not utilized enough for breeding of the *koji* molds because of the difficulty in functional analysis from low HR frequency. Exogenous DNAs were integrated into the genome mainly through two pathways, HR and NHEJ. In *A. oryzae* and *A. sojae*, the frequency of gene targeting was very low because most DNA fragments were randomly integrated into the genome through the NHEJ pathway. However, by disrupting genes related to NHEJ, *ku70*, *ku80*, or *ligD* (Takahashi et al. 2006; Mizutani et al. 2008), increase in gene-targeting frequency in the molds became possible. Using this technique, we constructed a comprehensive disruption strain library for a transcription factor gene of *A. oryzae* to analyze the function of the transcription factor. Further, we developed techniques for the construction of large chromosomal deletion strains and tried to generate a minimum genome factory of *A. oryzae* by deleting unessential chromosomal regions not only for cell growth but also for use in industrial processes.

Aspergillus oryzae and *Aspergillus sojae* have approximately 80 kb of homologous gene cluster for biosynthesis of aflatoxin, a potential carcinogen causing liver cancer. The past studies confirmed that the aflatoxin gene clusters of *A. oryzae* and *A. sojae* were genetically inactive and aflatoxins were not produced by the *koji* molds (Chang et al. 2007; Tominaga et al. 2006; Watson et al. 1999). Therefore, the cluster was considered an unnecessary chromosomal region, and we chose this cluster as a target for a large chromosomal deletion. Deletion of aflatoxin biosynthesis gene cluster was performed using two recombination methods, replacement-type

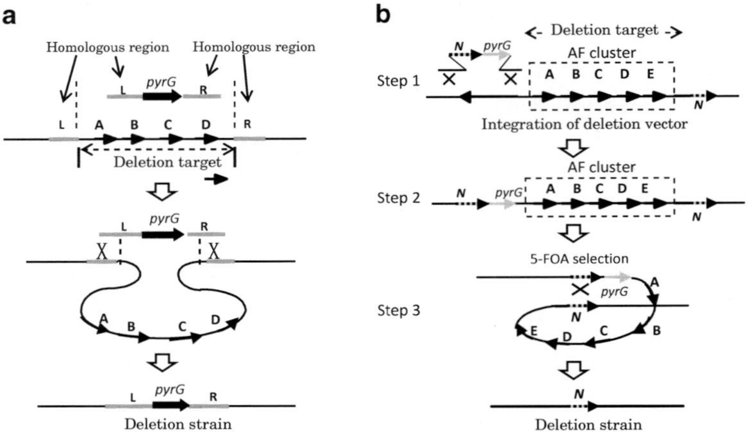

Fig. 8.1 Construction of large chromosomal deletions. **(a)** Replacement-type recombination. Genes *A–D* represent the deletion targets. A deletion vector comprising a *pyrG* marker and the recombination arms located at both ends of the deletion target were transformed into the *koji* mold cells. Both arms of the vector recombined with the corresponding homologous regions adjacent to the deletion target. Vector integration was completed and the deletion target was excised and replaced with *pyrG*. **(b)** Loop-out type recombination. Step 1: Genes *A–E* are the deletion targets (AF cluster). The deletion vector consisting of the *pyrG* marker and a DNA fragment homologous to the adjacent distal region (N) of the deletion target was transformed into cells of the *koji* mold. Step 2: The deletion vector was integrated into the proximal adjacent region of the deletion target. Step 3: The cells from which *pyrG* was lost as a result of the recombination between distal and proximal homologous regions (N) adjacent to the deletion target were selected by 5-FOA resistance. Deletion of genes A–E (AF cluster) was achieved

recombination (Takahashi et al. 2009) and loop-out recombination (Takahashi et al. 2008) (Fig. 8.1). Chromosomal deletion by replacement-type recombination was simply obtained by transforming *ku70*- or *ligD*-disrupted strains with a deletion vector comprising recombination arms located at both ends of the target chromosomal region and a transformation marker (Takahashi et al. 2009) (Fig. 8.1a). Deletion of large chromosomal region by loop-out recombination was achieved in three steps (Takahashi et al. 2008) (Fig. 8.1b). First, the deletion vector for the AF cluster (A–E in Fig. 8.1b) consisting of the *pyrG* marker and a DNA fragment homologous to the adjacent distal region (N in Fig. 8.1b) of the AF cluster was transformed into the cells of the *koji* mold (Step 1). The deletion vector was integrated into the proximal adjacent region of the AF cluster (Step 2). The cells that lost the *pyrG* gene as a result of recombination between distal and proximal homologous regions (N) adjacent to the AF cluster were selected by 5-FOA resistance; thus, AF cluster deletion was achieved (Step 3).

The replacement-type recombination can eliminate large chromosomal regions in both *ku70* and *ligD* mutant strains with high frequency. However, to accomplish another round of elimination, a transformation to remove the marker gene was necessary. In contrast, the loop-out recombination can eliminate chromosomal region without leaving a marker gene. Therefore, it is useful for introducing multiple

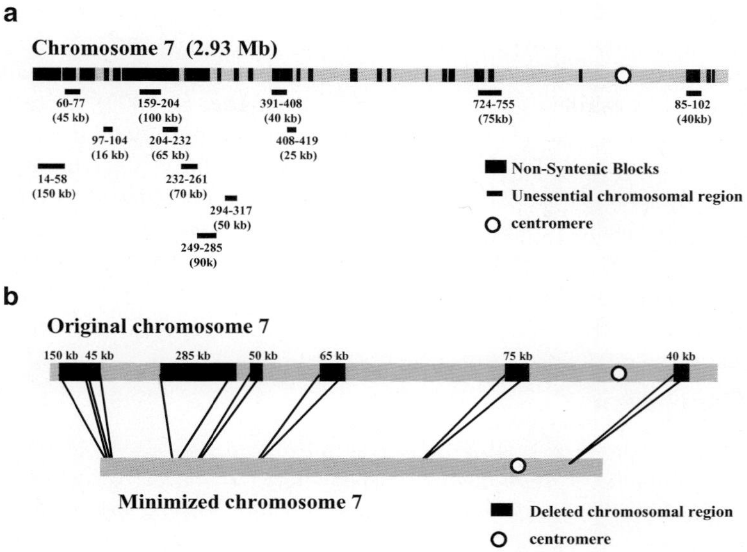

Fig. 8.2 Minimization of chromosome 7 of *A. oryzae*. (**a**) Identification of unessential chromosomal regions by systematic deletion of chromosome segments. The *top part* of the figure shows the site of putative nonsyntenic blocks (*black bars*). The *bottom part* of the figure shows the corresponding region that can be deleted using replacement-type recombination. (**b**) Minimization of chromosome 7 by introducing multiple deletions. Approximately 705 kb of chromosome 7 was deleted by seven rounds of multiple deletions

chromosomal deletions. Elimination of chromosomal region by loop-out recombination worked concisely in the *ku70*-disrupted strain. However, in the *ligD*-disrupted strain, unintended chromosomal deletions were frequently observed during the elimination of near-telomeric regions (Takahashi et al. 2012). The replacement-type recombination efficiently eliminated more than 470 kb of the chromosomal region in a single round of chromosomal deletion so long as the target was in a dispensable chromosomal region. In contrast, the loop-out recombination eliminated approximately a 200-kb length of chromosomal region concisely, because the efficiency of the loop-out recombination seemed to be affected by the structure of the target chromosome (Takahashi et al. 2008, 2012).

Aspergillus oryzae has eight chromosomes. Of these, chromosome number 7 is the smallest (2.9 Mb) and contains many chromosomal regions with low homology to the *A. nidulans* or *A. fumigatus* genome. These regions are expected to be dispensable for cell growth. Therefore, we tried to minimize the chromosome 7 of *A. oryzae* by deleting these unessential chromosomal regions.

The essential or unessential chromosomal regions were decided by systematically deleting parts of the chromosome 7 using replacement-type recombination (Jin et al. 2009). The results showed that an approximately 740-kb region of chromosome 7 was unessential for cell growth (Fig. 8.2a). Next, using loop-out recombination, we constructed a 705-kb (24 % of chromosome 7) deletion strain of chromosome 7 by seven rounds of multiple deletions (Jin et al. 2010) (Fig. 8.2b). The deletions of these regions were confirmed by array CGH analysis and pulsed-field gel electrophoresis. A strain with reduced chromosome 8 was obtained by deleting a 470-kb region of chromosome 8 (Takahashi et al. 2009). Moreover, we tried to combine the strains having a minimized chromosome 7 and those having a reduced chromosome 8. The fused strains were successfully generated bythe cell fusion method (Hara et al. 2012). The fused strain with the reduced chromosomes did not show beneficial properties such as increased cell growth or protein production. However, an increase in amylase activity and decrease in metabolite levels was observed in this strain (Jin et al. 2010).

8.5 Conclusion

Since the completion of whole-genome sequencing of *A. oryzae*, research on molecular biology and breeding techniques in the *koji* molds using biotechnology has progressed remarkably. At present, it is difficult to directly apply the findings from these studies in industrial production. However, generation of strains with novel functions using these techniques is expected to create a new market for fermented food. The strains constructed by chromosome minimization showed a decrease in metabolite levels, suggesting that they may be useful in the pharmaceutical industry as well as the fermentation industry.

Acknowledgments This study was supported by a program for the promotion of basic research activities for innovative biosciences of the Bio-Oriented Technology Research Advancement Institution.

References

Chang PK, Matsushima K, Takahashi T et al (2007) Understanding nonaflatoxigenecity of *Aspergillus sojae*: a windfall of aflatoxin biosynthesis research. Appl Microbiol Biotechnol 76:977–984

Galagan JE, Calvo SE, Cuomo C et al (2005) Sequencing of *Aspergillus nidulans* and comparative analysis with *A. fumigatus* and *A. oryzae*. Nature (Lond) 438:1105–1115

Hara S, Jin FJ, Takahashi T et al (2012) A further study on chromosome minimization by protoplast fusion in *Aspergillus oryzae*. Mol Genet Genomics 287:177–187

Ishibashi K, Suzuki K, Ando Y et al (2006) Nonhomologous chromosomal integration of foreign DNA is completely dependent on MUS-53 (human Lig4 homolog) in *Neurospora*. Proc Natl Acad Sci USA 103:14871–14876

Ito K, Umitsuki G, Oguma T et al (2011) Salt-tolerant and thermostable glutaminases of *Cryptococcus* species form a new glutaminase family. Biosci Biotechnol Biochem 75:1317–1324

Ito K, Matsushima K, Koyama Y (2012) Gene cloning, purification, and characterization of a novel peptidoglutaminase-asparaginase from *Aspergillus sojae*. Appl Environ Microbiol 78:5182–5188

Jin FJ, Takahashi T, Machida M et al (2009) Identification of a basic helix-loop-helix-type transcription regulator gene in *Aspergillus oryzae* by systematically deleting large chromosomal segments. Appl Environ Microbiol 75:5943–5951

Jin FJ, Takahashi T, Utsushikawa M et al (2010) A trial of minimization of chromosome 7 in *Aspergillus oryzae* by multiple chromosomal deletions. Mol Genet Genomics 283:1–12

Jin FJ, Takahashi T, Matsushima K et al (2011) SclR, a basic helix-loop-helix transcription factor, regulates hyphal morphology and promotes sclerotial formation in *Aspergillus oryzae* by systematically deleting large chromosomal segments. Eukaryot Cell 10:945–955

Koibuchi K, Nagasaki H, Yuasa A et al (2000) Molecular cloning and characterization of a gene encoding glutaminase from *Aspergillus oryzae*. Appl Microbiol Biotechnol 54:59–68

Machida M, Asai K, Sano M et al (2005) Genome sequencing and analysis of *Aspergillus oryzae*. Nature (Lond) 438:1157–1161

Masuo N, Yoshimune K, Ito K et al (2005) *Micrococcus luteus* K-3-type glutaminase from *Aspergillus oryzae* RIB40 is salt-tolerant. J Biosci Bioeng 100:576–578

Mizutani O, Kudo Y, Saito A et al (2008) A defect of *LigD* (human Lig4 homolog) for nonhomologous end joining significantly improves efficiency of gene-targeting in *Aspergillus oryzae*. Fungal Genet Biol 45:878–889

Nierman WC, Pain A, Anderson MJ et al (2005) Genomic sequence of the pathogenic and allergenic filamentous fungus *Aspergillus fumigatus*. Nature (Lond) 438:1151–1156

Ninomiya Y, Suzuki K, Ishii C et al (2004) Highly efficient gene replacements in *Neurospora* strains deficient for nonhomologous endjoining. Proc Natl Acad Sci USA 101:12248–12253

Ogawa M, Tokuoka M, Jin FJ et al (2010) Genetic analysis of conidiation regulatory pathways in koji-mold *Aspergillus oryzae*. Fungal Genet Biol 47:10–18

Sato A, Oshima K, Noguchi H et al (2011) Draft genome sequencing and comparative analysis of *Aspergillus sojae* NBRC4239. DNA Res 18:165–176

Takahashi T, Masuda T, Koyama Y (2006) Enhanced gene targeting frequency in *ku70* and *ku80* disruption mutants of *Aspergillus sojae* and *Aspergillus oryzae*. Mol Genet Genomics 275:460–470

Takahashi T, Jin FJ, Sunagawa M et al (2008) Generation of large chromosomal deletions in *koji* molds *Aspergillus oryzae* and *Aspergillus sojae* via a loop-out recombination. Appl Environ Microbiol 74:7684–7693

Takahashi T, Jin FJ, Koyama Y (2009) Nonhomologous end-joining deficiency allows large chromosomal deletions by replacement-type recombination in *Aspergillus oryzae*. Fungal Genet Biol 46:815–824

Takahashi T, Ogawa M, Koyama Y (2012) Analysis of the functions of recombination-related genes in the generation of large chromosomal deletions by loop-out recombination in *Aspergillus oryzae*. Eukaryot Cell 11:507–517

Tominaga M, Lee YH, Hayashi Y et al (2006) Molecular analysis of an inactive aflatoxin biosynthesis gene cluster in *Aspergillus oryzae* RIB strains. Appl Environ Microbiol 72:484–490

Watson AJ, Fuller LJ, Jeens DJ et al (1999) Homologs of aflatoxin biosynthesis genes and sequence of *aflR* in *Aspergillus oryzae* and *Aspergillus sojae*. Appl Environ Microbiol 65:307–310

Yamada O, Lee BR, Gomi K et al (1999) Cloning and functional analysis of the *Aspergillus oryzae* conidiation regulator gene *brlA* by its disruption and misscheduled expression. J Biosci Bioeng 87:424–429

Chapter 9
Comprehensive Libraries of *Escherichia coli* K-12 and Their Application

Hirotada Mori, Rikiya Takeuchi, Yuta Otsuka, Yong Han Tek, Wataru Nomura, and Barry L. Wanner

Abstract *Escherichia coli* K-12 was reported more than 100 years ago in Germany, and after the discovery of conjugative genetic material transfer in this bacterium, this organism quickly became widely used for biological research. Considering the extensive research in genetics, biochemistry, and molecular biology using this organism in the next half-century, it is not too much to say that the concept of genes and current genetic technology could not have been established without this organism. After genome sequencing biology, the new approach, systems biology, was launched to elucidate the cell at the systems level. In this new approach in biology, *E. coli* is one of the most appropriate unicellular organisms for this approach because of the huge amount of biological knowledge. Thus, we started to construct comprehensively prepared experimental resources, such as ORF clones and deletion strain libraries. We introduce here these resources and their application in the omics approach in systems biology.

Keywords ASKA • Barcode • *Escherichia coli* K-12 • Keio • Resource

9.1 Prologue

In the past two decades, technology innovation to unveil biological systems, such as genome sequencing, information technology, and high-throughput analysis tools, has been occurring far beyond our predictions and is still ongoing.

H. Mori (✉) • R. Takeuchi • Y. Otsuka • Y.H. Tek • W. Nomura
Graduate School of Biological Sciences, Nara Institute of Science and Technology, Takayama, Ikoma, Nara 630-0101, Japan
e-mail: hmori@gtc.naist.jp

B.L. Wanner
Department of Biological Sciences, Purdue University, West Lafayette, IN 47907, USA

H. Anazawa and S. Shimizu (eds.), *Microbial Production: From Genome Design to Cell Engineering*, DOI 10.1007/978-4-431-54607-8_9, © Springer Japan 2014

In the first year of the twenty-first century, a draft genome of the human had been determined, and many thought that was the end of the genome sequencing race. Ten years later, everybody realized that the end of one era is now the start of a new one. The sequencing project has now been directed to the personal genome and this may come true earlier than we expected.

On the other hand, even though more than a couple thousand complete microbial genomes have been determined, and also a living microorganism carrying a chemically synthesized genome has been reported, still we cannot design a living organism as we desire, even a small bacterium.

It is true that we have obtained blueprints of many organisms at the "AGCT" sequence level. We, however, know now the limitations of this knowledge are those are at the parts level, not the systems level, although many biological studies have been done using this organism for more than half a century.

Although in the field of molecular biology, which is more focused on an individual target gene(s), is still important, the approach to unveiling physiological activities at the system level has now been opened up based on the accumulation of past knowledge, innovation of systematic technologies, and comprehensive biological resources.

I introduce here our systems approach using comprehensive resources of *Escherichia coli* and discuss the perspectives in this field.

9.2 What and Why Is *E. coli*?

Escherichia coli K-12 is the gram-negative rod-shaped enteric bacterium commonly found in the intestine of animals and humans. Undoubtedly, *E. coli* is one of the best studied organisms on the Earth. Because of extensive analysis of this organism, many analysis tools for this microbe are available. Comprehensive experimental resources have been now well prepared and are available, such as open reading frame (ORF) clones and mutation libraries, and this organism is now one of the ideal organisms in systems biology.

Escherichia coli was discovered in 1885 by a German pediatrician, Theodor Escherich (1857–1911), in the feces of healthy individuals, and it was then named *Bacterium coli*. After reclassification in 1919, this bacterium was named *E. coli* after the discoverer (Castellani and Chalmers 1919). Currently, two types of *E. coli* strains are commonly used in molecular biology, the K-12 and B strains. K-12 was isolated in 1922 from a fecal sample of a patient convalescing from diphtheria and was labeled K-12 at Stanford University (Bachmann 1972). K-12 was not from the antigen but the registration number at Stanford. In 1997, entire genome sequences of two *E. coli* K-12 strains, W3110 (Yamamoto et al. 1997) and MG1655 (Blattner et al. 1997), had been determined.

The B strain goes back to the analysis of bacteriophages in Pasteur Institute around 1918. The strain name "B" was originated from the analyses of bacteriophages T1 and T7 by Delbruck in 1942. The genome sequence (Jeong et al. 2009) and pedigree (Daegelen et al. 2009) of B strain was analyzed in 2009.

After discovery of conjugation and recombination of *E. coli* by Tatum and Laderberg (1947), analyses using *E. coli* have increased explosively. Also, we hardly need to mention the contribution of this organism to clarifying the universal molecular mechanisms of fundamental biological processes, such as DNA replication, transcription, and translation.

Escherichia coli is also an industrially important organism. The first product of genetic engineering was human insulin, using *E. coli* in 1982 industrially by the clone of the human gene within 5 years of its first report.

9.3 Importance of Omics Approach

We think that systematic and comprehensive analyses can provide biological information in two important aspects: one is to discover unpredictable results and another is to obtain a global view of targets of interest. Although there is a huge accumulation of biological knowledge, there are still many unpredictable physiological connections between genes in a cell. The systematic approach is one of the powerful ways to identify such unpredictable relationships.

Global analyses may function as a compass showing the direction to the real goal. Even though the research is about an individual target gene or protein, this is the reason why a bird's-eye view by global analysis is so important.

No genes function entirely independent to other genes in a cell system. It is clear that collaboration between the vertical direction of individual research targets and the horizontal approach by systematic analysis may provide synergistic effects and accelerate biological science research.

9.4 Construction of Comprehensive Libraries

To perform a systematic analysis, comprehensively prepared libraries are one of the essential tools. In *E. coli*, the first comprehensive resource, the ordered phage clone library of *E. coli* W3110 genome, was established by Kohara et al. to construct a restriction enzyme map of the genome (Kohara et al. 1987). The Japanese *E. coli* genome project, launched in 1989, was the first systematic sequencing project in the world, based on this ordered phage clone library (Yura et al. 1992). During the genome project era, Japanese groups made great contributions to comprehensive libraries, such as full-length cDNA libraries of *Caenorhabditis elegans*, the mouse, the human, and *Oryza sative japonica* (rice).

Since 1997, with completion of the sequencing project of *E. coli*, we started to construct the entire set of ORF plasmid clone library (Fig. 9.1a). All the amino-acid coding regions except the initiation codon were amplified by polymerase chain reaction (PCR) and cloned into our original multicopy vector (Kitagawa et al. 2005). Using this library, cloned ORF regions were amplified by PCR, then immobilized on glass slides to make the full-length cDNA-type DNA microarray, which was done in

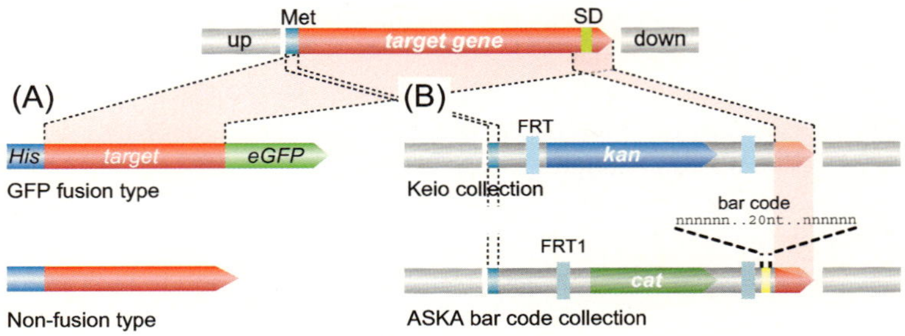

Fig. 9.1 (**A**) Open reading frame (ORF) plasmid clone libraries and (**B**) single-gene deletion collections. Plasmid clones carry the whole amino-acid coding region except the initiation codon. Single-deletion collections of coding region from the second to seventh from the C-terminus. Molecular barcode is 20 nt sequence (Yamamoto et al., in preparation)

collaboration with Takara Bio Company, and many transcriptome analyses were achieved.

Initially, we started to construct a plasmid clone library as His-tagged at the N-terminus and GFP-fused at the C-terminus. Since then, plasmid clone libraries have been expanding to the His-tagged but not the GFP fusion clone (Kitagawa et al. 2005), the gateway entry clone (Rajagopala et al. 2010), and now construction of the low-copy expression and self-transmissible clone library is underway (Yamamoto et al., in preparation). These resources have cited about 400 publications that contributed to this field.

A mutation library is also an important resource. *E. coli*, however, had long been thought to be a very difficult organism in which to generate mutants by homologous recombination, as compared with *Bacillus subtilis* or *Saccharomyces cerevisiae*, in which efficient homologous recombination can occur with less than about 50 bp. Even though using the exonuclease-deficient mutant, normally more than a couple hundred base pair homologous regions are required at both ends of a linear DNA fragment for practical recombination. For this reason, we first considered constructing a comprehensive mutant library by random transposon insertion (Miki et al. 2008). In 2000, however, Datzenko and Wanner developed an efficient one-step recombination system with a 40- to 50-bp homology region using λ-RED recombinase(Datsenko and Wanner 2000). PCR fragments were designed for each of the target ORF genes to create in-frame deletions of the second through seventh amino acids codons from the C-terminus, leaving the initiation codon and the translational signal for the downstream gene. Each fragment carries a Km resistance cassette between FRT sites, which are the target sites for site-specific recombination by FLP in the yeast *S. cerevisiae*. The removal of the Km resistance gene by FLP recombinase generates the in-frame deletion mutant strain (Baba et al. 2006). The final product, 34 amino acids in total length, consists of methionine and the 6 amino acids from the C-terminus including the translational product from the FRT site. These peptides, so far as we have tested, have not shown any phenotypic effects on the host strain.

Each of the primers for homologous recombination has 50-mer for the target site with 20-mer to amplify the antibiotic-resistant cassette.

The construction had been done in a gene-by-gene manner systematically, and an entire 2 years were needed to try the whole set of genes and another year for further confirmation and rearray as a stock.

The schematic illustration is shown in Fig. 9.1b; the second deletion library with molecular barcode is now almost completed and the library set for distribution is now under preparation. Currently, the National Bio-resource Center of the National Institute of Genetics, Mishima, Japan is the official distribution site of our resources.

9.5 Quality Control of Resources

Construction of the library is the most difficult and time-consuming step; however, quality control is another important task to keep libraries reliable. We actually have reconstructed more than 1,000 plasmid clones in response to annotation changes since 1997.

Since the completion of the genome project in 1997, we continued resequencing to check discrepancies between MG1655 and W3110 to produce a highly accurate sequence. Based on the most updated and accurate genome sequence (Hayashi et al. 2006), annotation jamborees had been held two times (Riley et al. 2006). Not only the correction of sequencing errors, but annotations of genes, especially the starting sites of CDS, were carefully considered. Finally more than 1,000 ORF genes had been reannotated for their start sites. Still, at present, revision of annotation is underway and about 200 gene annotations have been changed or added. Thus, reconstruction of resources should be always under consideration. For deletion strain collection, identification of partial duplication is also important for quality control (Yamamoto et al. 2009). The addition of new annotations of small ORFs and RNAs has accelerated recently, and two types of sRNA deletion libraries are almost completed.

9.6 Application of Experimental Resources

Our comprehensive resources are applicable for any type of systematic analyses. Some examples reflect our experiences.

The single-gene deletion library, the so-called Keio collection (Baba et al. 2006), has contributed more than 1,700 publications up to March 2013.

Thanks to technical innovation, high-throughput experiments have now become much easier than previously and are more and more popular. A variety of high-throughput dispensing machines by tip, capillary, or pin are now available, although still expensive. The 96-pinning system with manual operation is also available. Currently, we use the RoTor stamping system developed by Singer Instruments (UK) (http://www.singerinstruments.com/). This machine had been developed

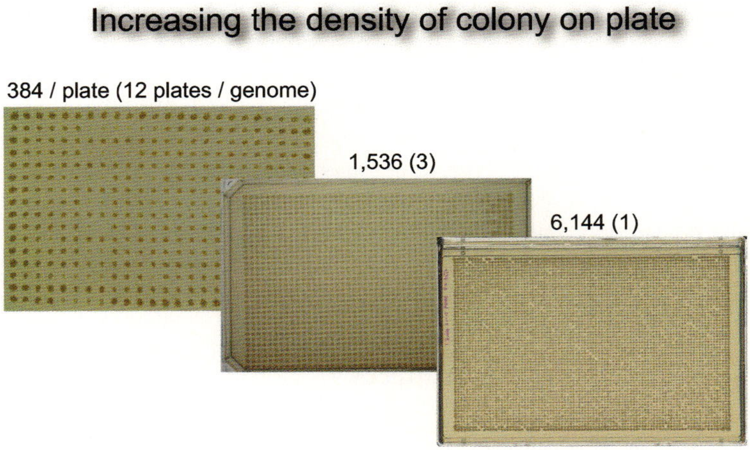

Fig. 9.2 High density of colony image on plate. Colonies were stamped by Singer RoTor robotic system by different pin densities supplied by the manufacturer. *Numbers* above the plate images are colony numbers of one plate; *numbers* in parentheses are the number required to cover the whole set of *Escherichia coli* genes

originally for picking and stamping of yeast colonies by collaboration between C. Boone's laboratory in Toronto University and Singer Instruments.

Escherichia coli K-12 has more than 4,000 genes including sRNA genes in current annotation, and the Keio collection contains two independent isolates of each of the protein-coding genes, in a total number of about 9,000 individual strains (Baba et al. 2006). The RoTor system provides from 96 to 1,536 density pins on a disposable plastic pad, and currently a 6,144-pin pad is also available from the company (Fig. 9.2). Keio collection are stored as glycerol frozen stock in 96- or 384-well microtiter plates and are easily transferred from 384-well liquid stock to 1,536 colonies on agar plates by the pinning robotic system. Once the 1,536-colony plate is made as a seed plate, the 1,536-pin pad is easily replicated onto a variety of plates, such as antibiotics, nutrients, or salts. Only three plates of the 1,536 density can cover the entire set of *E. coli* genes.

Another series of libraries are ORF plasmid clone libraries, so-called ASKA libraries(Kitagawa et al. 2005). Two types of N-terminus His-tagged clones have been established, fusion and non-fusion with enhanced green fluorescent protein (eGFP) at the C-terminus of target ORFs under the regulation of the isopropyl β-d-1-thiogalactopyranoside (IPTG) inducible promoter. These clones were constructed using amplified fragments of the amino-acid coding regions, except the first methionine codon by PCR. Initially we used the annotation registered in 1997 for the first complete genome sequence. As mentioned earlier, about 1,000 ORFs have been changed in their initiation position by annotation in 2005.

Research applications of these plasmid clone libraries already count more than 400 by Google Scholar. Plasmid clone and deletion libraries provide systems to analyze oversupplied condition (Pathania et al. 2009) and loss of function (Nichols et al. 2011), respectively. Plasmid clones also support systems to purify target gene products (Arifuzzaman et al. 2006). Once such comprehensive resources have been established, almost unlimited possible gates are opened.

Although a number of resources had been already established, such as luciferase fusion (Van Dyk et al. 2001), GFP fusion with promoters (Zaslaver et al. 2006), ORF plasmid clone (Kitagawa et al. 2005), and single-gene knockout (Baba et al. 2006), new series of comprehensively designed libraries are still needed for novel purposes. Another set of the single-gene deletion library for systematic genetic interaction analysis (Typas et al. 2008; Butland et al. 2008), the deletion library with barcode (manuscript in preparation), and the low-copy self-transmissible plasmid expression library by conjugation (manuscript in preparation) are examples.

9.7 Points to Note for Libraries

Plasmid clone library: Efforts to add new annotation of genes are continuing as are corrections of wrong annotations. The first comprehensive annotation of the complete genome was done in 1997 (Blattner et al. 1997; Yamamoto et al. 1997). The annotation was done basically with maximum information and a longer coding region was chosen when ambiguous multiple initiation sites exist. After acquiring a more accurate genome sequence (Hayashi et al. 2006), reannotation efforts accelerated (Riley et al. 2006). Finally about 1,000 genes had been corrected as to their start sites. This type of correction affects the plasmid clone library more seriously than the deletion library. We actually observed, at least in some cases, failure of expression when cloned with the wrong starting site. So, about 1,000 ASKA plasmid clones have been reconstructed based on the most reliable starting site (manuscript in preparation). For this reason, the ASKA library consists of more than 5,000 clones. Current distribution from the National Institute of Genetics includes updated plasmids (http://www.shigen.nig.ac.jp/ecoli/strain/top/top.jsp).

9.7.1 *Single Deletion Collection*

During establishing a single-gene deletion strain after homologous recombination between antibiotic cassette and the target gene, additional suppressive mutations might be introduced somewhere on the chromosome. It is clearly not practical to identify such suppressor mutations in each of the deletion strains. Two series of validations were done during the Keio collection construction, antibiotic resistance of the cassette and chromosomal structure by genomic PCR to check the predicted structure after the recombination. In some cases, two independent strains of this library showed different phenotypes, and we have validated the whole set of strains to identify partial duplications (Yamamoto et al. 2009). As a result, we have corrected the number of essential gene candidates in LB condition from 302 to 328 genes, excluding sRNA genes. The official distribution site of the library, the National Institute of Genetics Mishima Japan, is now distributing only the validated ones.

Reconstruction of the Keio collection based on the new annotation has not been started because of funding limitations. For practical points, a few cases have been found after the genome sequence correction and reannotation, such as fusion and fission of previously annotated genes. All this information was compiled and is now undergoing preparation to be opened (manuscript in preparation).

Although many points should be noted for usage of these libraries, these are still undoubtedly valuable resources.

9.8 Toward Genome Design

Although technologies to create genome-size clones (Itaya et al. 2008) or to chemically synthesize genomes (Gibson et al. 2010) are available, how to design a genome is still a large issue to be solved, and breeding of industrially valuable organisms is still a major research target. It is clear that one of the molecular mechanisms of these difficulties is caused by "robustness." In a cell system, there are flexible metabolic circuits to compensate for a perturbation, such as deletion of a certain enzymatic step. Isozyme or alternative pathway steps are one of those mechanisms. The central metabolic pathway steps are shown in Fig. 9.3a; this is a well-known pathway and many studies about these steps in biochemistry, enzymology, genetics,

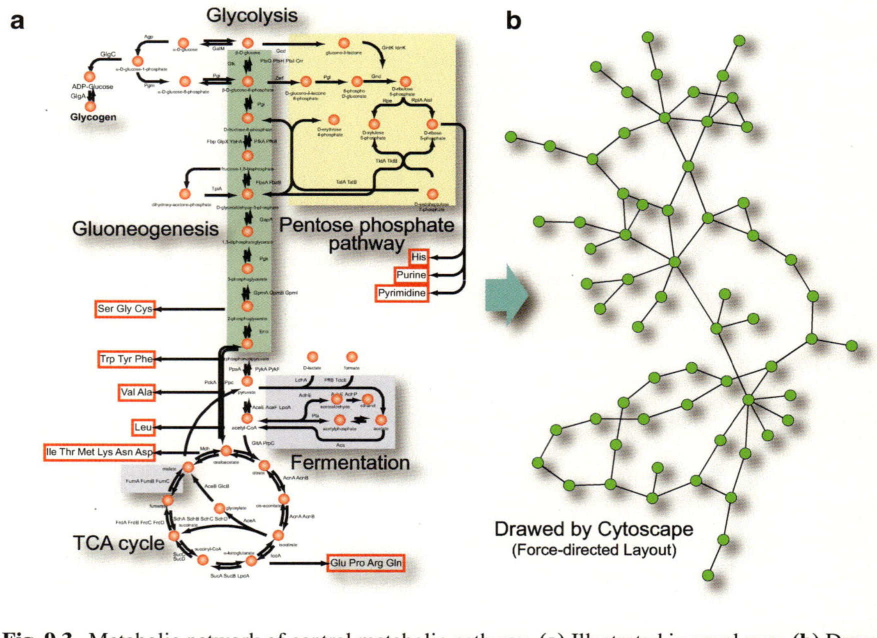

Fig. 9.3 Metabolic network of central metabolic pathway. **(a)** Illustrated in usual way. **(b)** Drawn by Cytoscape using force-directed layout algorithm

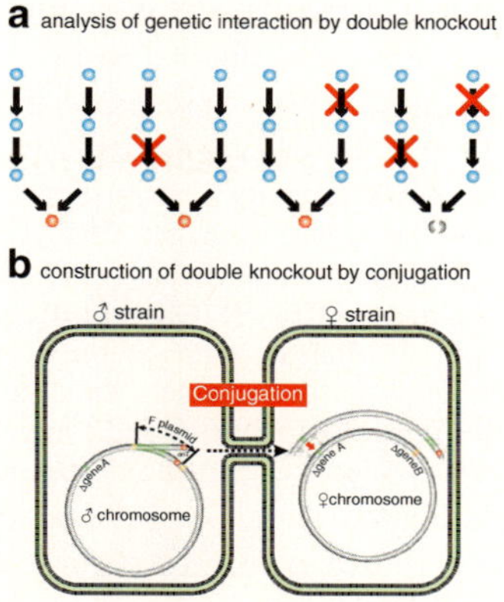

Fig. 9.4 Genetic interaction analysis by conjugation. **(a)** Concept of double-knockout method. **(b)** Construction method of double-knockout strain by conjugation. Gene deletion of gene À is transferred by conjugation to deletion strain of gene B. Homologous recombination generates double-knockout strain

physiology, molecular biology, and systems biology have been done. Redrawing using Cytoscape (Noor et al. 2009) (Fig. 9.3b) shows many loops that can be easily recognized, which may function as alternative pathways to compensate for disruption of a gene.

We had quantitatively and systematically analyzed the transcriptome, proteome, and metabolome of the central metabolic pathway (Ishii et al. 2007). The fluctuations of concentration of metabolites are much smaller in single-gene deletion strains than we expected. Change of growth rate showed more impact on the physiological state of a cell. The interpretation of this observation was that the cellular system tries to keep optimal concentration of each metabolite at a certain growth condition. Perturbation by gene deletion might be overcome by dynamic reconnection of metabolic circuits. To access this problem, we have been performing systematic genetic interaction analysis by double-gene knockout combining two types of single-gene knockout by conjugation as shown in Fig. 9.4 (Butland et al. 2008; Typas et al. 2008). Currently, the construction of the second single-gene deletion collection has been almost finished and a distribution copy is now under preparation. The new collection features a different antibiotic marker (Cm^R) and a 20-nt molecular barcode.

A quantification system is also our goal for improvement, and now time-series measurement is available using a transparent lighting system.

Currently, systematic genetic interaction analysis as just described is underway. Also, more high-throughput methods to identify genetic interaction are now under development for testing in various conditions.

9.9 Epilogue

Comprehensive resources are now one of the essential tools in the systems approach; however, construction including quality control itself is not easily funded. It is clear that the contribution of a biological database, especially a sequence database, is very large for modern biology. Contribution of comprehensively prepared experimental resources is also enormous, without question. In this aspect, Japanese groups have made great contributions in a variety of model organisms.

Acknowledgments Construction and quality control of our experimental resources have been done by support from the CREST funding program of Japan Science and Technology Agency, NEDO program of Ministry of Economy, Trade and Industry, Grant-in-Aid for Scientific Research (A), and KAKENHI (Grant-in-Aid for Scientific Research) on Priority Areas "System Genomics" from the Ministry of Education, Culture, Sports, Science and Technology of Japan. Part of the construction of the single-gene knockout library has been supported by local governmental funds from the Yamagata Prefecture and Tsuruoka City. Barry L. Wanner was supported by NIH GM62662.

References

Arifuzzaman M, Maeda M, Itoh A, Nishikata K, Takita C, Saito R, Ara T, Nakahigashi K, Huang HC, Hirai A, Tsuzuki K, Nakamura S, Altaf-Ul-Amin M, Oshima T, Baba T, Yamamoto N, Kawamura T, Ioka-Nakamichi T, Kitagawa M, Tomita M, Kanaya S, Wada C, Mori H (2006) Large-scale identification of protein–protein interaction of *Escherichia coli* K-12. Genome Res 16(5):686–691

Baba T, Ara T, Hasegawa M, Takai Y, Okumura Y, Baba M, Datsenko KA, Tomita M, Wanner BL, Mori H (2006) Construction of *Escherichia coli* K-12 in-frame, single-gene knockout mutants: the Keio collection. Mol Syst Biol 2(2006):0008

Bachmann BJ (1972) Pedigrees of some mutant strains of *Escherichia coli* K-12. Bacteriol Rev 36(4):525–557

Blattner FR, Plunkett G 3rd, Bloch CA, Perna NT, Burland V, Riley M, Collado-Vides J, Glasner JD, Rode CK, Mayhew GF, Gregor J, Davis NW, Kirkpatrick HA, Goeden MA, Rose DJ, Mau B, Shao Y (1997) The complete genome sequence of *Escherichia coli* K-12. Science 277(5331):1453–1474

Butland G, Babu M, Diaz-Mejia JJ, Bohdana F, Phanse S, Gold B, Yang W, Li J, Gagarinova AG, Pogoutse O, Mori H, Wanner BL, Lo H, Wasniewski J, Christopolous C, Ali M, Venn P, Safavi-Naini A, Sourour N, Caron S, Choi JY, Laigle L, Nazarians-Armavil A, Deshpande A, Joe S, Datsenko KA, Yamamoto N, Andrews BJ, Boone C, Ding H, Sheikh B, Moreno-Hagelseib G, Greenblatt JF, Emili A (2008) eSGA: *E. coli* synthetic genetic array analysis. Nat Methods 5(9):789–795.

Castellani A, Chalmers AJ (1919) Manual of tropical medicine, 3rd edn. Williams Wood, New York

Daegelen P, Studier FW, Lenski RE, Cure S, Kim JF (2009) Tracing ancestors and relatives of *Escherichia coli* B, and the derivation of B strains REL606 and BL21(DE3). J Mol Biol 394(4):634–643.

Datsenko KA, Wanner BL (2000) One-step inactivation of chromosomal genes in *Escherichia coli* K-12 using PCR products. Proc Natl Acad Sci USA 97(12):6640–6645

Gibson DG, Glass JI, Lartigue C, Noskov VN, Chuang RY, Algire MA, Benders GA, Montague MG, Ma L, Moodie MM, Merryman C, Vashee S, Krishnakumar R, Assad-Garcia N, Andrews-Pfannkoch C, Denisova EA, Young L, Qi ZQ, Segall-Shapiro TH, Calvey CH, Parmar PP, Hutchison CA 3rd, Smith HO, Venter JC (2010) Creation of a bacterial cell controlled by a chemically synthesized genome. Science 329(5987):52–56.

Hayashi K, Morooka N, Yamamoto Y, Fujita K, Isono K, Choi S, Ohtsubo E, Baba T, Wanner BL, Mori H, Horiuchi T (2006) Highly accurate genome sequences of *Escherichia coli* K-12 strains MG1655 and W3110. Mol Syst Biol 2(2006):0007

Ishii N, Nakahigashi K, Baba T, Robert M, Soga T, Kanai A, Hirasawa T, Naba M, Hirai K, Hoque A, Ho PY, Kakazu Y, Sugawara K, Igarashi S, Harada S, Masuda T, Sugiyama N, Togashi T, Hasegawa M, Takai Y, Yugi K, Arakawa K, Iwata N, Toya Y, Nakayama Y, Nishioka T, Shimizu K, Mori H, Tomita M (2007) Multiple high-throughput analyses monitor the response of *E. coli* to perturbations. Science 316(5824):593–597

Itaya M, Fujita K, Kuroki A, Tsuge K (2008) Bottom-up genome assembly using the *Bacillus subtilis* genome vector. Nat Methods 5(1):41–43.

Jeong H, Barbe V, Lee CH, Vallenet D, Yu DS, Choi SH, Couloux A, Lee SW, Yoon SH, Cattolico L, Hur CG, Park HS, Segurens B, Kim SC, Oh TK, Lenski RE, Studier FW, Daegelen P, Kim JF (2009) Genome sequences of *Escherichia coli* B strains REL606 and BL21(DE3). J Mol Biol 394(4):644–652.

Kitagawa M, Ara T, Arifuzzaman M, Ioka-Nakamichi T, Inamoto E, Toyonaga H, Mori H (2005) Complete set of ORF clones of *Escherichia coli* ASKA library (a complete set of *E. coli* K-12 ORF archive): unique resources for biological research. DNA Res 12(5):291–299.

Kohara Y, Akiyama K, Isono K (1987) The physical map of the whole *E. coli* chromosome: application of a new strategy for rapid analysis and sorting of a large genomic library. Cell 50(3):495–508

Miki T, Yamamoto Y, Matsuda H (2008) A novel, simple, high-throughput method for isolation of genome-wide transposon insertion mutants of *Escherichia coli* K-12. Methods Mol Biol 416:195–204.

Nichols RJ, Sen S, Choo YJ, Beltrao P, Zietek M, Chaba R, Lee S, Kazmierczak KM, Lee KJ, Wong A, Shales M, Lovett S, Winkler ME, Krogan NJ, Typas A, Gross CA (2011) Phenotypic landscape of a bacterial cell. Cell 144(1):143–156.

Noor R, Murata M, Yamada M (2009) Oxidative stress as a trigger for growth phase-specific sigmaE-dependent cell lysis in *Escherichia coli*. J Mol Microbiol Biotechnol 17(4): 177–187.

Pathania R, Zlitni S, Barker C, Das R, Gerritsma DA, Lebert J, Awuah E, Melacini G, Capretta FA, Brown ED (2009) Chemical genomics in *Escherichia coli* identifies an inhibitor of bacterial lipoprotein targeting. Nat Chem Biol 5(11):849–856.

Rajagopala SV, Yamamoto N, Zweifel AE, Nakamichi T, Huang HK, Mendez-Rios JD, Franca-Koh J, Boorgula MP, Fujita K, Suzuki K, Hu JC, Wanner BL, Mori H, Uetz P (2010) The *Escherichia coli* K-12 ORFeome: a resource for comparative molecular microbiology. BMC Genomics 11:470–477.

Riley M, Abe T, Arnaud MB, Berlyn MK, Blattner FR, Chaudhuri RR, Glasner JD, Horiuchi T, Keseler IM, Kosuge T, Mori H, Perna NT, Plunkett G 3rd, Rudd KE, Serres MH, Thomas GH, Thomson NR, Wishart D, Wanner BL (2006) *Escherichia coli* K-12: a cooperatively developed annotation snapshot–2005. Nucleic Acids Res 34(1):1–9

Tatum EL, Lederberg J (1947) Gene recombination in the bacterium *Escherichia coli*. J Bacteriol 53(6):673–684

Typas A, Nichols RJ, Siegele DA, Shales M, Collins SR, Lim B, Braberg H, Yamamoto N, Takeuchi R, Wanner BL, Mori H, Weissman JS, Krogan NJ, Gross CA (2008) High-throughput, quantitative analyses of genetic interactions in *E. coli*. Nat Methods 5(9):781–787

Van Dyk TK, Wei Y, Hanafey MK, Dolan M, Reeve MJ, Rafalski JA, Rothman-Denes LB, LaRossa RA (2001) A genomic approach to gene fusion technology. Proc Natl Acad Sci USA 98(5):2555–2560

Yamamoto Y, Aiba H, Baba T, Hayashi K, Inada T, Isono K, Itoh T, Kimura S, Kitagawa M, Makino K, Miki T, Mitsuhashi N, Mizobuchi K, Mori H, Nakade S, Nakamura Y, Nashimoto H, Oshima T, Oyama S, Saito N, Sampei G, Satoh Y, Sivasundaram S, Tagami H, Horiuchi T et al (1997) Construction of a contiguous 874-kb sequence of the *Escherichia coli* -K12 genome corresponding to 50.0–68.8 min on the linkage map and analysis of its sequence features. DNA Res 4(2):91–113

Yamamoto N, Nakahigashi K, Nakamichi T, Yoshino M, Takai Y, Touda Y, Furubayashi A, Kinjyo S, Dose H, Hasegawa M, Datsenko KA, Nakayashiki T, Tomita M, Wanner BL, Mori H (2009)

Update on the Keio collection of *Escherichia coli* single-gene deletion mutants. Mol Syst Biol 5:335.

Yura T, Mori H, Nagai H, Nagata T, Ishihama A, Fujita N, Isono K, Mizobuchi K, Nakata A (1992) Systematic sequencing of the *Escherichia coli* genome: analysis of the 0–2.4 min region. Nucleic Acids Res 20(13):3305–3308

Zaslaver A, Bren A, Ronen M, Itzkovitz S, Kikoin I, Shavit S, Liebermeister W, Surette MG, Alon U (2006) A comprehensive library of fluorescent transcriptional reporters for *Escherichia coli*. Nat Methods 3(8):623–628

Chapter 10
Insights into Metabolism and the Galactose Recognition System from Microarray Analysis in the Fission Yeast *Schizosaccharomyces pombe*

Kaoru Takegawa and Tomohiko Matsuzawa

Abstract The fission yeast *Schizosaccharomyces pombe* is a promising host for production of heterologous proteins. However, the oligosaccharide structures of yeasts, including *S. pombe*, differ significantly from those of mammalian cells and humans. In *S. pombe*, galactose residues are transferred to oligosaccharide moieties of glycoproteins by galactosyltransferases in the lumen of the Golgi apparatus. Therefore, UDP-galactose, a substrate for galactosyltransferases, should be synthesized in the cytosol, and transported into the Golgi apparatus by a UDP-galactose transporter. Because *S. pombe* cannot use galactose as a carbon or energy source, little is known about galactose metabolism in this species. A galactose-assimilating mutant of *S. pombe* that was able to grow in minimal galactose medium was isolated. Through DNA microarray analysis of gene expression profiles in the wild type and the mutant, three *gal* genes (*gal7*$^{+}$, *gal10*$^{+}$, and *gal1*$^{+}$) involved in galactose utilization were found to be highly expressed in the mutant. Although galactose residues are not essential for growth of *S. pombe*, galactosylation of protein is required for maintenance of normal cell shape, tolerance toward various drugs, and nonsexual flocculation. We identified fission yeast *gsf2*$^{+}$, encoding a flocculin that binds galactose residues located on cell-surface glycoconjugates by DNA microarray analysis. *S. pombe* appears to have a unique galactose-specific recognition system in which Gsf2/flocculin plays an essential role in mediating cell–cell interactions.

Keywords DNA microarray • Fission yeast • Flocculin • Galactosylation

K. Takegawa (✉) • T. Matsuzawa
Department of Bioscience and Biotechnology, Faculty of Agriculture, Kyushu University, Hakozaki 6-10-1, Fukuoka, 812-8581, Japan
e-mail: takegawa@agr.kyushu-u.ac.jp

H. Anazawa and S. Shimizu (eds.), *Microbial Production: From Genome Design to Cell Engineering*, DOI 10.1007/978-4-431-54607-8_10, © Springer Japan 2014

10.1 *Schizosaccharomyces pombe* Is an Attractive Host for Heterologous Protein Production

The fission yeast *Schizosaccharomyces pombe* is taxonomically and evolutionarily distant from the budding yeast *Saccharomyces cerevisiae. Schizosaccharomyces pombe* shares greater similarity to higher animals than the budding yeast *Saccharomyce cerevisiae* with respect to splicing mechanisms, cell division control, transcription-initiation mechanisms, and posttranslational modifications (Zhao and Lieberman 1995). Therefore, *Schizosaccharomyces pombe* has now become an attractive host for the expression of complex molecules, including glycosylated proteins derived from higher animals. We have developed *S. pombe*-based protein production systems that have been useful for producing many types of heterologous proteins from various organisms including humans (Giga-Hama et al. 2007; Takegawa et al. 2009; Idiris et al. 2010).

In the case of glycoproteins, especially for human therapeutic use, precise structural information on the oligosaccharide portion of the glycoprotein is extremely important. However, the oligosaccharide structures of yeasts, including *S. pombe*, differ significantly from those of mammalian cells and humans, and the difference between yeast and mammalian glycan structures of glycoproteins is a major problem that restricts use of yeasts as production hosts (Gemmill and Trimble 1999). Yeast-specific outer-chain structures may cause rapid removal of the corresponding glycoproteins from the bloodstream or may provoke an immune response in humans. For the production of therapeutic glycoproteins intended for human use, yeasts including *S. pombe* are currently less useful because of their inability to modify proteins with human-compatible glycan structures. Therefore, we have been analyzing the precise glycan structures in *S. pombe* glycosylation mutants to allow use of this organism as an alternative glycoprotein-producing host (Tabuchi et al. 1997; Ikeda et al. 2009; Ohashi and Takegawa 2010; Ohashi et al. 2009, 2010, 2011, 2012).

Schizosaccharomyces pombe was the sixth eukaryotic organism in which the genome was fully sequenced (Wood et al. 2002), and DNA microarray, proteome, and transcriptome analyses have subsequently been carried out (Lackner and Bähler 2008; Marguerat et al. 2012). We created *S. pombe* DNA microarrays using chemically synthesized DNA as probes, and have used them for analyses of the transcriptome in gene disruptants, as well as analyses of the transcriptome when heterologous proteins are expressed or when culture conditions are changed (Giga-Hama et al. 2007). This microarray system was also used to identify a promoter that can be used for the expression of heterologous proteins, and a new expression vector was successfully constructed (Fujita et al. 2006).

In the present review, we discuss current understanding of the biosynthetic pathway of galactose-containing sugar chains in *S. pombe* by use of DNA microarray analysis. We also review transcriptome analysis as a tool to advance metabolic engineering and glycotechnology efforts in *S. pombe*.

10.2 Galactosylation in Fission Yeast

Galactosylation, which occurs in the Golgi apparatus, is one of the unique characteristics of glycoprotein synthesis in *S. pombe*. Glycoproteins of *S. pombe* contain a large amount of galactose in addition to mannose (galactomannan), indicating that *S. pombe*, similar to animal cells, is equipped with mechanisms for the galactosylation of glycoproteins. Ascomycetous yeast species such as *S. cerevisiae* do not produce galactose-containing glycans. Although some basidiomycetous yeasts have been reported to contain Gal residues by monosaccharide composition analysis, their detailed overall glycan structures remain unknown. Furthermore, galactosylation is also rarely found in invertebrates and plants.

Schizosaccharomyces pombe *N*- and *O*-linked glycans contain large amounts of α-linked D-Gal in addition to α-linked D-mannose (Man), in contrast to other yeasts. *N*-linked oligosaccharides in *S. pombe* comprise a $Man_9GlcNAc_2$ inner core to which several α1,2-linked Gal residues and α1,3-linked Gal residues are added (Ziegler et al. 1999) and an outer chain composed of an α1,6-linked Man backbone decorated with α1,2-linked Gal (Gemmill and Trimble 1999). Some of these α1,2-linked Gal residues are further modified with pyruvylated β1,3-linked Gal residues (Gemmill and Trimble 1996; Andreishcheva et al. 2004). *O*-linked oligosaccharides in *S. pombe* consist of $Gal_{0-2}Man_{1-3}$ structures, which have an α1,2-linked Man backbone decorated with α1,2-linked or α1,3-linked Gal residues (Gemmill and Trimble 1999).

Galactose residues of glycoproteins are enzymatically attached by galactosyltransferases, which use UDP-galactose as substrate. In *S. pombe*, seven α1,2-galactosyltransferase-related genes (*gmh1*$^+$–*gmh6*$^+$ and *gma12*$^+$) have been identified. The encoded α1,2-galactosyltransferase-related enzymes have been partially characterized and their substrate specificities determined (Yoko-o et al. 1998; Ohashi et al. 2011). Gma12p is a general α1,2-galactosyltransferase able to transfer α1,2-Gal onto every terminal α1,2-Man residue on *N*- and *O*-linked glycans. The Gmh proteins seem to have different substrate preferences. Gma12p, Gmh2p, Gmh3p, and Gmh6p were also found to be involved in the galactosylation of *N*-linked glycans by acid phosphatase mobility analysis. Furthermore Gmh3p was reported to be involved in the galactosylation of core glycans by an in vitro galactosyltransferase assay using a solubilized microsomal fraction from a *gmh3*Δ mutant. Gmh2p and Gmh6p are involved in the galactosylation of *O*-linked glycan with Gmh2p appearing to have a preference for shorter glycans and Gmh6p for longer glycans. Furthermore, *gmh1*Δ, *gmh4*Δ, and *gmh5*Δ mutants have not been reported to have glycosylation defects. Recently, we searched for unidentified putative glycosyltransferases in the *S. pombe* genome sequence and identified three novel genes, named *otg1*$^+$–*otg3*$^+$ (α-*o*ne, *t*hree-*g*alactosyltransferase) (Ohashi et al. 2012). Gal-recognizing lectin blotting and HPLC analyses of pyridylaminated oligosaccharides after deletion of these three additional genes from a septuple α1,2-galactosyltransferase disruptant (7GalTΔ) strain demonstrated that the resultant disruptant missing 10 α-galactosyltransferase genes, 10GalTΔ, exhibited a complete loss of galactosylation (Ohashi et al. 2012).

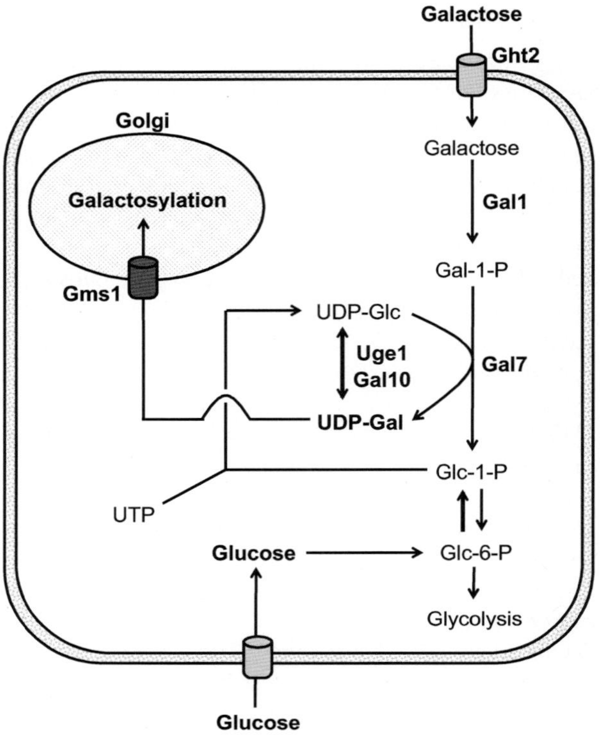

Fig. 10.1 UDP-galactose synthesis from galactose in *Schizosaccharomyces pombe* grown in galactose-containing glucose-rich medium. UDP-galactose is synthesized from extracellular glucose in a Uge1/Gal10-dependent manner, or from extracellular galactose in a Ght2-Gal7-dependent manner in the presence of glucose. UDP-galactose is transported by Gms1, the UDP-galactose transporter, from the cytosol into the Golgi lumen and is then utilized as a substrate for galactosylation

In the cytosol, the key enzyme for UDP synthesis is UDP-galactose/-glucose 4-epimerase, which catalyzes the interconversion of UDP-galactose and UDP-glucose. Interestingly, *S. pombe* has two types of genes encoding UDP-glucose/-galactose 4-epimerase, *uge1*$^{+}$ and *gal10*$^{+}$. Deletion of *uge1*$^{+}$ causes a defect in synthesis of UDP-galactose, and deletion mutants are defective in galactosylation when grown in glucose medium. Gal10 is involved in galactosylation of cell-surface proteins under conditions of glucose starvation (Suzuki et al. 2010). The galactosylation defect of *uge1*Δ mutants is suppressed in galactose-containing medium, indicating that galactose is transported from the medium to the cytosol and is then used for synthesis of UDP-galactose (Suzuki et al. 2010). The biosynthetic pathways for UDP-galactose in *S. pombe* are shown in Fig. 10.1.

Schizosaccharomyces pombe has eight hexose transporter genes, *ght1*$^{+}$–*ght8*$^{+}$. The galactosylation defect of a *uge1*Δ mutant defective in synthesis of UDP-galactose from glucose was suppressed in galactose-containing medium, but disruption of *ght2*$^{+}$ in the *uge1*Δ mutant reversed suppression of the galactosylation defect. Expression of *S. cerevisiae GAL2* in *uge1*Δ*ght2*Δ cells suppressed the defective galactosylation phenotype in galactose-containing medium. Therefore, *ght2*$^{+}$, which is highly expressed in the presence of glucose, is essential for UDP-galactose synthesis from extracellular galactose when cells are grown on glucose (Matsuzawa et al. 2013a).

10.3 Why Is *S. pombe* Unable to Grow on Galactose Medium?

In *S. cerevisiae*, the galactose metabolic pathway (Leloir pathway), which effects the complete transformation of galactose to glucose, has been extensively characterized. The galactokinase Gal1p (ATP-α-D-galactose-1-phosphotransferase), the transferase Gal7p [uridine diphosphoglucose (UDPG)-α-D-galactose-1-phosphouridylyltransferase], and epimerase Gal10p (UDPG-4-epimerase) function in this pathway whereas *GAL2* encodes the galactose permease (Rubio-Texeira 2005).

In contrast to *S. cerevisiae*, *S. pombe* does not grow in media containing galactose as the sole carbon source even though *S. pombe* has the galactose-assimilating genes (*gal1*$^+$, *gal7*$^+$, and *gal10*$^+$) required for galactose metabolism, and galactose is transported from the medium to the cytosol by hexose transporters (Suzuki et al. 2010). To clarify why *S. pombe* does not grow on galactose, we isolated gain-of-function *S. pombe* mutants (FG2-8) able to grow on galactose by using a proofreading-deficient DNA polymerase (Matsuzawa et al. 2011a). DNA microarray analysis was performed initially to examine differences between the wild-type strain and the FG2-8 mutant. SPBPB2B2.10c (*gal7*$^+$), SPBPB2B2.12c (*gal10*$^+$), and SPBPB2B2.13 (*gal1*$^+$), orthologues of *S. cerevisiae GAL10*, *GAL7*, and *GAL1*, respectively, were constitutively upregulated in the FG2-8 mutant (Matsuzawa et al. 2010) (Table 10.1). In addition, SPBPB2B2.07c, SPBPB2B2.08, and SPBPB2B2.09c located near the *gal* genes (between DUF999 and DUF999) were also upregulated (Fig. 10.2). The fission yeast *gal* genes are located at the end of the right arm of chromosome 2. Telomere heterochromatin, which is subject to gene silencing (Kanoh et al. 2005), extends approximately 50 kb from the chromosome end on the right arm of chromosome 2. Our results are consistent with the possibility that gene silencing represses expression of the *gal* genes in the wild-type strain, and that the silencing was diminished in FG2-8 (Matsuzawa et al. 2011a). These results suggest that the *gal* genes were repressed independently of telomeric gene silencing (Fig. 10.2). Identification of the mutation responsible for constitutive expression of the *gal* genes in FG2-8 is the subject of ongoing work based on cloning of the wild-type allele. Its discovery will provide new information about galactose metabolism and gene silencing in *S. pombe* and may lead to development of a galactose-based gene expression system for heterologous protein production.

Table 10.1 Upregulated genes in FG2-8 cells grown in glucose medium

Systematic name	Annotation	Ratio
SPBPB2B2.08	Sequence orphan	14.98
SPBPB2B2.12c	Orthologue of *Saccharomyces cerevisiae GAL10*	14.05
SPBPB2B2.09c	2-Dehydropantoate 2-redutase	9.86
SPBC1348.07	DUF999, SPAC1348.07	7.24
SPBPB2B2.10c	Orthologue of *S. cerevisiae GAL7*	7.09
SPBPB2B2.13	Orthologue of *S. cerevisiae GAL1*	5.98
SPBPB2B2.07c	DUF999	5.71

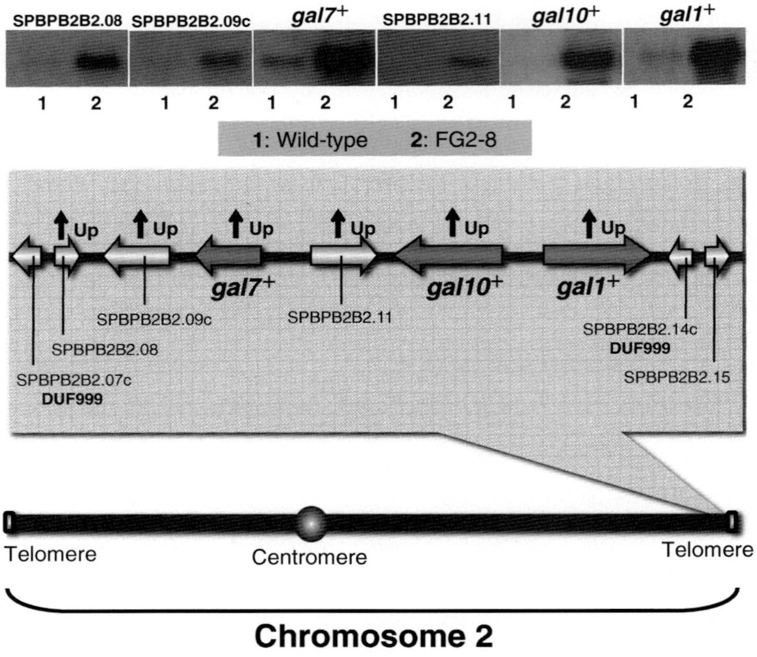

Fig. 10.2 Genes upregulated in FG2-8 cells. Wild-type (*1*) and FG2-8 (*2*) cells were cultured in glucose medium, after which total RNA was extracted. Genes between SPBPB2B2.08 and *gal1*$^+$ (SPBPB2B2.13) were upregulated in FG2-8 cells

10.4 Cell-Surface Galactosylation Is Essential for Cell–Cell Recognition in *S. pombe*

Heterothallic strains of fission yeast, h^+ and h^-, are usually nonflocculent when cultured separately. Cells of homothallic strains of *S. pombe* become intensely flocculent (sexual flocculation) after prolonged cultivation. In contrast to sexual flocculation, little attention has been given to flocculent mutants of heterothallic strains and to mechanisms of nonsexual flocculation in *S. pombe*. We have isolated fission yeast mutants that flocculate constitutively on growth in liquid media (Tanaka et al. 1999). One of the flocculent mutants, *gsf1* (*g*alactose-*s*pecific *f*locculation), exhibited a dominant, nonsexual, and calcium-dependent aggregation phenotype, which suggested that galactose residues rather than mannose residues in cell wall glycoproteins may be receptors that mediate cell–cell adhesion in *gsf1*-mediated flocculation (Tanaka et al. 1999). Although the *S. pombe* genome has been sequenced, a candidate protein involved in nonsexual flocculation had not been found.

Flocculation of *S. cerevisiae* is a nonsexual form of cell aggregation that is calcium dependent and reversible. The mechanism by which *S. cerevisiae* cells flocculate has been studied extensively both biochemically and genetically.

Table 10.2 Upregulated genes in *ScFLO8-overexpressing Schizosaccharomyces pombe* cells

Systematic name	Annotation	Ratio
SPCC1742.01	Sequence orphan	4.52
SPCC1450.08c	*S. pombe*-specific families	4.34
SPAC977.14c	Aldo/keto reductase	4.33
SPBC16A3.01	Septin, Spn3	3.41
SPAC22A12.04c	40S ribosomal protein	3.38
SPBC211.03c	Guanyl-nucleotide exchange factor	3.20
SPBC31E1.06	GTP-binding protein, Bms1	3.09
SPAC5H10.06c	Alcohol dehydrogenase, Adh4	2.71

This phenomenon is thought to involve cell-surface components and to result from a lectin-like interaction between a cell-wall sugar-binding protein and cell-surface mannan. Several dominant flocculation genes including *FLO1*, *FLO5*, and *FLO10* have been identified in *S. cerevisiae* (Goossens and Willaert 2010). The N-terminal regions of Flo1p, Flo5p, and Flo10p are important for sugar recognition (Goossens and Willaert 2010). Recently, the domain structure of Flo5p was resolved, and the N-terminal Flo5A domain (S23 to H271) was found to be required for calcium-dependent flocculation (Veelders et al. 2010).

We found that nonsexual flocculation in *S. pombe* was induced by expression of *S. cerevisiae FLO8* (Matsuzawa et al. 2011b). The *FLO8* gene, encoding a transcriptional activator of the dominant flocculation genes *FLO1* and *FLO11*, induces nonsexual flocculation in *S. cerevisiae* (Kobayashi et al. 1996). To determine why *FLO8* expression induced flocculation in *S. pombe*, microarray analysis was performed in wild-type and *FLO8*-overexpressing cells (Table 10.2). In the *FLO8-overexpressing* strain, a gene designated *gsf2*$^+$ was specifically induced. The Gsf2 protein consists of 43.9 % serine and threonine residues and contains an N-terminal signal sequence, a C-terminal GPI anchor signal sequence, eight repeated sequences consisting of 78 and 44 amino acids, respectively, and 16 putative *N*-glycosylation NX(S/T) sites. Although Gsf2p does not share sequence similarity with Flo1p, Flo11p, or Flo5p, the protein structures (GPI anchor, high Ser/Thr content, and repeated sequences) resemble budding yeast flocculins. We conclude that Gsf2p aggregates cells into 'flocs' by binding to galactose-containing cell-surface oligosaccharides (Matsuzawa et al. 2011b).

The *pvg4*$^+$ gene originally was isolated as a multicopy suppressor of a *pvg4* mutant defective in pyruvylation of *N*-linked oligosaccharides (Andreishcheva et al. 2004). However, we did not detect a defect in cell-surface pyruvylation in a *pvg4/mbx2* deletion mutant, as assessed by alcian blue staining and a Q-Sepharose binding assay. Instead, the deletion reduced the adhesion and flocculation of otherwise flocculent mutants by reducing *gsf2*$^+$ expression. *mbx2*$^+$-overexpressing strains exhibited nonsexual and calcium-dependent aggregation, which was inhibited in the presence of galactose but mediated by the induction of *gsf2*$^+$. These findings indicate that the MADS box transcription factor Mbx2 mediates flocculation via the transcriptional activation of *gsf2*$^+$ in fission yeast (Fig. 10.3) (Matsuzawa et al. 2012). In addition, we found that Gsf1, a newly identified transcription factor containing an

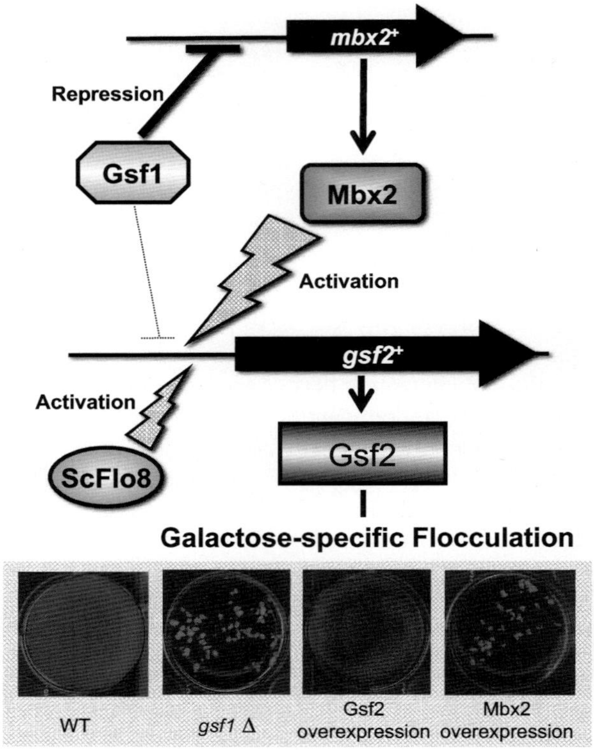

Fig. 10.3 Regulation of nonsexual flocculation in *S. pombe*. Nonsexual flocculation and expression of *gsf2*$^+$ encoding galactose-specific flocculin are activated by Mbx2 and ScFlo8, whereas nonsexual flocculation is repressed by Gsf1 in *S. pombe*. Deletion of *gsf1*$^+$ or overexpression of *mbx2*$^+$ or *gsf2*$^+$ causes galactose-specific nonsexual flocculation

N-terminal Zn-finger-type DNA-binding domain, represses expression of *mbx2*$^+$ and/or *gsf2*$^+$ and, therefore, negatively regulates flocculation in wild-type cells (Fig. 10.3) (Matsuzawa et al. 2013b).

Flocculation and sedimentation of yeast cells to the bottom of fermentation vessels is a desirable industrial property. We found that overexpression of *gsf2*$^+$ or *mbx2*$^+$ strongly induces flocculation of wild-type *S. pombe*, suggesting that secreted heterologous proteins from culture broth can easily be purified without centrifugation.

10.5 Conclusion

The glycan moieties associated with these glycoproteins provide the cell surface with a net negative charge, which plays a significant role in the regulation of influx/efflux processes and in cell–cell interactions such as cell aggregation and pathogenic adhesion. In general, sialylated and sulfated glycans are associated with vertebrate glycoproteins. *S. pombe N*-linked polysaccharides contain pyruvic acid 4,6-ketal-linked to Galβ1,3-(PvGal) (Gemmill and Trimble 1996). Few reports have documented the presence of pyruvylated sugar residues in eukaryotic polysaccharides and glycoprotein oligosaccharides.

Mutants of *S. pombe* lacking the pyruvyl group in PvGal have been isolated, leading to the identification of five genes (*pvg1–5*$^{+}$) involved in PvGal biosynthesis (Andreishcheva et al. 2004). BLAST searches established that the amino-acid sequence of Pvg1p shares weak similarity with predicted bacterial pyruvyltransferases. Recently, we confirmed that Pvg1p transferred a pyruvyl residue from PEP to *p*NP-β-Gal, resulting in the synthesis of 4,6-ketal-linked PvGal, indicating that Pvg1p has pyruvyltransferase activity (Yoritune et al. 2013). Furthermore, we succeeded in determining the three-dimensional crystal structure of recombinant Pvg1p (our unpublished results). In contrast to observations noted in a previous report (Andreishcheva et al. 2004), we found that Pvg1p is located in the Golgi apparatus and therefore proposed that pyruvylation of oligosaccharides occurs in the Golgi apparatus in *S. pombe* (Yoritune et al. 2013). Sialic acids are synthesized in vitro by linking pyruvate to mannosamine. It is interesting to note the close structural similarity between the 4,6-ketal-linked pyruvate of the PvGal epitope and α2,6-linked sialic acid common in glycoproteins of higher eukaryotes (Gemmill and Trimble 1996). We are now testing whether 4,6-ketal-linked pyruvate behaves similarly to α2,6-linked sialic acid in mammalian cells.

References

Andreishcheva EN, Kunkel JP, Gemmill TR et al (2004) Five genes involved in biosynthesis of the pyruvylated Galbeta1,3-epitope in *Schizosaccharomyces pombe* N-linked glycans. J Biol Chem 279:35644–35655

Fujita Y, Tohda H, Giga-Hama Y et al (2006) Heat shock-inducible expression vectors for use in *Schizosaccharomyces pombe*. FEMS Yeast Res 6:883–887

Gemmill TR, Trimble RB (1996) *Schizosaccharomyces pombe* produces novel pyruvate-containing N-linked oligosaccharides. J Biol Chem 271:25945–25949

Gemmill TR, Trimble RB (1999) Overview of N- and O-linked oligosaccharide structures found in various yeast species. Biochim Biophys Acta 1426:227–237

Giga-Hama Y, Tohda H, Takegawa K et al (2007) *Schizosaccharomyces pombe* minimum genome factory. Biotechnol Appl Biochem 46:147–155

Goossens K, Willaert R (2010) Flocculation protein structure and cell–cell adhesion mechanism in *Saccharomyces cerevisiae*. Biotechnol Lett 32:1571–1585

Idiris A, Tohda H, Kumagai H et al (2010) Engineering of protein secretion in yeast: strategies and impact on protein production. Appl Microbiol Biotechnol 86:403–417

Ikeda Y, Ohashi T, Tanaka N et al (2009) Identification and characterization of a gene required for α1,2-mannose extension in the O-linked glycan synthesis pathway *Schizosaccharomyces pombe*. FEMS Yeast Res 9:115–125

Kanoh J, Sadaie M, Urano T et al (2005) Telomere binding protein Taz1 establishes Swi6 heterochromatin independently of RNAi at telomeres. Curr Biol 25:1808–1819

Kobayashi O, Suda H, Ohtani T et al (1996) Molecular cloning and analysis of the dominant flocculation gene FLO8 from *Saccharomyces cerevisiae*. Mol Gen Genet 251:707–715

Lackner DH, Bähler J (2008) Translational control of gene expression from transcripts to transcriptomes. Int Rev Cell Mol Biol 271:199–251

Marguerat S, Schmidt A, Codlin S et al (2012) Quantitative analysis of fission yeast transcriptomes and proteomes in proliferating and quiescent cells. Cell 151:671–683

Matsuzawa T, Fujita Y, Tanaka N et al (2011a) New insights into galactose metabolism by *Schizosaccharomyces pombe*: isolation of a galactose-assimilating mutant. J Biosci Bioeng 111:158–166

Matsuzawa T, Morita T, Tanaka N et al (2011b) Identification of a galactose-specific flocculin essential for nonsexual flocculation and hyphal growth in *Schizosaccharomyces pombe*. Mol Microbiol 82:1531–1544

Matsuzawa T, Yoritsune K, Takegawa K (2012) MADS box transcription factor Mbx2/Pvg4 regulates invasive growth and flocculation by inducing *gsf2*$^{+}$ expression in fission yeast. Eukaryot Cell 11:151–158

Matsuzawa T, Hara F, Tanaka N et al (2013a) *ght2*$^{+}$ is required for UDP-galactose synthesis from extracellular galactose by *Schizosaccharomyces pombe*. Appl Microbiol Biotechnol 97:4957–4964

Matsuzawa T, Kageyama Y, Ooishi K et al (2013b) The zinc finger protein Gsf1 regulates Gsf2-dependent flocculation in fission yeast. FEMS Yeast Res 13:259–266

Ohashi T, Takegawa K (2010) N- and O-linked oligosaccharides completely lack galactose residues in the gms1och1 mutant of *Schizosaccharomyces pombe*. Appl Microbiol Biotechnol 86:263–272

Ohashi T, Ikeda Y, Tanaka N et al (2009) The och1 mutant of *Schizosaccharomyces pombe* produces galactosylated core structures of N-linked oligosaccharides. Biosci Biotechnol Biochem 73:407–414

Ohashi T, Nakakita S, Sumiyoshi W et al (2010) Production of heterologous glycoproteins by a glycosylation-defective *alg3och1* mutant of *Schizosaccharomyces pombe*. J Biotechnol 150:348–356

Ohashi T, Nakakita S, Sumiyoshi W et al (2011) Structural analysis of α1,3-linked galactose-containing oligosaccharides in *Schizosaccharomyces pombe* mutants harboring single and multiple α-galactosyltransferase genes disruptions. Glycobiology 21:340–351

Ohashi T, Fujiyama K, Takegawa K (2012) Identification of novel α1,3-galactosyltransferase and elimination of α-galactose-containing glycans by disruption of multiple α-galactosyltransferase genes in *Schizosaccharomyces pombe*. J Biol Chem 287:38866–38875

Rubio-Texeira M (2005) A comparative analysis of the GAL genetic switch between not-so-distant cousins: *Saccharomyces cerevisiae* versus *Kluyveromyces lactis*. FEMS Yeast Res 5:1115–1128

Suzuki S, Matsuzawa T, Nukigi Y et al (2010) Characterization of two different types of UDP-glucose/-galactose 4-epimerase involved in galactosylation in fission yeast. Microbiology 156:708–718

Tabuchi M, Tanaka N, Iwahara S et al (1997) The *Schizosaccharomyces pombe gms1*$^{+}$ gene encodes an UDP-galactose transporter homologue required for protein galactosylation. Biochem Biophys Res Commun 232:121–125

Takegawa K, Tohda H, Sasaki M et al (2009) Production of heterologous proteins using the fission-yeast (*Schizosaccharomyces pombe*) expression system. Biotechnol Appl Biochem 53:227–235

Tanaka N, Awai A, Bhuiyan MSA et al (1999) Cell surface galactosylation is essential for nonsexual flocculation in *Schizosaccharomyces pombe*. J Bacteriol 181:1356–1359

Veelders M, Bruckner S, Ott D et al (2010) Structural basis of flocculin-mediated social behavior in yeast. Proc Natl Acad Sci USA 107:22511–22516

Wood V, Gwilliam R, Rajandream MA et al (2002) The genome sequence of *Schizosaccharomyces pombe*. Nature (Lond) 415:871–880

Yoko-o T, Roy SK, Jigami Y (1998) Differences in in vivo acceptor specificity of two galactosyltransferases, the gmh3^{+} and gma12$_{+}$ gene products from *Schizosaccharomyces pombe*. Eur J Biochem 257:630–637

Yoritune K, Matsuzawa T, Ohashi T et al (2013) The fission yeast Pvg1p shows galactose-specific pyruvyltransferase activity. FEBS Lett 587:917–921

Zhao Y, Lieberman HB (1995) *Schizosaccharomyces pombe*: a model for molecular studies of eukaryotic genes. DNA Cell Biol 14:359–371

Ziegler FD, Cavanagh J, Lubowski C et al (1999) Novel *Schizosaccharomyces pombe N*-linked GalMan9GlcNac isomers. Role of the Golgi GMA12 galactosyltransferase in core glycan galactosylation. Glycobiology 9:497–505

Part IV
Applications of Advanced Technologies for Production

Chapter 11
Multi-enzymatic Systems for the Production of Chiral Compounds

Akira Iwasaki, Noriyuki Ito, and Yoshihiko Yasohara

Abstract Two multi-enzymatic systems for the production of optically active compounds are discussed. One is a system for deracemization via enzymatic stereo-inversion useful for chiral 1,2-diol synthesis, which involves stereoselective oxidation and stereoselective reduction using an alcohol dehydrogenase and reductase with opposite stereospecificities. The recycling of NAD^+ was carried out by a water-forming NADH oxidase. The recycling of NADPH was carried out by an NADPH-dependent glucose dehydrogenase. The recombinant *Escherichia coli* capable of overproducing the four enzymes was constructed for purposes of generating catalysts useful for the deracemization of racemic diols. Optically active 3-chloro-1,2-propanediol and (*S*)-1,2-pentanediol could be prepared by using the recombinant *E. coli* cells in a one-pot reaction with high yield. Second is a transamination system for chiral secondary amine synthesis. Two novel bacterial ω-transaminases capable of catalyzing a stereoselective transamination between a ketone and amine were discovered by enrichment culturing. An (*R*)-specific ω-transaminase was isolated from a microorganism and characterized, and the relevant gene was cloned. Addition of lactate dehydrogenase into the transaminase reaction mixture was effective at increasing the yield by ensuring the reversibility of the reaction. A recombinant *E. coli* system capable of overproducing the ω-transaminase, lactate dehydrogenase, and glucose dehydrogenase used for regenerating NADH was constructed. By using the recombinant enzyme systems, various kinds of useful optically active amines could be prepared in good yields and high optical purity without using high-pressure and high-temperature conditions.

Keywords Bioconversion • Chiral synthesis • Stereoinversion • Transaminase

---

A. Iwasaki • N. Ito • Y. Yasohara (✉)
Kaneka Corporation, 1-8 Miyamae, Takasago, Hyogo 676-8688, Japan
e-mail: Yoshihiko_Yasohara@kn.kaneka.co.jp

H. Anazawa and S. Shimizu (eds.), *Microbial Production: From Genome Design to Cell Engineering*, DOI 10.1007/978-4-431-54607-8_11, © Springer Japan 2014

11.1 Introduction

Biocatalysts such as microorganisms or enzymes can catalyze highly selective reactions at ambient temperature and pressure, which are difficult to achieve through standard organic synthetic techniques. Manufacturing optically active compounds, which are important raw materials for pharmaceuticals and agrochemicals, can be carried out using biocatalysts. However, biocatalysts generally have such low productivity that their applications have been limited because of their high cost. Most industrially useful enzymatic reactions are hydrolytic reactions, with a few exceptions. For biocatalytic reactions to be widely adopted in manufacturing, it is necessary to overcome their low productivity and broaden their applicability to a larger number of reaction types.

There are two important issues in the next-generation development of bioprocesses. One is the replacement of chemical processes with bioprocesses while considering the preservation of the environment and conservation of energy. The other is the development of novel industrial processes by maximizing the use of biocatalysts. Creation of a multi-enzymatic system that contains coupling enzymes able to function in a one-pot reaction is one of the most important challenges. In this chapter, some examples of chemical processes using multi-enzymatic systems are described by detailing the identification of suitable enzymes, creation of a recombinant microorganism as a catalytic system, and establishment of the processes.

11.2 Synthesis of Optically Active Alcohols by Enzymatic Stereoinversion

Asymmetrical reduction is important in organic chemistry. Use of an enzymatic coupling reaction containing alcohol dehydrogenase and a coenzyme (NAD(P)H) regenerating enzyme are powerful industrial tools similar to the reaction catalyzed by an organometallic complex (Nanba et al. 2007). However, it is difficult to apply this reaction when the ketone substrate is expensive or unstable. We investigated an economical process to synthesize optically active alcohols from a racemic alcohol substrate.

11.2.1 An Enzymatic Stereoinversion

Optically active 3-chloro-1,2-propanediol (CPD) is an important chiral building block. A manufacturing process using the racemate as a starting material is not cost effective because the yield is at most only 50 %, and the chemical catalyst is expensive.

1-Chloro-3-hydroxyacetone (CHA) is an unstable compound for purposes of industrial handling. A reductase can convert CHA to optically active CPD. The configuration of one enantiomer can be reversed by coupling of the asymmetrical oxidation of CPD and asymmetric reduction of the produced CHA. The racemic CPD can then be converged to chiral CPD as the other enantiomer is not changed (Fig. 11.1).

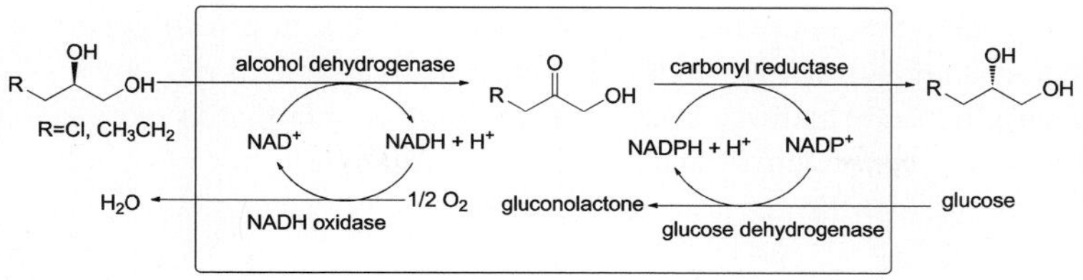

Fig. 11.1 An enzymatic stereoinversion for chiral diol synthesis

Table 11.1 Effect of coenzyme dependency on enzymatic stereoinversion of 3-chloro-1,2-propanediol (CPD)

Recycle of NAD^+	Recycle of NADPH	Optical purity of CPD after the reaction [% enantiomeric excess (e.e.)]
None	None	0
None	NADPH-dependent GDH	34
NADH oxidase	NADPH-dependent GDH	100
NADH oxidase	Nonselective GDH (*Bacillus*)	5
NADH oxidase	Nonselective GDH (*Bacillus*)	19

Coenzyme regeneration via reduction and oxidation is necessary to achieve a more efficient reaction. Glucose dehydrogenase (GDH) has been successfully used on an industrial scale for the regeneration of the reduced coenzyme formed during a reaction. NADH oxidase was selected for the regeneration of the oxidized form of the coenzyme because it catalyzes an irreversible reaction with no harmful side products. Thus, a water-forming NADH oxidase is more useful than a peroxide-forming NADH oxidase because the peroxides produced may have a negative effect on the system enzymes and compounds. For industrial applications, a more stable enzyme could be prepared by random mutations.

Table 11.1 shows the effect of the coenzyme species on a stereoinversion reaction where the oxidation and the reduction occur at the same time. The stereoinversion reaction succeeded only when the NADPH-dependent glucose dehydrogenase was used in the presence of the four enzymes as shown in Fig. 11.1.

11.2.2 NADPH-Dependent Glucose Dehydrogenase

Glucose dehydrogenase derived from *Bacillus* sp. is used for the industrial regeneration of NADPH and NADH. It can be a powerful force to drive the reduction reaction to completion because the reaction irreversibly rapidly converts the produced gluconolactone into gluconic acid. As already described, an NADPH-dependent enzyme is required to complete the stereoinversion reaction. We researched the coenzyme dependency of microbial glucose dehydrogenases.

The glucose dehydrogenases derived from lactic acid bacteria such as *Lactobacillus* and *Pediococcus* spp. show high selectivity. In addition, these novel NADPH-dependent glucose dehydrogenases have an optimum pH that is more acidic than other reported glucose dehydrogenases.

11.2.3 Synthesis of Optically Active (R)-3-Chloro-1,2-Propanediol

A glycerol dehydrogenase derived from *Cellulomonas* sp. was discovered as a (*S*)-CPD-oxidizing enzyme by microorganism screening. (*R*)-CPD could be produced from 10 g/l racemic CPD with 50 % yield and 99.8 % enantiomeric excess (e.e.) by *Escherichia coli* overexpressing the gene of this glycerol dehydrogenase. The intermediate CHA was not accumulated because of its chemical instability. NAD^+ was regenerated as a result of the cell's own regenerative machinery. Synthesis of (*R*)-CPD from racemic CPD was carried out using a system containing four enzymes. The recombinant *E. coli* cells harboring the plasmid vector where the genes for the four enzymes were inserted downstream of the *lac* promoter provided the catalysts for this reaction. The four enzymes were carbonyl reductase S1 derived from *Candida magnoliae* AKU4643 (Wada et al. 1998) as the CHA-reducing enzyme, a NADPH-dependent glucose dehydrogenase from a lactic acid bacterium as the NADPH-regenerating enzyme, the glycerol dehydrogenase described previously as the (*S*)-CPD-oxidizing enzyme, and the NADH oxidase derived from *Streptococcus mutans* NCINB11723 (Matsumoto et al. 1996) as the NAD^+-regenerating enzyme. A total of 100 g/l of optically pure (*R*)-CPD was quantitatively accumulated after 45 h of reaction using the cultivated broth from these recombinant *E. coli* cells (Fig. 11.2). In the same way, optically active (*S*)-1,2-pentanediol could be synthesized from its racemic substrate (Fig. 11.3).

11.2.4 Synthesis of Optically Active (S)-3-Chloro-1,2-Propanediol

Ochrobactrum sp. KNKc71-3 was isolated from a soil sample and identified as the producer of (*R*)-CPD-oxidizing activity by enrichment culturing. The CDP-oxidizing enzyme may be 2,3-butandiol dehydrogenase based on its substrate specificity and amino-acid sequence homology analysis. The recombinant *E. coli* producing four enzymes for the synthesis of (*S*)-CPD from racemic CPD was prepared by the same method as that described for (*R*)-CPD synthesis. The four enzymes were *Ochrobactrum* dehydrogenase, NADH oxidase, NADPH-dependent glucose dehydrogenase, and the reported carbonyl reductase from *Rhodotorula glutinis* var. *dairenensis* IFO415 (Kizaki et al. 2005) as the CHA-reducing enzyme. A total of 150 g/l (*S*)-CPD with more than 95 % e.e. was accumulated in the reaction mixture in the presence of glucose and catalytic amounts of NAD^+ and $NADP^+$ by adding the substrate in two steps (Fig. 11.4).

Fig. 11.2 (*R*)-3-Chloro-1,3-propanediol production by an enzymatic stereoinversion reaction

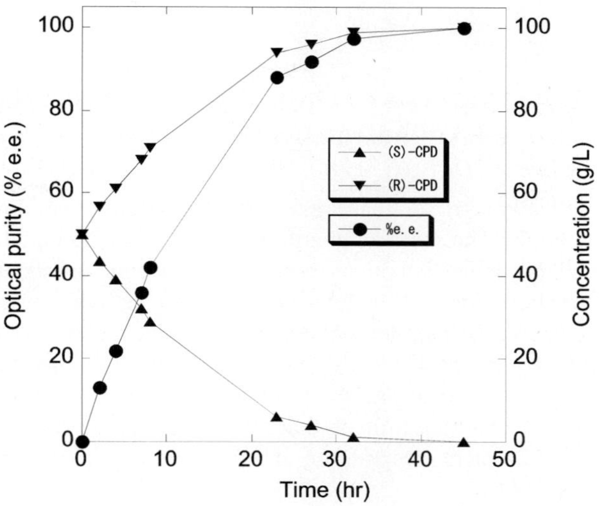

Fig. 11.3 (*S*)-1,2-Pentanediol production by an enzymatic stereoinversion reaction

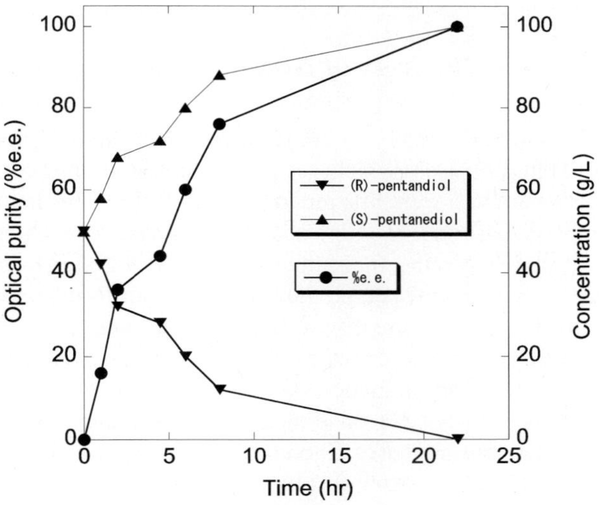

Fig. 11.4 (*S*)-3-Chloro-1,3-propanediol production by an enzymatic stereoinversion reaction

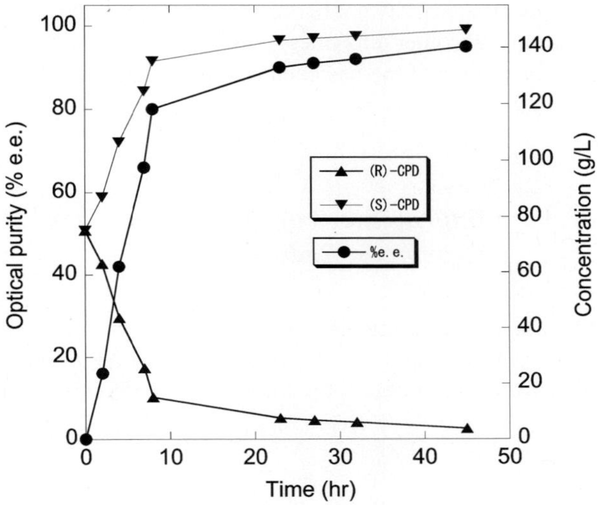

11.3 Synthesis of Optically Active Secondary Amines by Aminotransferase

Optically active amines are as important as chiral building blocks as optically active alcohols. The main synthetic methodology for production of optically active secondary amines is optical resolution of racemic compounds by optically active carboxylic acids. The racemic amines can be synthesized from the corresponding ketone under high temperature and pressure. There is also a multistep process that involves the conversion of an optically active alcohol starting material. Regardless, more economical and environmentally friendly synthetic processes are required. The use of a transaminase that can catalyze the stereoselective amination of a ketone under ambient temperature and pressure is one such promising way (Turner and Carr 2006).

11.3.1 Discovery of Novel Transaminases

Transaminase catalyzes the transfer of an amino group into a carbonyl group. ω-Transaminases, which act on secondary amines for the synthesis of optically active amines, were first reported in the 1990s (Matcham and Bowen 1996). Several reports exist on L-α-amino acid transaminases (Leuchtenberger 1995).

An (*R*)-specific transaminase from *Arthrobacter* sp. KNK168 (Iwasaki et al. 2012) and an (*S*)-specific transaminase from *Pseudomonas fluorescens* KNK08-18 (Ito et al. 2011) were discovered from soil samples by enrichment culturing experiments. Both these enzymes are pyridoxal phosphate dependent and catalyze the reversible transamination between α-phenetylamine and pyruvic acid; these enzymes do not act on β-alanine and 4-aminobutyric acid, in contrast to other reported transaminases (Table 11.2). The encoding gene of each enzyme was cloned and successfully expressed in *E. coli*, respectively. This is the first report to isolate an (*R*)-specific transaminase from a microorganism where previously (*R*)-specific transaminase activity could only be obtained by in silico screening and protein modifications of other enzymes (Höhne et al. 2010). The substrate specificity of *Arthrobacter* enzyme has been changed to include its range of applications (Savile et al. 2010).

11.3.2 Improvement of the Transaminase Reaction Using Enzymatic Reduction

The transaminase reaction is reversible. It is industrially difficult to separate the product from the four compounds in the reaction mixture after the reaction. We solved this problem by adding lactate dehydrogenase to the reaction mixture.

Table 11.2 Comparison of enzymatic properties of typical **ω**-amino-acid transaminases

	RTA[a]	TPF[b]	ω-Amino acid: pyruvate transaminase[c]	4-Aminobutyrate 2-ketoglutarate transaminase[d]
Origin	*Arthrobacter* sp. KNK168	*Pseudomonas fluorescens* KNK08-18	*Pseudomonas* sp. F-126	*Pseudomonas* sp. F-126
Subunit MW	37,000	53,000	43,000	44,000
Native MW	148,000	120,000	172,000	176,000
Number of subunits	4	2	4	4
Optimal pH	8.0–9.0	7.0–9.0	8.5–10.5	8.5–9.0
Optimal temperature	30 °C	40 °C	60 °C	60 °C
Specific activity[e]				
(*S*)-α-Phenetylamine: pyruvate	0 U/mg	40 U/mg	Not tested	Not tested
(*R*)-α-Phenetylamine: pyruvate	19	0	Not tested	Not tested
β-Alanine: pyruvate	0	0	4 U/mg	Not tested
4-Aminobutyrate: 2-ketoglutarate	0	0	Not tested	47 U/mg

[a]Iwasaki et al. 2012
[b]Ito et al. 2011
[c]Yonaha et al. 1977
[d]Yonaha and Toyama 1980
[e]One unit of enzyme activity is defined as the amount that catalyzes the formation of 1 μmol of the product for 1 min

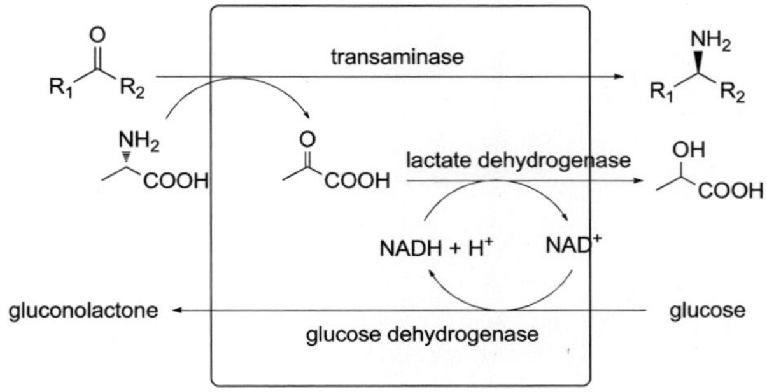

Fig. 11.5 Improvement of a transaminase reaction using enzymatic reduction

The transamination reaction could be carried out in almost stoichiometric yield by converting pyruvic acid to lactic acid using lactate dehydrogenase (Fig. 11.5). This improved reaction, which requires three enzymes, transaminase, lactate dehydrogenase, and glucose dehydrogenase, functions to regenerate the NADH required by lactate dehydrogenase. The recombinant *E. coli* co-overproducing these three enzymes was constructed to prepare the biocatalyst system and simplify the process using the same method as already described. The pharmaceutical intermediates described in Tables 11.3 and 11.4 were synthesized in an effective manner by using the recombinant *E. coli* cells as a catalyst.

Table 11.3 Production of optically active secondary amines by an (*R*)-specific transaminase

Product	Concentration (mM)	Reaction period (h)	Yield (%)	Optical purity (% e.e.)
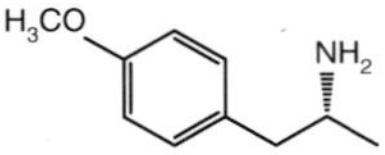	306	68	98	99.6
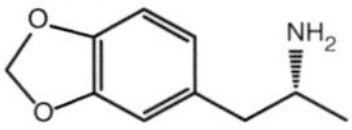	561	27	80	99.9
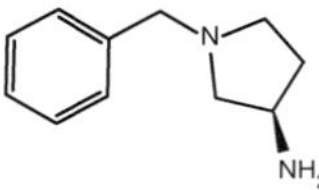	171	23	71	99.9
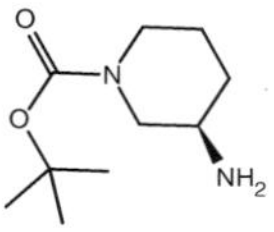	239	20	75	99.9
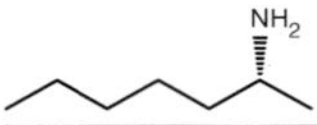	175	20	97.5	99.9

Table 11.4 Production of optically active secondary amines by an (*S*)-specific transaminase

Product	Concentration (mM)	Reaction period (h)	Yield (%)	Optical purity (% e.e.)
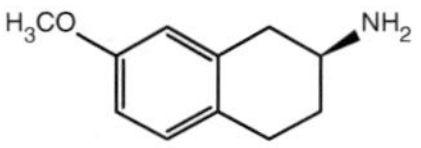	57	20	93	97.0
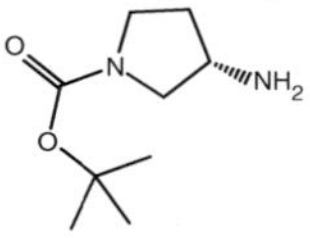	270	23	100	99.4
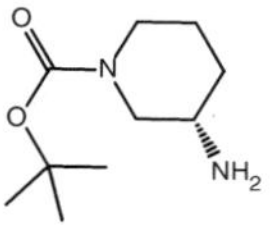	151	23	100	99.9
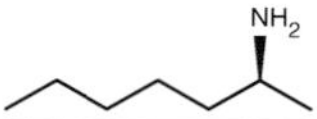	175	20	98	98.8

References

Höhne M, Schätzle S, Jochens H et al (2010) Rational assignment of key motifs for function guides *in silico* enzyme identification. Nat Chem Biol 6:807–813

Ito N, Kawano S, Hasegawa J et al (2011) Purification and characterization of a novel (*S*)-enantioselective transaminase from *Pseudomonas fluorescens* KNK18-08 for the synthesis of optically active amines. Biosci Biotechnol Biochem 75:2093–2098

Iwasaki A, Matsumoto K, Hasegawa J et al (2012) A novel transaminase, (*R*)-amine:pyruvate aminotransferase, from *Arthrobacter* sp. KNK168 (FERM BP-5228): purification, characterization, and gene cloning. Appl Microbiol Biotechnol 93:1563–1573

Kizaki N, Sawa I, Yano M et al (2005) Purification and characterization of a yeast carbonyl reductase for synthesis of optically active (*R*)-styrene oxide derivatives. Biosci Biotechnol Biochem 69:79–86

Leuchtenberger W (1995) Tranaminations and miscellaneous reactions. In: Drauz K, Waldmann H (eds) Enzyme catalysis in organic synthesis. VCH, New York, pp 470–480

Matcham GW, Bowen ARS (1996) Biocatalysis for chiral intermediates: meeting commercial and technical challenges. Chim Oggi 14:20–24

Matsumoto J, Higuchi M, Shimada M et al (1996) Molecular cloning and sequence analysis of the gene encoding the H_2O-forming NADH oxidase from *Streptococcus mutans*. Biosci Biotechnol Biochem 60:39–43

Nanba H, Yasohara Y, Hasegawa J et al (2007) Bioreactor systems for the production of optically active amino acids and alcohols. Org Process Res Dev 11:503–508

Savile CK, Janey JM, Mundorff EC et al (2010) Biocatalytic asymmetric synthesis of chiral amines from ketones applied to Sitagliptin manufacture. Science 329:305–309

Turner NJ, Carr R (2006) Biocatalytic routes to nonracemic chiral amines. In: Patel RN (ed) Biocatalysis in the pharmaceutical and biotechnology industries. CRC, Boca Raton, pp 743–755

Wada M, Kataoka M, Kawabata H et al (1998) Purification and characterization of NADPH-dependent carbonyl reductase, involved in stereoselective reduction of ethyl 4-chloro-3-oxobutanoate, from *Candida magnoliae*. Biosci Biotechnol Biochem 62:208–285

Yonaha K, Toyama S (1980) γ-Aminobutyrate:α-ketoglutarate aminotransferase from *Pseudomonas* sp. F-126: purification, crystallization, and enzymologic properties. Arch Biochem Biophys 200:156–164

Yonaha K, Toyana S, Yasuda M et al (1977) Properties of crystalline ω-amino acid: pyruvate aminotransferase of *Pseudomonas* sp. F-126. Agric Biol Chem 41:1701–1706

Chapter 12
Use of Organic Solvent-Tolerant Microorganisms in Bioconversion

Akinobu Matsuyama

Abstract In the chemical industry, there is a need for innovative manufacturing processes that are not dependent on fossil resources and have reduced environmental impact. Bioprocesses are advantageous as environmentally friendly or green processes and are also superior to chemical processes in terms of reaction specificity and substrate selectivity. However, the aqueous reaction fields used in bioprocesses are not suitable for the conversion, reaction, or production of the hydrophobic materials that account for the majority of chemicals. Therefore, to enable more general use of chemical production technologies employing bioprocesses, the author has screened for organic solvent-tolerant microorganisms, that is, microorganisms with the ability to maintain cellular structure in organic solvents, and created biocatalysts using such microorganisms as host cells.

Keywords Organic solvent-tolerant microorganism • Nonaqueous reaction • LogPow • Schreenig • *Kocuria rhizophila* • *Rhodococcus* spp. • Gene expression system • Mandelic acid • Ethyl- Chloro-Hydroxybutyrate

12.1 Introduction

To improve sustainability, there is a strong need for a major transformation in the chemical industry toward manufacturing processes that are not dependent on fossil resources and have reduced environmental impact. For that reason, there is a need

A. Matsuyama (✉)
Green Product Development Center, Daicel Corporation,
1-1 Shinko-cho, Myoko, Niigata 944-8550, Japan
e-mail: ak-matsu@jp.daicel.com

H. Anazawa and S. Shimizu (eds.), *Microbial Production: From Genome Design to Cell Engineering*, DOI 10.1007/978-4-431-54607-8_12, © Springer Japan 2014

to advance green chemistry and achieve a number of technical breakthroughs, and one powerful method is to broaden the range of use of bioprocesses that are suitable for resource recycling. Microbial conversion (biocatalytic reaction) is one type of bioprocess, and it is superior to chemical processes in terms of reaction specificity and substrate selectivity. However, the aqueous reaction fields that are frequently used in microbial conversion are not suitable for conversion or production of hydrophobic materials, and thus there are limits on raw materials and products. Therefore, to produce a wide variety of chemicals, there is a need for microbial conversion technology for nonaqueous reaction fields (e.g., organic solvent, biphasic organic solvent/aqueous, aqueous solution of organic solvent, supercritical fluid). The advantages of producing materials in nonaqueous reaction fields include the following. (1) It is possible to increase the concentration of raw materials and products that are difficult to dissolve in water, and thus manufacturing costs and fixed costs can be reduced; (2) it is possible to avoid inhibition of the biocatalyst by the raw materials and products; (3) it is possible to prevent hydrolysis of raw materials and products; (4) it is easy to recover products; and (5) it is easy to integrate with chemical processes. The disadvantages include (1) organic solvents are toxic to and inhibit biocatalysts; (2) materials that are insoluble in cell-derived water, and in organic solvents, emerge onto the interface; and (3) production systems become more complex (Salter et al. 1995). In terms of conventional technology, attempts have been made to improve stability of enzymes through chemical modification and fixation onto carriers to enable use in organic solvents. However, there are major issues relating to microbial conversion in nonaqueous reaction fields. Essentially, microorganisms themselves are susceptible to materials such as organic solvents, and there are many cases where use as a biocatalyst is difficult in nonaqueous reaction fields. Therefore, there is a need to develop new microorganisms that can be used even in nonaqueous reaction fields.

There have been various reports on the organic solvent tolerance of microorganisms, such as a report in which the ability of various microorganisms to form colonies was examined after layering organic solvents onto agar medium inoculated with microorganisms (Aono and Nakajima 1997), and a report on selection of organic solvents for material conversion in organic solvent (Salter et al. 1995). However, there are almost no previous findings in which new organisms were screened by evaluating function relating to reactions in various organic solvents to achieve microbial conversion in nonaqueous reaction fields. Therefore, the author has separately considered function of the biocatalyst as a host cell, and catalytic function of enzymes, established an assembly technique concept that optimally combines host cells and enzymes (Matsuyama et al. 2004) (Fig. 12.1), and thereby aimed to develop a system enabling broad application in manufacturing a variety of chemicals. To achieve this, microorganisms with the ability to maintain cellular structure in organic solvents were screened, and biocatalysts were developed using those microorganisms as host cells. The biocatalysts were then used for microbial conversion in nonaqueous reaction fields.

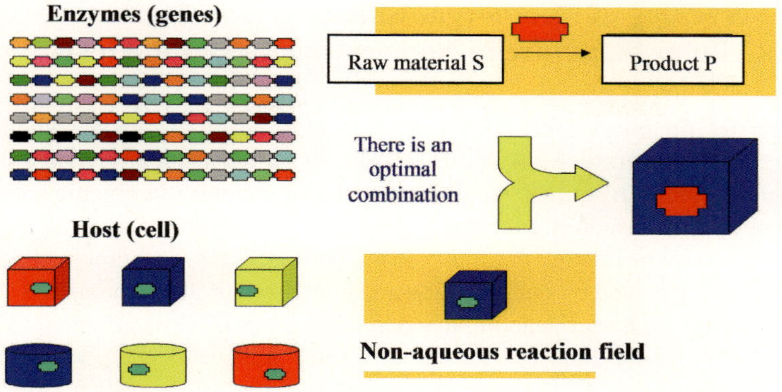

Fig. 12.1 Conceptual diagram of host and enzyme assembly

12.2 Analysis and Evaluation of the Function of Microorganisms in Nonaqueous Reaction Fields

When using biocatalysts in nonaqueous reaction fields such as organic solvents, the organic solvent will inhibit enzyme reactions, and inhibit reproduction and kill the cells themselves. Therefore, to develop biocatalysts that can be used in nonaqueous reaction fields, the author analyzed, and then established an evaluation method for, microorganism function under these conditions. First, the author examined what types of cell function resulted in superior performance of host cells as biocatalysts in the nonaqueous reaction fields of organic solvents with various polarity values. As models of nonaqueous reaction fields, a total of 28 types of organic solvents were selected from primary alcohols, *n*-alkanes, ester acetates, and ketones exhibiting various log P_{ow} values. These agents were each added directly into the medium, and then, as host cell models, the survival of *E. coli*, budding yeast, and *Rhodococcus* spp. microorganisms was evaluated in terms of inhibition of colony formation. Just as in previous findings (Aono et al. 1994; Weber and de Bont 1996), it was confirmed that there is a correlation between the log P_{ow} value and reproduction inhibition of organic solvents. On the other hand, the results also suggested that the relationship varies depending on factors such as differences in the functional groups of the organic solvents. Next, the author examined the effect of these organic solvents on microorganism morphology and comprehensive gene expression. Analysis was conducted, in the presence of each organic solvent, of inhibition of colony formation by cells (*E. coli*, *Rhodococcus* spp. microorganisms, budding yeast), changes in morphology detected using a scanning electron microscope and transmission electron microscope, changes in internal structure detected using flow cytometry, and induced gene clusters examined using cDNA microarrays. The results showed that cell function is greatly affected not only by the log P_{ow} value of the organic solvent, which is an indicator of hydrophilicity and hydrophobicity, but also by the types of compounds and their functional groups (Fujita et al. 2004).

12.3 Screening of Microorganisms That Can Maintain Cellular Structure in Nonaqueous Reaction Fields

Screening for organic solvent tolerance was not done by evaluating whether the microorganism can grow in an organic solvent. From the standpoint of enabling use in organic synthesis reactions, maintaining cellular structure in organic solvents is the most important capability for use as a biocatalyst, and thus screening was done using a system for evaluating the ability to maintain cellular structure, from a microorganism library owned by our company and composed of 200 genera, 709 species, and 1,477 strains of bacteria, actinomycetes, and yeasts. First, as primary screening, 93 genera, 206 species, and 294 strains were selected because there was almost no change in turbidity of the cell suspension when exposed to a 50 % solution of 1-pentanol. The microbe strains selected in primary screening were successively evaluated in 1-octanol and 1-nonanol, which have even higher log P_{ow} values than 1-pentanol, and pared down to 12 genera, 18 species, and 23 strains (Fig. 12.2). These microbes were evaluated using 28 types of organic solvent, and in the end there was almost no change in turbidity of the cell suspension over the full range of log P_{ow} values. *Kocuria rhizophila* DC2201 (referred to below as DC2210) was selected as the organic solvent-tolerant microorganism with the greatest ability to maintain cellular structure (Fujita et al. 2006) (Fig. 12.3).

12.4 Comparison of Organic Solvent Tolerance of DC2201 and *E. coli*

The states of solutions were compared when DC2201 and *E. coli* microbes were incubated in biphasic aqueous/organic solvent solutions using seven types of organic solvents with different log P_{ow} values (butanol, pentane, heptane, octane, ethyl

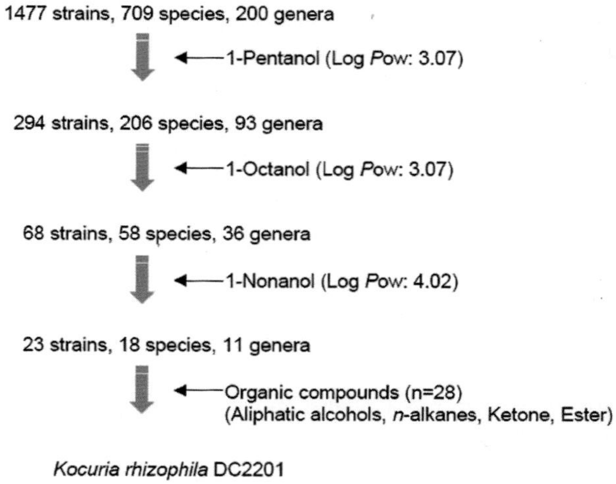

Fig. 12.2 Screening *Kocuria rhizophila* DC2201

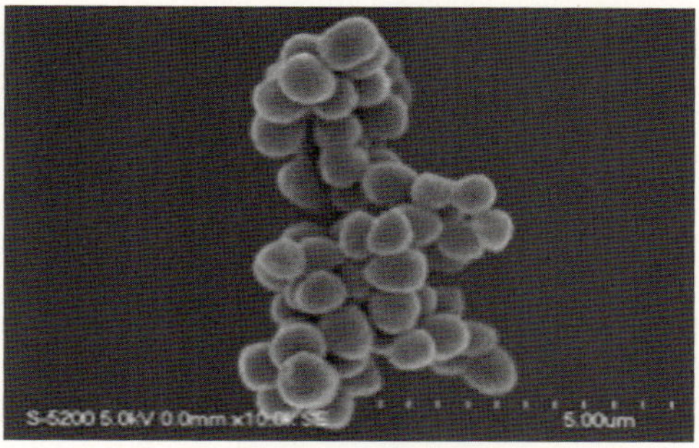

Fig. 12.3 Electron micrograph of *Kocuria rhizophila* DC2201. [Photograph courtesy of National Institute of Technology and Evaluation (NITE)]

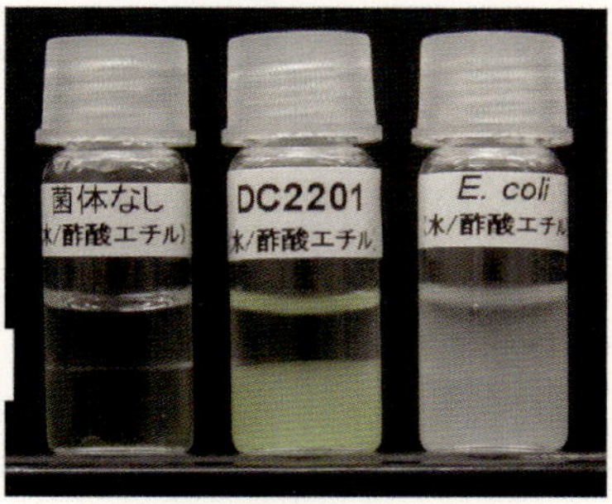

Fig. 12.4 State of interface with mocrobes in ethyl acetate/aqueous solution

acetate, butyl acetate, and toluene). In the case of *E. coli*, viscosity of the solution increased, and there was formation of an intermediate layer thought to be caused by cytoplasmic components that leaked out from lysis. With DC2201, on the other hand, there was almost no evident formation of an intermediate layer, and little leakage of cytoplasmic components caused by lysis. These results suggest that proteins such as enzymes were stably retained inside the cells (Fig. 12.4). When maintenance of the cellular structure of *E. coli* and DC2201 was evaluated using various primary alcohols and *n*-alkanes, it was found that although *E. coli* is dissolved when the log P_{ow} value is in the interval 2–4, there is almost no impact on DC2201 over the entire range of log P_{ow} values (Fig. 12.5).

12.5 Development of a Gene Expression System for DC2201

To develop microbial conversion using DC2201 as the biocatalyst, it was essential to establish gene recombination technology for high-level expression of heterologous genes, taking DC2201 as the host cell, but there have been almost no findings

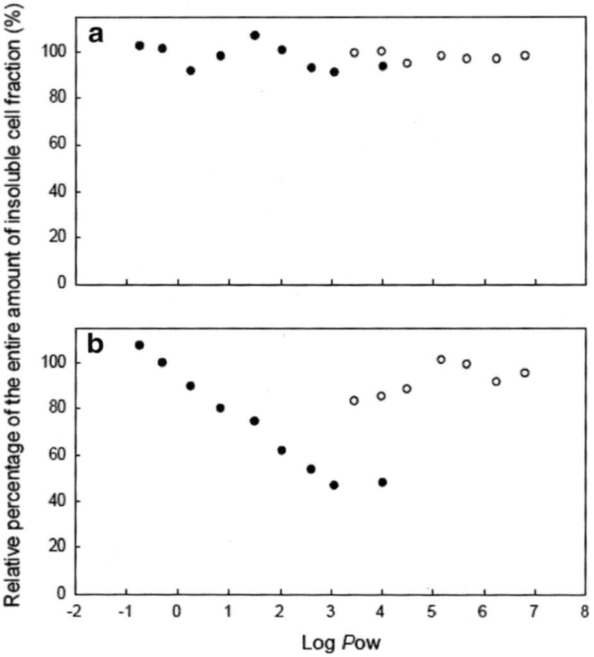

Fig. 12.5 Comparison of cellular structure maintenance between DC2201 (**a**) and *Escherichia coli* (**b**) in response to various types of primary alcohol (*filled circles*) and *n*-alkane (*open circles*)

thus far in this area. Therefore, the author found a usable replication origin from a closely related species of DC2201, acquired a promoter for high-level expression, and developed a host vector-type gene expression system for DC2201.

Next, it was decided to show the superiority of DC2201 by examining reactions in an organic-aqueous biphasic medium using DC2201 (Fig. 12.6).

12.6 Production of (*R*)-Mandelic Acid Using DC2201

(*R*)-Mandelic acid (RMA) is an industrially useful chiral compound used as a resolving reagent for chiral amine and in other applications. Figure 12.7 shows production of RMA from mandelonitrile. A comparison was conducted to determine whether DC2201 or *E. coli* is superior as the host cell in RMA production, using transformants of DC2201 and *E. coli* expressing organic solvent-tolerant nitrilase (F73Nit) (Matsuda 2004). In the aqueous/ethyl acetate biphasic reaction field used as a model of nonaqueous reaction fields, the reaction almost did not proceed at all with the enzyme alone (cell free), but with the DC2201 transformant, 210 g/l RMA accumulated in the water phase in a reaction time of 48 h, and the reaction yield was 91 %. On the other hand, the reaction yield for *E. coli* was 50 % or less, and in the

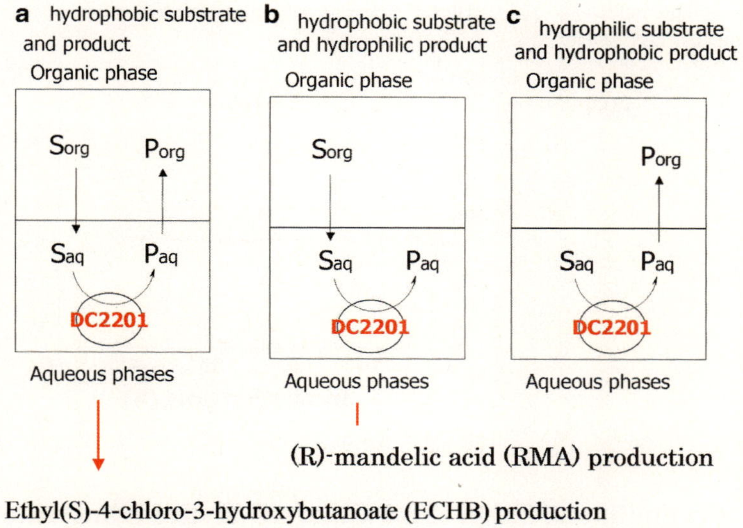

Fig. 12.6 Example reaction of organic-aqueous biphasic medium

Fig. 12.7 Production of (*R*)-mandelic acid (RMA) from mandelonitrile

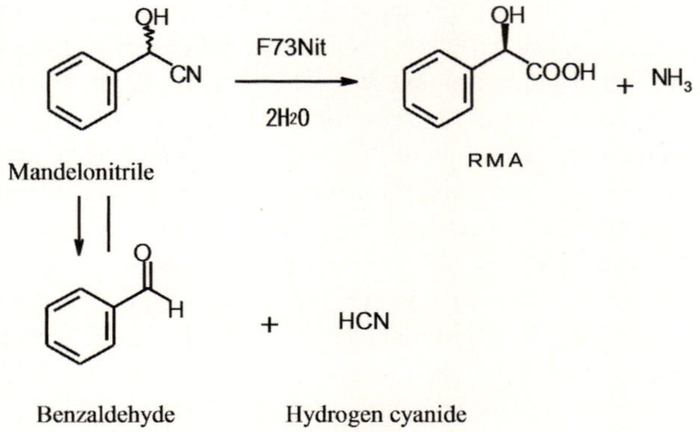

aqueous/ethyl acetate biphasic reaction field, it was shown that DC2201 is far superior to *E. coli* as a host cell (Fig. 12.8). RMA was obtained as a crystal with a refining yield of 81 % from the reaction solution of the DC2201 transformant, and the consistent yield was 74 %. Thus, it was found that RMA can be produced with productivity at a world-class level by using DC2201.

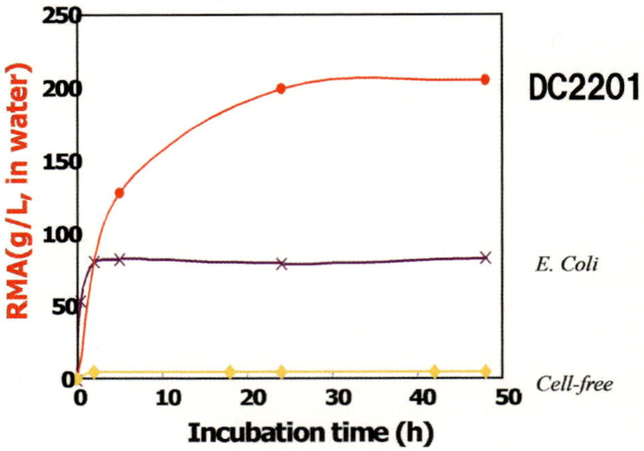

Fig. 12.8 Production of RMA using transformants of DC2201 and *E. coli* expressing organic solvent-tolerant nitrilase (F73Nit)

12.7 Production of Ethyl (*S*)-4-Chloro-3-Hydroxybutyrate Using DC2201

Ethyl (*S*)-4-chloro-3-hydroxybutyrate (ECHB) is a chiral compound that is useful as an intermediate for drugs to treat hyperlipidemia. Therefore, a comparison was conducted to determine whether DC2201 or *E. coli* is superior as a host cell in ECHB production using transformants coexpressing (*R*)-2-octanol dehydrogenase (PfODH) and formate dehydrogenase (McFDH) (Fig. 12.9). ECHB production was performed in an aqueous/butyl acetate biphasic reaction field, using transformants of DC2201 and *E. coli*. Although addition of NAD^+ was essential in the reaction using the *E. coli* transformant, in the reaction using the DC2201 transformant, almost the same amount of ECHB was produced when NAD^+ was added (35.5 % w/v) as when it was not added (34.5 % w/v). Thus, there was no change in the amount of ECHB produced regardless of whether NAD^+ was added. In systems where NAD^+ was not added, ECHB accumulated at a far higher concentration when using a DC2201 transformant than when using an *E. coli* transformant. Therefore, if DC2201 is used as the host cell in an aqueous/butyl acetate biphasic reaction field, it is possible to efficiently use the coenzymes in the cell because cellular structure is maintained, and there is no need to add expensive NAD^+. High productivity is achieved, and the approach is markedly superior to using *E. coli* as the host cell (Fig. 12.10).

12.8 Conclusion

In the future, using organic solvent-tolerant microorganisms such as DC2201 as biocatalysts will enable reactions previously impossible in aqueous reaction fields to be carried out in nonaqueous reaction fields, and this should enable use of various

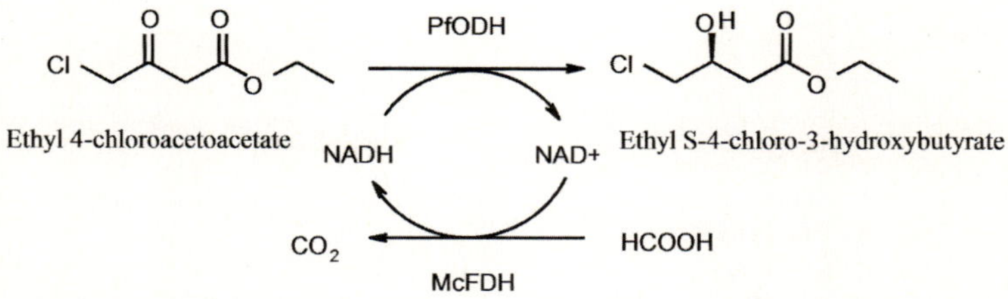

Fig. 12.9 Ethyl (*S*)-4-chloro-3-hydroxybutyrate (ECHB) production using PfODH/McFDH

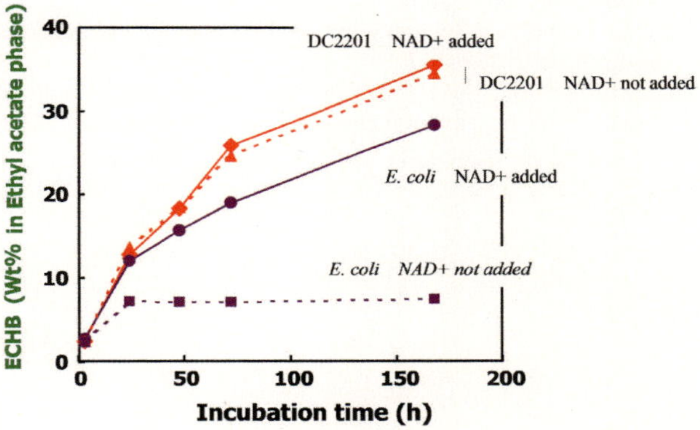

Fig. 12.10 ECHB production using transformants of DC2201 and *E. coli* coexpressing PfODH and McFDH

hydrophobic raw materials that previously could not be used. There will also likely be a dramatic increase in the number of chemicals that can be produced using microbial conversion, which has previously been limited by aqueous reaction fields. Further extension of the technique should enable processes using ionic liquid reaction fields where cellulose biomass is dissolved, and various substances are produced from that material. By using not only *E. coli*, which has been the primary microorganism used thus far, but also organic solvent-tolerant microorganisms such as DC2201, and combining biocatalysts that appropriately assemble cellular function and enzyme function with nonaqueous reaction fields such as organic solvents and ionic liquids, it should be possible to infinitely expand the possibilities for producing chemicals and materials using microbial conversion.

This research was commissioned by NEDO (New Energy and Industrial Technology Development Organization) and conducted as one part of "Development of Basic Technology for Environmentally-friendly Manufacturing using Microorganism Function."

References

Aono R, Nakajima H (1997) Tolerance of *E. coli* for hydrophobic organic solvents. Protein Nucleic Acid Enzyme 42(15):2532–2541 (in Japanese)

Aono R et al (1994) Mapping of organic solvent tolerance gene *ostA* in *Escherichia coli* K12. Biosci Biotechnol Biochem 58:1231–1235

Fujita K et al (2004) Comprehensive gene expression analysis of the response to straight-chain alcohols in *Saccharomyces cerevisiae* using cDNA microarray. J Appl Microbiol 97:57–67

Fujita K et al (2006) The cell structural properties of *Kocuria rhizophila* for aliphatic alcohol exposure. Enzyme Microb Technol 39(3):511–551

Matsuda S (2004) Screening of nitrilases showing tolerance against organic solvent. Proc Soc Biotech Jpn 113 (in Japanese)

Matsuyama A, Kobayashi Y, Nagasawa T (2004) Achieving more general use of green bioprocesses using the "assembly technique". Biosci Industry 62(1):23–26 (in Japanese)

Salter GJ et al (1995) Solvent selection for whole cell biotransformation in organic media. Crit Rev Biotechnol 15:139–177

Weber FJ, de Bont JA (1996) Adaptation mechanisms of microorganisms to the toxic effects of organic solvents on membranes. Biochim Biophys Acta 1286(3):225–245

Chapter 13
Approaches for Improving Protein Production by Cell Surface Engineering

Takeko Kodama, Kenji Manabe, Katsutoshi Ara, and Junichi Sekiguchi

Abstract *Bacillus subtilis* is an attractive host organism because it has a naturally high secretory capacity and exports proteins directly into the extracellular medium. Secreted proteins emerge from translocation to the compartment between the cell wall and cytoplasmic membrane. It has been reported that the increased net negative charge of the cell wall is involved in protein folding and stability at the cytoplasmic membrane–cell wall interface of *B. subtilis*. In *Escherichia coli*, Sec translocase requires a functional interaction with the membrane acidic lipids such as phosphatidylglycerol (PG) and cardiolipin (CL). In Sect. 13.1, we describe altered compositions of anionic polymers on the cell surface and cell membrane lipid of *B. subtilis*, which is improved for protein secretion. Inactivation of extracellular proteases is essential for improvement of secreted proteins with *B. subtilis* as a host. Previously we reported that the decrease of extracellular protease amounts in *B. subtilis* mutants led to the stabilization of autolysins, making the cells more prone to lysis (Kodama et al., J Biosci Bioeng 103:13–21, 2007). In Sect. 13.2, we describe the improvement process for protein production from the aspect of prevention of cell lysis.

Keywords Autolysin • *Bacillus subtilis* • Lipid membrane • Protein production • Teichuronic acid

---

T. Kodama • K. Manabe • K. Ara
Biological Science Laboratories, Kao Corporation, 2606 Akabane, Ichikai, Haga-gun, Tochigi 321-3497, Japan

J. Sekiguchi (✉)
Faculty of Textile Science and Technology, Shinshu University, 3-15-1 Tokida, Ueda 386-8567, Japan
e-mail: jsekigu@shinshu-u.ac.jp

H. Anazawa and S. Shimizu (eds.), *Microbial Production: From Genome Design to Cell Engineering*, DOI 10.1007/978-4-431-54607-8_13, © Springer Japan 2014

13.1 Improvement of Cell Wall and Membrane

13.1.1 Change of Anionic Polymer on Cell Surface

The anionic polymer that constitutes the cell wall of *Bacillus subtilis* 168 consists of a major teichoic acid (WTA), lipoteichoic acid (LTA), and a minor teichoic acid. The former two components contain poly(glycerol phosphate) as the main ingredient. The latter contains poly(glucosyl *N*-acetylgalactosamine 1-phosphate) as the main ingredient. Substitution of phosphate-rich teichoic acids with phosphate-free teichuronic acid [poly(*N*-acetyl galactosamine glucuronic acid)] occurs during the transition to phosphate-limited growth (Ellwood and Tempest 1969). It is likely that the negative charge of these polymers maintains cation homeostasis and assists in the assimilation of metal cations for cellular function (Hughes et al. 1973). On the other hand, the *dltABCDE* operon is responsible for the modification of both WTA and LTA with D-alanine, which decreases the density of the negative charge in the cell wall (Neuhaus and Baddiley 2003). The increased net negative charge of the cell wall in *dlt* mutants may increase the binding of divalent cations to the cell wall, which in turn may improve the posttranslocational folding of exported proteins (Hyyryläinen et al. 2000, 2007).

We examined whether the altered charge of the cell wall by the modulation of teichuronic acid affects protein productivity. We constructed mutants with deletion and overexpression (expressed by tandem promoter of *rplU* and *tuaABCDEFGH*) of the *tuaABCDEFGH* operon. The cell walls of the wild type, and deletion and overexpression strains of *tuaABCDEFGH* in the exponential growth phase, gave 0.15, 0.14, and 0.51 mmol g^{-1} uronic acid, respectively. In the stationary phase, the amounts of uronic acid in the wild type and the deletion and overexpression strains were 0.24, 0.16, and 0.42 mmol g^{-1} uronic acid, respectively. The amount of teichuronic acid in the wild type in the stationary phase was 1.5 times higher than that in the exponential phase because the *tuaABCDEFGH* operon was induced in the stationary phase by the depletion of phosphate in the medium. The amount of uronic acid in the *tuaABCDEFGH* deletion mutant at the stationary phase was almost the same as that of the wild type at the exponential phase. In both phases, the cell wall in the *tuaABCDEFGH* overexpression strain contained 2.8 times higher uronic acid than the wild type in the exponential phase. There was no marked difference in the amount of phosphate per cell weight among the wild type and the deletion and overexpression strains. In the overexpression strain, both Egl-237 cellulase (Hakamada et al. 2000) and M-protease (Kobayashi et al. 1995) were 1.2 times or higher than those in the wild-type strain (Fig. 13.1a, b). AmyK38 amylase (Hagihara et al. 2001) and KP-43 protease (Nonaka et al. 2004) produced by the deletion strain were about 1.2 and 1.4 times higher than those in the wild type, respectively (Fig. 13.1c, d). Thus, both the deletion and overexpression of *tuaABCDEFGH* significantly affected protein productivity. These results suggest that the proportion of teichuronic acid optimal for enzyme productivity depends on numerous enzymatic properties.

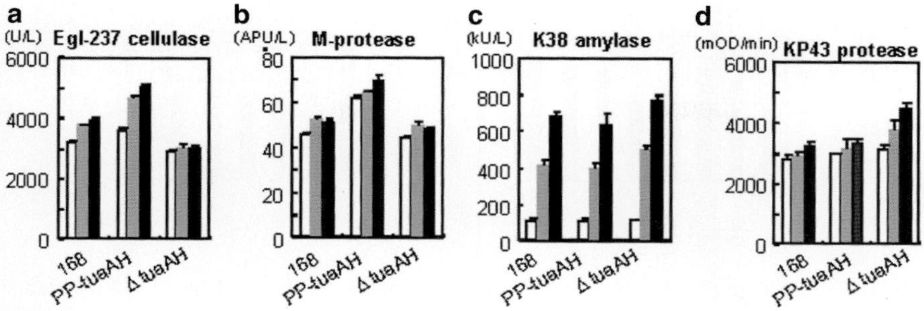

Fig. 13.1 Effects of productivity of various enzymes in the absence and in the presence of the overexpression of teichuronic acid. Cells were cultured in the modified 2xL broth at 30°C. The accumulation of Egl-237 cellulase (**a**), M-protease (**b**), K38 amylase (**c**) and KP43 protease (**d**) in the culture medium was measured at 24 (white bars), 48 (gray bars) and 72 (black bars) h. The results presented are the average of three individual experiments. Error bars correspond to the standard deviation of the means (STDEV)

13.1.2 Change of Cell Membrane Lipid Composition

Escherichia coli cells from exponentially grown cultures have an unusually simple phospholipid composition, generally reported to be 75 % zwitterionic phosphatidylethanolamine (PE), 20 % anionic phospholipid phosphatidylglycerol (PG), and 5 % anionic cardiolipin (CL) (Cronan 2003). The protein translocation reaction, driven by SecA ATPase through the conserved SecYEG complex, has a known requirement for acidic phospholipids in *E. coli* (de Vrije et al. 1988). It has been proposed that FtsY requires a functional interaction with inner membrane lipids at a late stage of the signal recognition particle pathway (de Leeuw et al. 2000). It has been reported that PE is required for the proper conformation and active transport function of LacY (Bogdanov et al. 2002).

The membrane of *B. subtilis* is a complex structure composed mainly of PG and PE. Other components include a relatively large amount (<30 %) of neutral glycolipids (GL), a variable amount of positively charged lysylphosphatidylglycerol (LPG), and a small amount of CL. There are almost no reports of how protein secretion plays a role in the cell membrane lipid in *B. subtilis*. To elucidate the membrane lipid involved in protein secretion, we examined the protein production of *B. subtilis* strains with altered membrane compositions. We inactivated genes involved in the biosynthesis of complex lipids, including the genes necessary for the synthesis of PE (*psd*), PS (*pssA*), GL (*ugtP*), LPG (*mprF*), and CL (*ywnE* and *ywjE*), and evaluated the productivity of Egl-237 cellulase (Hakamada et al. 2000) and AmyK38 amylase (Hagihara et al. 2001) in the mutants. Interestingly, bacteriolysis did not occur in any mutants as a result of the deletion of membrane lipid synthetic genes. The productivity of Egl-237 cellulase increased 1.2 times or more by single deletion of *mprF*, *ywnE*, *ywjE*, and *ugtP* from the wild type (Fig. 13.2a). Moreover, strains with deletions of *ywnE* and *ywjE* (CL synthetic genes) exhibited 1.5 times higher cellulase productivity than the wild type (Fig. 13.2a). The amylase increased by 1.4

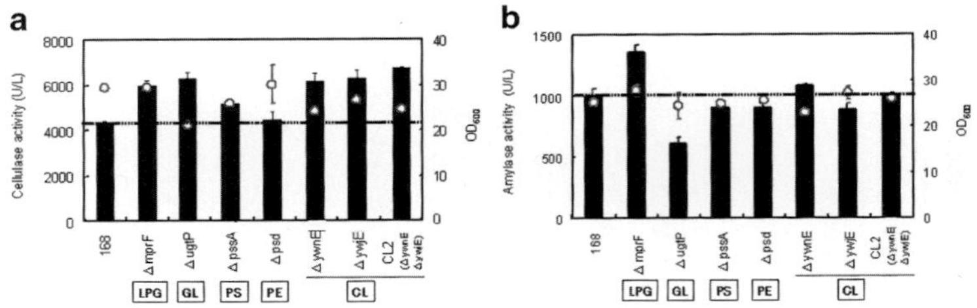

Fig. 13.2 Effects of productivity of various enzymes in the deletion mutants of cell membrane lipid. Cells were cultured in the modified 2xL-mal broth at 30°C. The accumulation of Egl-237 cellulase (**a**) and K38 amylase (**b**) in the culture medium was measured at 72 h (black bars). Open circles indicates cell density of OD_{600nm}. The results presented are the average of three individual experiments. Error bars correspond to the standard deviation of the means (STDEV)

times as a result of the deletion of *mprF* (Fig. 13.2b). The results confirmed that increased enzyme productivity could be obtained by deleting the synthetic genes of LPG and GL that do not exist in *E. coli*. These observations suggest that altered compositions of the cell membrane lipid of *B. subtilis* improve protein secretion.

13.2 Prevention Method of Cell Lysis

13.2.1 Inactivation of the Major Autolysins

The *spo0A* mutant with very low protease activities and the multiple protease-deficient mutants were better hosts for extracellular protein production (Fahnestock and Fisher 1987). However, extensive cell lysis of the *spo0A* mutant during the late stationary phase caused a serious problem for enzyme production (Kodama et al. 2007). The Dpr8 (eight-extracellular-protease-deficient) strain also exhibited cell lysis during the late stationary phase (Kodama et al. 2007). Zymography of the cell wall lytic proteins indicated that the *spo0A* and Dpr8 mutants contained high amounts of four cell wall lytic enzymes [*N*-acetylmuramoyl-L-alanine amidase of LytC (CwlB; Kuroda and Sekiguchi 1991), endo-β-*N*-acetylglucosamidase of LytD (CwlG; Rashid et al. 1995), D,L-endopeptidases of LytE (CwlF; Ishikawa et al. 1998) and LytF (CwlE; Ohnishi et al. 1999)] on the cell surface (Kodama et al. 2007). Because SigD-RNA polymerase transcribes many cell wall lytic enzyme genes (i.e., from the promoters of *lytD* and *lytF*, and also from one of the dual promoters of *lytC*), we constructed the *spo0A/sigD/lytC* mutant to prevent cell lysis in the *spo0A* mutant. In addition, we constructed the *spo0A/lytC* mutant because LytC is a major autolysin localized on the entire vegetative cell surface (Yamamoto et al. 2003) and its mutant becomes resistant to cell lysis (Kuroda and Sekiguchi 1991). The resultant mutants did not show any significant decrease in cell density until 75 h of cultivation (Fig. 13.3a). These results suggested that inactivation of autolysins in the *spo0A* mutant was effective to prevent cell lysis.

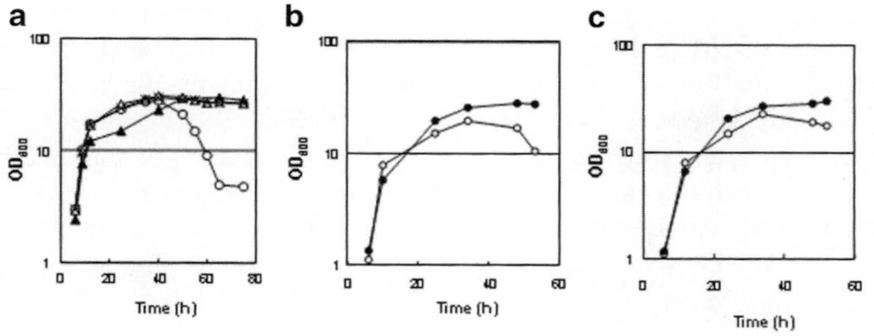

Fig. 13.3 Prevention of the cell lysis by inactivation of autolysins (**a**), inhibition of cell wall lytic enzyme with IseA (**b**) and modification of peptidoglycan with PdaC deacetylase (**c**). (**a**) Closed triangles, wild type strain; open circles, spo0A mutant; open triangles, spo0A/lytC mutant; crosses, spo0A/sigD/lytC mutant. (**b** and **c**) Closed symbol, addition of 1 mM IPTG; open symbol, no addition of IPTG. Cells were cultured in the modified 2xL broth at 30°C

13.2.2 Inhibition of the Autolysins

Bacillus subtilis has four major D,L-endopeptidases [LytE, LytF, CwlS (YojL; Fukushima et al. 2006) and CwlO (YvcE; Yamaguchi et al. 2004)], which cleave the linkage of D-γ-glutamyl-meso-diaminopimelic acid in peptidoglycan. IseA was identified as a inhibitor of the cell wall lytic activities of these D,L-endopeptidases in vitro (Yamamoto et al. 2008). We examined the prevention of cell lysis by inhibition of these D,L-endopeptidases with IseA. Hence, we constructed the overexpression of IseA in the *spo0A* mutant. The mutant harboring pDG-IseA was grown in a medium containing 1 mM isopropyl β-D-1-thiogalactopyranoside (IPTG). The overexpression of IseA in the *spo0A* mutants did not lead to decreasing optical cell density at 54 h of the late stationary phase culture (Fig. 13.3b). The results showed that the IseA is effective to prevent cell lysis by inhibition of the D,L-endopeptidases such as LytE and LytF, CwlS and CwlO.

13.2.3 Modification of Peptidoglycan

In *B. subtilis*, the *N*-terminal domain of CwlT (Fukushima et al. 2008) and the SLT domain of CwlP (Sudiarta et al. 2010a) are muramidases (for example, lysozyme) that cleave the linkage between *N*-acetylmuramic acid (MurNAc) and GlcNAc. CwlQ (Sudiarta et al. 2010b) is a bifunctional enzyme expressing both soluble-lytic transglycosylase and muramidase activities. However, the muramidase could not hydrolyze Mur-GlcNAc, MurNAc-Glc, or Mur-Glc. Therefore, the *N*-acetyl groups of the peptidoglycan hexosamine residues are responsible for the muramidase activity. *N*-Acetylglucosaminidase hydrolyzes the glycosidic bond between GlcNAc and MurNAc, and the enzyme activity is thought to depend on acetylation of the peptidoglycan hexosamine residues. *B. subtilis* PdaC was identified as a deacetylase of

MurNAc in peptidoglycan, and its deletion mutant showed sensitivity to lysozyme (Kobayashi et al. 2012). Therefore, we considered that PdaC protects the cell wall from muramidase and *N*-acetylglucosaminidase by deacetylating peptidoglycan. Hence, we constructed the overexpression of PdaC in the *spo0A* mutant harboring pDG-PdaC, and then grew the strain in a medium containing 1 mM IPTG. The strain did not exhibit decreasing optical cell density at 54 h of the late stationary phase culture (Fig. 13.3c). The results showed that PdaC was effective to prevent cell lysis by a high rate of deacetylation in peptidoglycan, leading to resistance to muramidase and *N*-acetylglucosaminidase.

References

Bogdanov M, Heacock PN, Dowhan W (2002) A polytopic membrane protein displays a reversible topology dependent on membrane lipid composition. EMBO J 21:2107–2116

Cronan JE (2003) Bacterial membrane lipids: where do we stand? Annu Rev Microbiol 57:203–224

de Leeuw E, te Kaat K, Moser C et al (2000) Anionic phospholipids are involved in membrane association of FtsY and stimulate its GTPase activity. EMBO J 19:531–541

de Vrije T, de Swart RL, Dowhan W et al (1988) Phosphatidylglycerol is involved in protein translocation across *Escherichia coli* inner membranes. Nature (Lond) 334:173–175

Ellwood DC, Tempest DW (1969) Control of teichoic acid and teichuronic acid biosynthesis in chemostat cultures of *Bacillus subtilis* var. *niger*. Biochem J 111:1–5

Fahnestock SR, Fisher KE (1987) Protease-deficient *Bacillus subtilis* host strains for production of staphylococcal protein A. Appl Environ Microbiol 53:379–384

Fukushima T, Afkham A, Kurosawa S et al (2006) A new D,L-endopeptidase gene product, YojL (renamed CwlS), plays a role in cell separation with LytE and LytF in *Bacillus subtilis*. J Bacteriol 188:5541–5550

Fukushima T, Kitajima T, Yamaguchi H et al (2008) Identification and characterization of novel cell wall hydrolase CwlT: a two-domain autolysin exhibiting *N*-acetylmuramidase and D,L-ENDOPEPTIDASE activities. J Biol Chem 283:11117–11125

Hagihara H, Hayashi Y, Endo K et al (2001) Novel alpha-amylase that is highly resistant to chelating reagents and chemical oxidants from the alkaliphilic *Bacillus* isolate KSM-K38. Appl Environ Microbiol 67:1744–1750

Hakamada Y, Hatada Y, Koike K et al (2000) Deduced amino acid sequence and possible catalytic residues of a thermostable, alkaline cellulase from an alkaliphilic *Bacillus* strain. Biosci Biotechnol Biochem 64:2281–2289

Hughes AH, Hancock IC, Baddiley J (1973) The function of teichoic acids in cation control in bacterial membranes. Biochem J 132:83–93

Hyyryläinen HL, Vitikainen M, Thwaite J et al (2000) D-Alanine substitution of teichoic acids as a modulator of protein folding and stability at the cytoplasmic membrane/cell wall interface of *Bacillus subtilis*. J Biol Chem 275:26696–26703

Hyyryläinen HL, Pietiäinen M, Lundén T et al (2007) The density of negative charge in the cell wall influences two-component signal transduction in *Bacillus subtilis*. Microbiology 153:2126–2136

Ishikawa S, Hara Y, Ohnishi R et al (1998) Regulation of a new cell wall hydrolase gene, *cwlF*, which affects cell separation in *Bacillus subtilis*. J Bacteriol 180:549–2555

Kobayashi T, Hatada Y, Adachi S et al (1995) Purification and properties of an alkaline protease from alkalophilic *Bacillus* sp. KSM-K16. Appl Microbiol Biotechnol 43:473–481

Kobayashi K, Sudiarta IP, Kodama T et al (2012) Identification and characterization of a novel polysaccharide deacetylase C (PdaC) from *Bacillus subtilis*. J Biol Chem 287:9765–9776
Kodama T, Endo K, Ara K et al (2007) Effect of *Bacillus subtilis spo0A* mutation on cell wall lytic enzymes and extracellular proteases, and prevention of cell lysis. J Biosci Bioeng 103:13–21
Kuroda A, Sekiguchi J (1991) Molecular cloning and sequencing of a major *Bacillus subtilis* autolysin gene. J Bacteriol 173:304–312
Neuhaus FC, Baddiley J (2003) A continuum of anionic charge: structures and functions of D-ALANYL-TEICHOIC acids in gram-positive bacteria. J Microbiol Mol Biol Rev 67:686–723
Nonaka T, Fujihashi M, Kita A et al (2004) The crystal structure of an oxidatively stable subtilisin-like alkaline serine protease, KP-43, with a C-terminal beta-barrel domain. Biol Chem 279:47344–47351
Ohnishi R, Ishikawa S, Sekiguchi J (1999) Peptidoglycan hydrolase LytF plays a role in cell separation with CwlF during vegetative growth of *Bacillus subtilis*. J Bacteriol 181:3178–3184
Rashid MH, Mori M, Sekiguchi J (1995) Glucosaminidase of *Bacillus subtilis*: cloning, regulation, primary structure and biochemical characterization. Microbiology 141:2391–2404
Sudiarta IP, Fukushima T, Sekiguchi J (2010a) Bacillus subtilis CwlP of the SP-β prophage has two novel peptidoglycan hydrolase domains, muramidase and cross-linkage digesting D,D-ENDOPEPTIDASE. J Biol Chem 285:41232–41243
Sudiarta IP, Fukushima T, Sekiguchi J (2010b) *Bacillus subtilis* CwlQ (previous YjbJ) is a bifunctional enzyme exhibiting muramidase and soluble-lytic transglycosylase activities. Biochem Biophys Res Commun 398:606–612
Yamaguchi H, Furuhata K, Fukushima T et al (2004) Characterization of a new *Bacillus subtilis* peptidoglycan hydrolase gene, *yvcE* (named *cwlO*), and the enzymatic properties of its encoded protein. J Biosci Bioeng 98:174–181
Yamamoto H, Kurosawa S, Sekiguchi J (2003) Localization of the vegetative cell wall hydrolases LytC, LytE, and LytF on the *Bacillus subtilis* cell surface and stability of these enzymes to cell wall-bound or extracellular proteases. J Bacteriol 185:6666–6677
Yamamoto H, Hashimoto M, Higashitsuji Y et al (2008) Post-translational control of vegetative cell separation enzymes through a direct interaction with specific inhibitor IseA in *Bacillus subtilis*. Mol Microbiol 70:168–182

Chapter 14
Strategies for Increasing the Production Level of Heterologous Proteins in *Aspergillus oryzae*

Mizuki Tanaka and Katsuya Gomi

Abstract The filamentous fungus *Aspergillus oryzae*, which has long been used for the production of Japanese traditional fermented foods, has been attracting attention as a promising host for recombinant protein production because of its safety and a high capacity for protein secretion. The production levels in *A. oryzae* of most non-fungal proteins of higher eukaryote origin are generally low compared to those of homologous or other fungal proteins. The reduced yields of heterologous proteins are also a common problem in other filamentous fungal species. Thus, there have been many attempts to improve the production yield of heterologous proteins, particularly those derived from higher eukaryotes. Consequently, several trials have successfully resulted in an increase in the production level of recombinant proteins. This review focuses on the various strategies for improving heterologous protein production in *A. oryzae*, mainly for improving transcription and translation efficiencies, mRNA stability, and protein secretion, as well as for preventing proteolytic degradation. Although these strategies are individually effective in improvement of protein yields, combination of these methods should be required to achieve the commercial production levels of heterologous proteins from *A. oryzae*. In addition to the efforts in productivity improvement from the molecular aspects, application of the solid-state fermentation process would be advantageous for heterologous protein production by *A. oryzae*.

Keywords *Aspergillus oryzae* protein production • mRNA stability • Protein secretion • Proteolytic degradation • Solid-state fermentation • Transcription and translation efficiencies

M. Tanaka • K. Gomi (✉)
Graduate School of Agricultural Science, Tohoku University,
1-1, Tsutsumidori-Amamiyamachi, Aoba-ku, Sendai 981-8555, Japan
e-mail: gomi@biochem.tohoku.ac.jp

H. Anazawa and S. Shimizu (eds.), *Microbial Production: From Genome Design to Cell Engineering*, DOI 10.1007/978-4-431-54607-8_14, © Springer Japan 2014

14.1 Introduction

The filamentous fungus *Aspergillus oryzae* secretes large amounts of enzymatic proteins represented by amylolytic and proteolytic enzymes. On the basis of this advantage, *A. oryzae* has been used in Japan for the production of traditional fermented foods, such as sake, soy sauce, and miso (soybean paste) for more than 1,000 years (Machida et al. 2008). This long history of food industrial use proves the safety of *A. oryzae*, and this fungus is classified as "generally recognized as safe (GRAS)" by the U.S. Food and Drug Administration (FDA) in the United States of America. Therefore, *A. oryzae* has recently become attractive as a host organism for recombinant protein production. For instance, it has been reported that *A. oryzae* secreted 3 g/L of aspartic proteinase derived from a filamentous fungus *Rhizomucor miehei* (Christensen et al. 1988). However, the secretory yields of heterologous proteins are generally low compared with those of homologous proteins or proteins from closely related fungal species. The reduced yield of heterologous proteins is also a common problem in other filamentous fungal species, such as *Aspergillus niger* and *Trichoderma reesei*. To improve the production level of heterologous proteins, various attempts have been conducted, and several trials resulted in an increase in heterologous protein level. In this chapter, we describe the various strategies to improve the production level of heterologous proteins in *A. oryzae*.

14.2 Strategy for Improving the mRNA Level of Heterologous Genes

14.2.1 Utility of High-Expression Promoter

The use of an expression promoter that has a high transcriptional activity is the most popular method for higher expression of the heterologous gene. In *A. oryzae*, the promoters of amylolytic genes, such as the α-amylase-encoding gene, *amyA* or *amyB*, and glucoamylase-encoding gene, *glaA*, are widely used for high expression of heterologous genes. The amylolytic gene expression is induced by the presence of maltooligosaccharides, such as starch and maltose, and the transcriptional activator, AmyR, regulates their gene expression (Gomi et al. 2000). The introduction of multiple tandem copies of the fragment comprising *cis*-element region [designated region III or SRE (starch-responsive element)], to which AmyR is specifically bound, into the α-glucosidase gene (*agdA*) promoter remarkably increased the gene expression activity (Minetoki et al. 1998). Moreover, the modified *glaA* promoter (*glaA*142 promoter) harboring the 12 tandem repeats of the region III has the highest promoter activity in *A. oryzae* (Minetoki 2000; Tamalanpudi et al. 2007).

In addition to the amylolytic genes promoter, some promoters were developed for constitutive expression of the heterologous genes in *A. oryzae*. A DNA fragment designated No. 8AN was isolated from the *A. niger* genome by screening the sequences that could direct high-level expression of the β-glucuronidase (GUS) encoding gene (*uidA*) of *Escherichia coli* as a reporter gene (Ozeki et al. 1996). The *uidA* gene expression level under the control of No. 8AN was approximately threefold higher than that expressed by the *amyB* promoter. A BLAST search based on the nucleotide sequence (787 bp) of No. 8AN suggested that this fragment contained the intergenic region between An08g05050 and An08g05060, both of which are divergently transcribed, with their translational start codons. Genome sequencing data of *A. niger* showed that the intergenic region between An08g05050 and An08g05060 is only 470 bp in length, including 5′-untranslated region (5′-UTR) of both genes, and thus both genes would share the common promoter that directs high-level gene expression. On the other hand, as a result of screening for highly expressed EST clones in submerged culture of *A. oryzae*, *sodM* gene encoding a manganese superoxide dismutase of *A. oryzae* was found to be abundantly expressed. Although the *sodM* promoter showed a high transcription activity under the standard submerged culture condition, its promoter activity could be enhanced approximately threefold by the addition of 0.01 % H_2O_2 in the standard liquid medium (Ishida et al. 2004).

Insertion of the multiple region IIIs into the *A. oryzae* enolase gene promoter (P*enoA*) and No. 8AN is also an effective way to improve the promoter activity (Minetoki 2000; Tsuboi et al. 2005). In particular, No. 8–142 promoter harboring 12 copies of region III in No. 8AN had significantly high transcription activity under the condition using glucose as a sole carbon source (Minetoki 2000; Tamalanpudi et al. 2007). Because the production of amylolytic enzymes is severely restricted in the glucose-containing medium caused by carbon catabolite repression, No. 8–142 promoter would be advantageous in the purification of secreted heterologous protein from culture supernatant that contains a small amount of amylolytic enzymes as contaminated proteins.

In Japan, solid-state fermentation is used for the production of fermented foods as well as for the production of industrially valuable enzymes. Because *A. oryzae* can secrete higher levels of hydrolytic enzymes in solid-state culture compared to that in submerged culture, it would be of great importance to construct an efficient protein expression system in *A. oryzae* grown in solid-state culture. Among the amylolytic enzymes produced by *A. oryzae*, glucoamylase (GlaB) is most important in sake brewing because it liberates glucose that is fermented to ethanol by yeast from limit dextrin or maltooligosaccharides, which are produced from starch by α-amylase. The *glaB* gene is highly and specifically expressed in solid-state fermentation (Hata et al. 1998; Ishida et al. 2000), and thus the *glaB* promoter seemed to be suitable for high-level gene expression in solid-state culture. In addition to the *glaB* promoter, the promoters of α-amylase (*amyA*) and *hlyA* (encoding hemolysin-like protein) were identified as the favorable promoter showing high transcription activity in solid-state culture (Bando et al. 2011). Particularly, the *hlyA* promoter had the highest transcription activity under solid-state culture condition among the promoters examined, including *amyA* and *glaA* promoters. In addition, the *hlyA* promoter was found to be more active than the *sodM* promoter in submerged culture, suggesting that the *hlyA* promoter is useful to direct high-level expression of the heterologous gene in both solid-state and submerged cultures.

14.2.2 Codon Optimization of Heterologous Gene

In heterologous protein production, codon optimization is one of the effective methods for improving production level in any organisms. In general, codon optimization is believed to improve translational efficiency by removing the rare codons that could interrupt the translation elongation. In filamentous fungi, however, some reports showed that codon optimization increased mRNA level of heterologous genes (Gouka et al. 1996; Te'o et al. 2000; Scholtmeijer et al. 2001; Koda et al. 2005; Li et al. 2007; Tokuoka et al. 2008; Leroch et al. 2011). In the case of the production of house dust mite allergen (Der f 7) derived from *Dermatophagoides farina*, the steady-state mRNA level in *A. oryzae* was increased more than threefold by codon optimization (Tokuoka et al. 2008). Further analysis revealed that the transcription products derived from Der f 7 were prematurely polyadenylated within the coding region when native cDNA was expressed in *A. oryzae* (Tokuoka et al. 2008). The prematurely polyadenylated transcripts were not detected when codon-optimized Der f 7 cDNA was expressed, indicating that codon optimization prevents premature polyadenylation within the coding region, which results in an increase in steady-state mRNA levels.

Because the premature polyadenylation within the coding region resulted in the formation of truncated mRNA without a translational termination codon, it was presumed that these transcription products are degraded by a nonstop mRNA degradation pathway, which is an mRNA quality control mechanism. Comparison of native and codon-optimized Der f 7 mRNA stabilities using transcription inhibitor revealed that native Der f 7 transcripts were remarkably unstable compared with codon-optimized Der f 7 transcripts (Tanaka et al. 2012). In addition, codon optimization of only the 3′-half region of native Der f 7 cDNA, where premature polyadenylation sites were exclusively situated, increased steady-state mRNA level, and half-life of this chimeric Der f 7 mRNA was similar to that of codon-optimized Der f 7 mRNA (Tanaka et al. 2012). These observations suggested that heterologous gene transcripts were significantly stabilized by codon optimization, which was probably caused by the prevention of premature polyadenylation within the coding region (Fig. 14.1).

In eukaryotes, polyadenylation sites are determined by several sequence elements, so-called 3′-end processing signals. In silico analysis using expressed sequence tags and genomic sequencing data showed that the *A. oryzae* 3′-end processing signals consisted of AU-rich sequence elements (Fig. 14.2, Tanaka et al. 2011). The GC content of the native Der f 7 gene was 37.8 %, and codon optimization of Der f 7 gene resulted in a GC content of 52.8 %. This result suggests that AU-rich sequences within the native Der f 7 coding region may be involved in premature polyadenylation, and codon optimization could result in eliminating the cryptic 3′-end processing signals within the coding region of Der f 7 gene. In support of this idea, the steady-state mRNA levels of heterologous genes were increased in most cases where codon optimization resulted in high GC contents of heterologous genes in filamentous fungi (Table 14.1). In mammals, it has been known that polyadenylation signal AAUAAA, which is located 10–35 nucleotides upstream of the poly(A) site, is highly

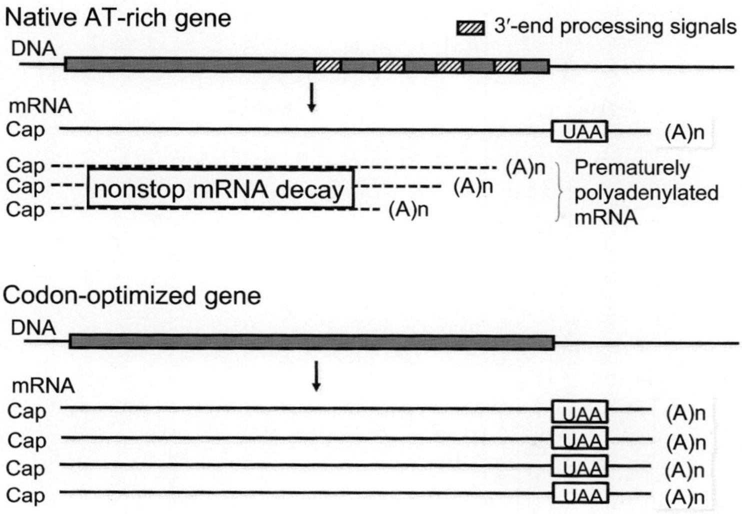

Fig. 14.1 A schematic model for the effect of codon optimization on improvement of the mRNA level of a heterologous gene

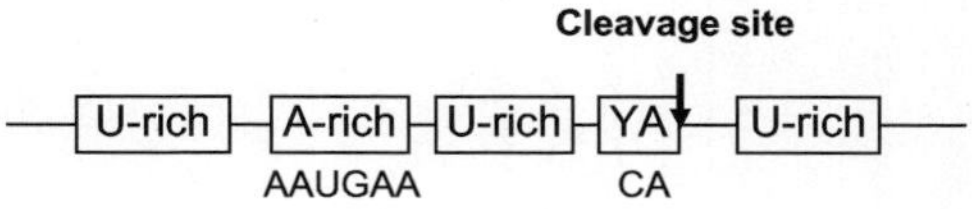

Fig. 14.2 Schematic representation of the 3′-end processing signals in *Aspergillus oryzae. Arrow* indicates the cleavage and polyadenylation site. The AAUGAA sequence is the most frequently found hexanucleotide sequence in A-rich region, although this sequence accounts for only 6 % of *A. oryzae* transcripts (Tanaka et al. 2011)

conserved. However, all the sequence elements that constituted the *A. oryzae* 3′-end processing signals are less well conserved. Therefore, because it would be difficult to effectively eliminate the cryptic 3′-end processing signals within the coding region, codon optimization (i.e., alteration of AT-biased codons) of the prematurely polyadenylated region within the heterologous gene is the most efficient method for prevention of premature polyadenylation in filamentous fungi.

14.2.3 Introduction of Intron

In eukaryotes, it is known that splicing of introns is coupled to transcription, mRNA stability, and mRNA transport from the nucleus to cytoplasm. In mammalian cells, it has been reported that splicing of introns results in increased steady-state mRNA levels. In the filamentous fungus *Schizophyllum commune*, existence of introns within the heterologous genes contributed to accumulation of steady-state mRNA

Table 14.1 Codon optimization of heterologous genes resulting in increase in steady-state mRNA levels in filamentous fungi

			GC contents		
Heterologous genes	Derived organisms	Host filamentous fungi	Native (%)	Codon-optimized	References
α-Galactosidase (*aglA*)	*Cyamopsis tetragonoloba* (guar)	*Aspergillus niger*	41	Codon optimization for yeast[a]	Gouka et al. (1996)
Xylanase (*xynB*)	*Dictyoglomus thermophilum*	*Trichoderma reesei*	39	60 %	Te'o et al. (2000)
Hygromycin B resistance gene (*hph*)	*Escherichia coli*	*Schizophyllum commune*	58	Partial codon optimization[b]	Scholtmeijer et al. (2001)
α-Glucan phosphorylase	*Solanum tuberosum* (potato)	*Aspergillus niger*	43	56 %	Koda et al. (2005)
Xylanase (*xynA*)	*Orpinomyces* PC-2	*Trichoderma reesei*	44	61 %	Li et al. (2007)
House dust mite allergen (Der f 7)	*Dermatophagoides farina*	*Aspergillus oryzae*	38	53 % Partial codon optimization[c]	Tokuoka et al. (2008) Tanaka et al. (2012)
Enhanced GFP (eGFP)	Codon optimized for human cell	*Botrytis cinerea*	62	45 %	Leroch et al. (2011)

[a]Although there is no information on GC content, some AT-rich regions were removed by codon optimization
[b]The AT-rich region constituted of 36 nucleotides was removed by codon optimization
[c]Only the 3′-half region, where premature polyadenylation sites were exclusively situated, was codon optimized

independently of transcription (Lugones et al. 1999), which suggests that the introduction of an intron into heterologous cDNA is an effective method for an increase in steady-state mRNA derived from heterologous cDNA in filamentous fungi.

14.2.4 Fusion of Homologous Gene

The fusion of a carrier protein is also a major strategy for improvement of the production level of heterologous proteins. In filamentous fungi, α-amylase and glucoamylase have been used as a carrier protein, and production levels of heterologous proteins were significantly increased by fusion to these secreted proteins (Tsuchiya et al. 1994; Gouka et al. 1997a; Gouka et al. 1997b; Tokuoka et al. 2008; Ohno et al. 2011; Hisada et al. 2013). In some cases, including Der f 7, fusion of a homologous gene could contribute to increased steady-state mRNA levels of the heterologous gene (Gouka et al. 1997a; Tokuoka et al. 2008). The steady-state mRNA levels of native and codon-optimized Der f 7 were increased more than fivefold by fusion of *glaA* gene (Tokuoka et al. 2008). However, the half-life of nonfused Der f 7 mRNA was almost equal to *glaA*-fused Der f 7 mRNA (Tanaka et al. 2012). This result suggested that improvement of the steady-state mRNA level of *glaA*-fused Der f 7 was independent of mRNA stability, probably because of the efficiency of transcription activity or mRNA transport from the nucleus to the cytoplasm. It is possible that the introns of a homologous gene fused to heterologous cDNA have an effect on the improved steady-state mRNA levels. Further studies are needed to understand the effect of homologous gene fusion on improving the mRNA levels of heterologous gene.

14.2.5 3′-Untranslated Region (3′-UTR)

In eukaryotes, it is known that the stability of some mRNAs is regulated by binding of specific proteins to the 3′-UTR of target mRNAs. In *Aspergillus nidulans*, the transcripts involved in nitrogen metabolism, including a transcription factor gene *areA*, are rapidly degraded in response to the presence of glutamine (Morozov et al. 2000; Caddick et al. 2006). The stabilities of these transcripts depend on their 3′-UTRs, and the fusion of these 3′-UTRs to heterologous genes triggers the degradation of the heterologous gene transcripts in response to glutamine or ammonium (Morozov et al. 2000; Caddick et al. 2006). In *Saccharomyces cerevisiae*, the effect of 5,302 terminator regions on GFP expression level was comprehensively evaluated, and fluorescent intensity of GFP fused to the terminator of ribosomal protein gene was approximately 2.5 fold higher compared to that of a *PGK1* terminator-fused GFP (Yamanishi et al. 2013). Although the terminator of *amyB* or α-glucosidase-encoding gene (*agdA*) is generally used for heterologous gene expression in *A. oryzae*, there is no information on the activity of terminator in filamentous fungi. Because too little attention has been paid to 3′-UTRs in terms of their effect on the transcript stability and transcription efficiency, it is possible that there is a certain terminator more suitable for heterologous gene expression in *A. oryzae*.

14.3 Strategy for Improving the Translation Efficiency of Heterologous Genes

14.3.1 5′-Untranslated Region (5′-UTR)

In eukaryotes, it has been known that the 5′-untranslated region (5′-UTR) has an important role in control of translation efficiency. Using the *E. coli* GUS gene (*uidA*) as the reporter, Koda et al. (2004) examined the effect of 5′-UTR on heterologous protein production in *A. oryzae*, and they revealed that alternation of 5′-UTR of No. 8–142 promoter with that of its own gene (An08g05060) or the *A. oryzae enoA* gene resulted in a significant increase in GUS activity. Steady-state mRNA levels of the *uidA* gene were similar to each other, suggesting that the increase in GUS activity was owing to enhanced translation efficiency but not to enhanced transcription efficiency or increased mRNA stability. In addition, 5′-UTR of heat-shock protein gene (*hsp12*) enhanced the translation efficiency more, especially under high-temperature condition (37 °C), and GUS activity in the construct with the improved promoter fused to the *hsp12* 5′-UTR was approximately 20 fold higher than that in the control construct (Koda et al. 2006).

14.3.2 Codon Optimization

As described here, codon optimization increases the steady-state mRNA level of heterologous gene in filamentous fungi. However, to the best of our knowledge, there are no known experimental data on the effect of codon optimization on translation efficiency of a heterologous gene in filamentous fungi. As it is well known that codon optimization or expression of rare codon tRNAs are powerful tools for heterologous protein expression in bacterial cells, further studies are required to understand the effect of rare codons on the translation efficiency in filamentous fungi.

14.4 Strategy for Removing the Bottlenecks at the Posttranslational Process

14.4.1 Improvement at Secretion Pathway

The nascent peptide of a secretory protein is translocated into the endoplasmic reticulum (ER) through the recognition of a signal sequence. It has been reported that a 28-amino-acid region (N28) within the propeptide of *Rhizopus oryzae* lipase has an important role in protein translocation into the ER (Hama et al. 2008). When N28 was fused to a llama-variable heavy-chain antibody fragment (V_{HH}), secretion yield of V_{HH} in *A. oryzae* was about 3.5 fold higher compared with the nonfused construct (Okazaki et al. 2012).

Overexpression of secreted heterologous protein genes may result in an accumulation of their unfolded or misfolded proteins in the ER lumen, triggering the unfolded protein response (UPR). On UPR activation, such misfolded proteins in the ER lumen are either refolded with the assistance of the ER chaperones and foldases, or degraded through the ER-associated degradation (ERAD) pathway and vacuole/lysosomal degradation pathway. In *A. niger*, production levels of heterologous proteins were increased by overproduction of ER chaperones and foldases (Conesa et al. 2002; Valkonen et al. 2003) or by disruption of ERAD components (Jacobs et al. 2009; Carvalho et al. 2010). Overexpression of the UPR-activated form of the transcription factor *hacA* resulted in a significant increase (more than twofold) in the expression level of the ER chaperone gene *bipA*. This enhanced transcription of *bipA* led to an increase in the production level of heterologous proteins, bovine chymosin, and laccase from a white rot fungus *Trametes versicolor*. Although the laccase transcript levels were lower in overexpression transformants of the activated *hacA* form than in the parental strain, a greater amount of secreted laccase was produced by the transformants compared to the parental strain (Valkonen et al. 2003). Similarly, overproduction of another ER chaperone, calnexin, was found to increase the production level (~fivefold) of a manganese peroxidase from the white rot fungus *Phanerochaete chrysosporium* in *A. niger* (Conesa et al. 2002). However, overproduction of the ER chaperones and foldases has not been always effective in improvement of production levels of heterologous proteins (Ward et al. 1990; van Gemeren et al. 1997; Wiebe et al. 2001). For example, overexpression of the *cypB* gene encoding ER foldase, cyclophilin, did not result in an increase in the production level of tissue plasminogen activator in *A. niger* (Wiebe et al. 2001).

On the other hand, several putative genes involved in the ERAD pathway were disrupted in *A. niger*, and the effect of disruption of those genes on heterologous protein production was examined. Overexpression of glucoamylase-GUS fusion (GlaGUS) gene in the disruptants of the *derA* and *hrdC*, which each encodes a homologue of *Saccharomyces cerevisiae* Der1 and Hrd3, respectively, resulted in a twofold increase in *bipA* mRNA level compared to the control and concomitantly led to the highest accumulation of GlaGUS protein intracellularly. This finding suggests that absence of the ERAD pathway proteins, DerA and HrdC, may result in the longer retention of overproduced heterologous proteins in ER, leading to a delay in protein degradation (Carvalho et al. 2010).

In addition to ERAD, vacuole/lysosomal degradation is another mechanism by which misfolded or aberrant proteins are removed and degraded from the secretory pathway. The aberrant proteins are sorted by this quality control mechanism from the Golgi apparatus to the endosomal system and finally degraded in a vacuole or lysosome. Based on the observations that *S. cerevisiae* Vps10, a transmembrane receptor protein, functions in recognition and sorting of the vacuole proteins such as carboxypeptidase Y and proteinase A to the vacuole, but also in targeting of recombinant proteins from the late Golgi to a vacuole for degradation, the effect of disruption of *Aovps10* gene, *S. cerevisiae VPS10* orthologue, on heterologous protein production in *A. oryzae* was examined. Similar to the *S. cerevisiae vps10* mutant, the *Aovps10* disruption mutant missorted and secreted a vacuolar carboxypeptidase AoCpyA. Furthermore, the yields of bovine chymosine and human lysozyme fused to α-amylase in the *A. oryzae Aovps10* disruption mutant were increased up to 3.0- and 2.2 fold, respectively, compared to the wild-type strain (Yoon et al. 2010). These results indicated that deletion of a sorting receptor for vacuolar protein targeting would be highly effective for improving the production level of heterologous proteins in *A. oryzae*. Observations in experiments in which the genes involved in ERAD and vacuole degradation were disrupted in *A. oryzae* suggest that heterologous proteins are degraded by ER and post-ER quality control pathways in filamentous fungi.

In heterologous protein production by other filamentous fungi, there are many reports of improvements at the secretion pathway, and these reports are included in review papers (Gouka et al. 1997b; Fleissner and Dersch 2010; Ward 2012).

14.4.2 Reduction of Proteolytic Activity

A. oryzae produces copious amounts of proteases and peptidases extracellularly, which is quite advantageous in its use in traditional fermentation industries, such as soy sauce and miso manufacturing in Japan. However, this characteristic adversely hampers efficient production of heterologous proteins in *A. oryzae*, because the heterologous proteins produced can be degraded by their own proteolytic enzymes. Therefore, proteolytic degradation of secreted heterologous protein is the one of the major problems for heterologous protein production in *A. oryzae*, as is also true for the closely related industrial filamentous fungus *A. niger*. In this context, inactivation of proteases and peptidases is essential for improvement of heterologous protein production in *Aspergillus* fungi. In earlier studies in *A. niger*, disruption of the *pepA* gene encoding a major extracellular aspartic protease resulted in a twofold increase in the production level of bovine chymosin (Berka et al. 1990; Davies 1994). Furthermore, disruption of additional two protease genes, *pepB* and *pepE*, which each encode an extracellular non-pepsin-type protease and intracellular (vacuolar) pepsin-type protease, respectively, was carried out. Consequently, culture supernatants from the *ΔpepA/ΔpepB* double disruptant and *ΔpepA/ΔpepB/ΔpepE* triple disruptant showed reduced proteolytic degradation of *A. niger* pectin lyase PelB (van den Hombergh et al. 1997).

Recently, similar attempts to disrupt the proteolytic genes have also been made in *A. oryzae*. In the beginning, single-gene disruptants for five proteolytic enzymes, PepA (aspartic protease; aspergillopepsin O), PepE (vacuolar aspartic protease), AlpA (alkaline protease), TppA (tripeptidyl peptidase), and PalB (calpain-like protease), were constructed, and the effect of their disruption on the secretory production of human lysozyme (HLY) was evaluated. As a result, disruption of the *tppA* gene was most effective in HLY production. Further disruption of the *pepE* gene in the *ΔtppA* disruptant resulted in a nearly 2-fold increase in the level of secreted HLY compared to the parental wild-type strain (Jin et al. 2007). The resultant *ΔtppA/ΔpepE* double disruptant also showed a dramatic reduction in proteolytic degradation of bovine chymosin, leading to a nearly 2-fold increase in the production level

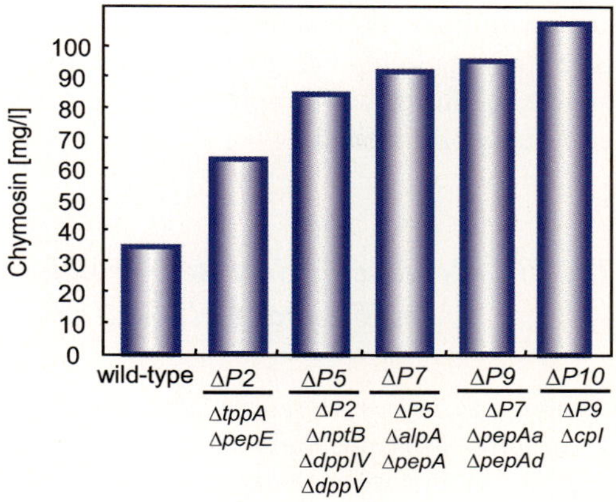

Fig. 14.3 Production levels of secreted bovine chymosin by disruption strains for multiple proteolytic enzymes in *A. oryzae*. Chymosin production levels of each disruptant strain and wild-type are derived from the reports by Yoon et al. (2009, 2011)

(Yoon et al. 2009). In addition, disruption of additional proteolytic genes was attempted using the *ΔtppA*/*ΔpepE* double disruptant as a host strain, and finally, by disruption of ten proteolytic genes, the production yield of secreted chymosin was increased up to 3.8 fold compared with that of the wild-type strain (Fig. 14.3, Yoon et al. 2011). Although the decoupled proteolytic gene disruptant showed highly improved productivity of secreted heterologous proteins, extracellular protease and peptidase activities still remained relatively high; protease activity at pH 5.5 and at 8.0 in the disruptant was approximately 50 % and 70 % of that in the wild type, respectively. Genome analysis revealed that *A. oryzae* possesses the 134 putative protease- or peptidase-encoding genes within its genomic sequence (Kobayashi et al. 2007). Therefore, further disruption of additional proteolytic genes might be required to reduce the extracellular protease and peptidase activities.

On the other hand, in *A. niger*, PrtT was identified as the transcription factor that regulates extracellular proteolytic genes (Punt et al. 2008). In *A. oryzae*, ten extracellular proteolytic genes were upregulated by overexpression of *prtR* (orthologue of *prtT*; Mizutani et al. 2008), and extracellular protease activities in the *prtR* disruptant were approximately threefold (3-day culture) and eightfold (5-day culture) lower as compared to those in the wild type. In agreement with the reduced extracellular protease activity, the production yield of HLY in the *prtR* disruptant was approximately twofold higher compared to that in wild-type strain (Matsuura and Gomi, unpublished data). In addition, double disruption of *prtR* and *pepE* resulted in a nearly threefold increase in HLY production compared to the wild-type strain (Matsuura and Gomi, unpublished data). Because extracellular proteolytic enzymes are produced under solid-state culture condition (Kitano et al. 2002), proteolytic degradation of heterologous proteins seems to be more problematic in solid-state fermentation. The *ΔprtR*/*ΔpepE* double disruptant showed a significantly lower protease activity even though in solid-state culture, thus sufficiently preventing proteolytic degradation of heterologous proteins in solid-state fermentation. In fact, when *Cryptococcus* lipase was produced by *A. oryzae* in solid-state fermentation, the amount of lipase protein produced began to decrease after 2-day culture in the wild-type strain, possibly being degraded by its own proteolytic enzymes, but it kept on increasing during a 4-day culture in the *ΔprtR/ΔpepE* double disruptant (Mizutani et al., unpublished data).

14.4.3 Release of Secreted Proteins Bound to the Fungal Cell Wall

Although *A. oryzae* secretes a large amount of α-amylase (Taka-amylase A, TAA), TAA activity disappeared in the later stage of submerged culture. Recently, it has been shown that disappearance of TAA activity was caused by adsorption of TAA on to a cell wall component, most likely chitin (Sato et al. 2011). In addition to TAA, bovine serum albumin, hen egg lysozyme, and human lysozyme were also adsorbed onto the cell wall of *A. oryzae* (Yabuki and Fukui 1970; Matsuura and Gomi, unpublished data). Therefore, it would be necessary to note that secreted heterologous proteins may adsorb to the cell wall, resulting in an apparently reduced production level of the protein of interest in the later stage of submerged culture. As the TAA bound to the cell wall was liberated by addition of 0.1 M phosphate buffer, adsorption of heterologous proteins to the cell wall could be inhibited by addition of phosphate buffer to the culture medium (Sato et al. 2011).

14.5 Conclusions

As described here, many strategies for improving heterologous protein production level have been reported. In most cases, however, the yield of nonfungal heterologous proteins did not exceed tens of milligrams per liter of culture broth. Therefore, combination of these various methods is needed to produce a sufficient amount of heterologous proteins from *A. oryzae*.

In the process of Japanese fermented food production, *A. oryzae* is grown in solid-state culture using rice and wheat bran as a base material. This solid-state culture is a unique feature of *A. oryzae*, and production level of secretory protein in solid-state culture is much higher than that in submerged culture. In addition, the traditional fermentation technologies in Japan provide a wealth of knowledge and facilities for solid-state fermentation of *A. oryzae*. Therefore, solid-state fermentation would also have an enormous advantage in heterologous protein production in *A. oryzae*.

The molecular biological analysis and functional analysis of individual genes have been dramatically developed after the publication of *A. oryzae* genome sequencing data (Machida et al. 2005) and the construction of a convenient host stain for gene knockouts (Mizutani et al. 2008). In addition, novel methods of selectable marker gene rescue were developed for disruptions of multiple genes

(Maruyama and Kitamoto 2008; Mizutani et al. 2012). The basic analysis of heterologous protein expression will be much promoted by using these applications, and the knowledge obtained from those analyses will contribute to improvement of the heterologous protein production level in *A. oryzae*.

References

Bando H, Hisada H, Ishida H, Hata Y, Katakura Y, Kondo A (2011) Isolation of a novel promoter for efficient protein expression by *Aspergillus oryzae* in solid-state culture. Appl Microbiol Biotechnol 92:561–569

Berka RM, Ward M, Wilson LJ, Hayenga KJ, Kodama KH, Carlomagno LP, Thompson SA (1990) Molecular cloning and deletion of the gene encoding aspergillopepsin A from *A. awamori*. Gene (Amst) 86:153–162

Caddick MX, Jones MG, van Tonder JM, Le Cordier H, Narendja F, Strauss J, Morozov IY (2006) Opposing signals differentially regulate transcript stability in *Aspergillus nidulans*. Mol Microbiol 62:509–519

Carvalho ND, Arentshorst M, Kooistra R, Stam H, Sagt CM, van den Hondel CA, Ram AF (2010) Effects of a defective ERAD pathway on growth and heterologous protein production in *Aspergillus niger*. Appl Microbiol Biotechnol 89:357–373

Christensen T, Woeldike H, Boel F, Mortensen SB, Hjortshoej K, Thim L, Hansen M (1988) High level expression of recombinant genes in *Aspergillus oryzae*. Biotechnology 6:1419–1422

Conesa A, Jeenes D, Archer DB, van den Hondel CA, Punt PJ (2002) Calnexin overexpression increases manganese peroxidase production in *Aspergillus niger*. Appl Environ Microbiol 68:846–851

Davies RW (1994) Heterologous gene expression and protein secretion in *Aspergillus*. Prog Ind Microbiol 29:521–560 (In: Martinelli SD, Kinghorn JR (1994) *Aspergillus*: 50 years on. Elsevier, Amsterdam)

Fleissner A, Dersch P (2010) Expression and export: recombinant protein production systems for *Aspergillus*. Appl Microbiol Biotechnol 87:1255–1270

Gomi K, Akeno T, Minetoki T, Ozeki K, Kumagai C, Okazaki N, Iimura Y (2000) Molecular cloning and characterization of a transcriptional activator gene, *amyR*, involved in the amylolytic gene expression in *Aspergillus oryzae*. Biosci Biotechnol Biochem 64:816–827

Gouka RJ, Punt PJ, Hessing JG, van den Hondel CAMJJ (1996) Analysis of heterologous protein production in defined recombinant *Aspergillus awamori* strains. Appl Environ Microbiol 62:1951–1957

Gouka RJ, Punt PJ, van den Hondel CAMJJ (1997a) Glucoamylase gene fusions alleviate limitations for protein production in *Aspergillus awamori* at the transcriptional and (post)transcriptional levels. Appl Environ Microbiol 63:488–497

Gouka RJ, Punt PJ, van den Hondel CAMJJ (1997b) Efficient production of secreted proteins by *Aspergillus*: progress, limitations and prospects. Appl Microbiol Biotechnol 47:1–11

Hama S, Tamalampudi S, Shindo N, Numata T, Yamaji H, Fukuda H, Kondo A (2008) Role of N-terminal 28-amino-acid region of *Rhizopus oryzae* lipase in directing proteins to secretory pathway of *Aspergillus oryzae*. Appl Microbiol Biotechnol 79:1009–1018

Hata Y, Ishida H, Ichikawa E, Kawato A, Suginami K, Imayasu S (1998) Nucleotide sequence of an alternative glucoamylase-encoding gene (glaB) expressed in solid-state culture of *Aspergillus oryzae*. Gene 207:127–134

Hisada H, Tsutsumi H, Ishida H, Hata Y (2013) High production of llama variable heavy-chain antibody fragment (VHH) fused to various reader proteins by *Aspergillus oryzae*. Appl Microbiol Biotechnol 97:761–766

Ishida H, Hata Y, Kawato A, Abe Y, Kashiwagi Y (2004) Isolation of a novel promoter for efficient protein production in *Aspergillus oryzae*. Biosci Biotechnol Biochem 68:1849–1857

Ishida H, Hata Y, Kawato A, Abe Y, Suginami K, Imayasu S (2000) Identification of functional elements that regulate the glucoamylase-encoding gene (glaB) expressed in solid-state culture of *Aspergillus oryzae*. Curr Genet 37:373–379

Jacobs DI, Olsthoorn MM, Maillet I, Akeroyd M, Breestraat S, Donkers S, van der Hoeven RA, van den Hondel CA, Kooistra R, Lapointe T, Menke H, Meulenberg R, Misset M, Müller WH, van Peij NN, Ram A, Rodriguez S, Roelofs MS, Roubos JA, van Tilborg MW, Verkleij AJ, Pel HJ, Stam H, Sagt CM (2009) Effective lead selection for improved protein production in *Aspergillus niger* based on integrated genomics. Fungal Genet Biol 46:S141–S152

Jin FJ, Watanabe T, Juvvadi PR, Maruyama J, Arioka M, Kitamoto K (2007) Double disruption of the proteinase genes, *tppA* and *pepE*, increases the production level of human lysozyme by *Aspergillus oryzae*. Appl Microbiol Biotechnol 76:1059–1068

Kitano H, Kataoka K, Furukawa K, Hara S (2002) Specific expression and temperature-dependent expression of the acid protease-encoding gene (*pepA*) in *Aspergillus oryzae* in solid-state culture (Rice-Koji). J Biosci Bioeng 93:563–567

Kobayashi T, Abe K, Asai K, Gomi K, Juvvadi PR, Kato M, Kitamoto K, Takeuchi M, Machida M (2007) Genomics of *Aspergillus oryzae*. Biosci Biotechnol Biochem 71:646–670

Koda A, Minetoki T, Ozeki K, Hirotsune M (2004) Translation efficiency mediated by the 5′ untranslated region greatly affects protein production in *Aspergillus oryzae*. Appl Microbiol Biotechnol 66:291–296

Koda A, Bogaki T, Minetoki T, Hirotsune M (2005) High expression of a synthetic gene encoding potato alpha-glucan phosphorylase in *Aspergillus niger*. J Biosci Bioeng 100:531–537

Koda A, Bogaki T, Minetoki T, Hirotsune M (2006) 5′ Untranslated region of the Hsp12 gene contributes to efficient translation in *Aspergillus oryzae*. Appl Microbiol Biotechnol 70:333–336

Leroch M, Mernke D, Koppenhoefer D, Schneider P, Mosbach A, Doehlemann G, Hahn M (2011) Living colors in the gray mold pathogen *Botrytis cinerea*: codon-optimized genes encoding green fluorescent protein and mCherry, which exhibit bright fluorescence. Appl Environ Microbiol 77:2887–2897

Li XL, Skory CD, Ximenes EA, Jordan DB, Dien BS, Hughes SR, Cotta MA (2007) Expression of an AT-rich xylanase gene from the anaerobic fungus *Orpinomyces* sp. strain PC-2 in and secretion of the heterologous enzyme by *Hypocrea jecorina*. Appl Microbiol Biotechnol 74:1264–1275

Lugones LG, Scholtmeijer K, Klootwijk R, Wessels JG (1999) Introns are necessary for mRNA accumulation in *Schizophyllum commune*. Mol Microbiol 32:681–689

Machida M, Asai K, Sano M, Tanaka T, Kumagai T, Terai G, Kusumoto K, Arima T, Akita O, Kashiwagi Y, Abe K, Gomi K, Horiuchi H, Kitamoto K, Kobayashi T, Takeuchi M, Denning DW, Galagan JE, Nierman WC, Yu J, Archer DB, Bennett JW, Bhatnagar D, Cleveland TE, Fedorova ND, Gotoh O, Horikawa H, Hosoyama A, Ichinomiya M, Igarashi R, Iwashita K, Juvvadi PR, Kato M, Kato Y, Kin T, Kokubun A, Maeda H, Maeyama N, Maruyama J, Nagasaki H, Nakajima T, Oda K, Okada K, Paulsen I, Sakamoto K, Sawano T, Takahashi M, Takase K, Terabayashi Y, Wortman JR, Yamada O, Yamagata Y, Anazawa H, Hata Y, Koide Y, Komori T, Koyama Y, Minetoki T, Suharnan S, Tanaka A, Isono K, Kuhara S, Ogasawara N, Kikuchi H (2005) Genome sequencing and analysis of *Aspergillus oryzae*. Nature (Lond) 438:1157–1161

Machida M, Yamada O, Gomi K (2008) Genomics of *Aspergillus oryzae*: learning from the history of koji mold and exploration of its future. DNA Res 15:173–183

Maruyama J, Kitamoto K (2008) Multiple gene disruptions by marker recycling with highly efficient gene-targeting background (DeltaligD) in *Aspergillus oryzae*. Biotechnol Lett 30:1811–1817

Minetoki T (2000) The expression system of heterologous genes in *Aspergilli* (in Japanese). Kagaku to Seibutsu 38:831–838

Minetoki T, Kumagai C, Gomi K, Kitamoto K, Takahashi K (1998) Improvement of promoter activity by the introduction of multiple copies of the conserved region III sequence, involved in the efficient expression of *Aspergillus oryzae* amylase-encoding genes. Appl Microbiol Biotechnol 50:459–467

Mizutani O, Kudo Y, Saito A, Matsuura T, Inoue H, Abe K, Gomi K (2008) A defect of LigD (human Lig4 homolog) for nonhomologous end joining significantly improves efficiency of gene-targeting in *Aspergillus oryzae*. Fungal Genet Biol 45:878–889

Mizutani O, Masaki K, Gomi K, Iefuji H (2012) Modified Cre-loxP recombination in *Aspergillus oryzae* by direct introduction of Cre recombinase for marker gene rescue. Appl Environ Microbiol 78:4126–4133

Morozov IY, Martinez MG, Jones MG, Caddick MX (2000) A defined sequence within the 3′ UTR of the *areA* transcript is sufficient to mediate nitrogen metabolite signalling via accelerated deadenylation. Mol Microbiol 37:1248–1257

Ohno A, Maruyama J, Nemoto T, Arioka M, Kitamoto K (2011) A carrier fusion significantly induces unfolded protein response in heterologous protein production by *Aspergillus oryzae*. Appl Microbiol Biotechnol 92:1197–1206

Okazaki F, Aoki JI, Tabuchi S, Tanaka T, Ogino C, Kondo A (2012) Efficient heterologous expression and secretion in *Aspergillus oryzae* of a llama variable heavy-chain antibody fragment V_{HH} against EGFR. Appl Microbiol Biotechnol 96:81–88

Ozeki K, Kanda A, Hamachi M, Nunokawa Y (1996) Construction of a promoter probe vector autonomously maintained in *Aspergillus* and characterization of promoter regions derived from *A. niger* and *A. oryzae* genomes. Biosci Biotechnol Biochem 60:383–389

Punt PJ, Schuren FH, Lehmbeck J, Christensen T, Hjort C, van den Hondel CA (2008) Characterization of the *Aspergillus niger prtT*, a unique regulator of extracellular protease encoding genes. Fungal Genet Biol 45:1591–1599

Sato H, Toyoshima Y, Shintani T, Gomi K (2011) Identification of potential cell wall component that allows Taka-amylase A adsorption in submerged cultures of *Aspergillus oryzae*. Appl Microbiol Biotechnol 92:961–969

Scholtmeijer J, Wösten HAB, Springer J, Wessels JGT (2001) Effect of introns and AT-rich sequences on expression of the bacterial hygromycin B resistance gene in the basidiomycete *Schizophyllum commune*. Appl Environ Microbiol 67:481–483

Tamalanpudi S, Talukder MM, Hama S, Tanino T, Suzuki Y, Kondo A, Fukuda H (2007) Development of recombinant *Aspergillus oryzae* whole-cell biocatalyst expressing lipase-encoding gene from *Candida antarctica*. Appl Microbiol Biotechnol 75:387–395

Tanaka M, Sakai Y, Yamada O, Shintani T, Gomi K (2011) *In silico* analysis of 3′-end-processing signals in *Aspergillus oryzae* using expressed sequence tags and genomic sequencing data. DNA Res 18:189–200

Tanaka M, Tokuoka M, Shintani T, Gomi K (2012) Transcripts of a heterologous gene encoding mite allergen Der f 7 are stabilized by codon optimization in *Aspergillus oryzae*. Appl Microbiol Biotechnol 96:1275–1282

Te'o VS, Cziferszky AE, Bergquist PL, Nevalainen KM (2000) Codon optimization of xylanase gene *xynB* from the thermophilic bacterium *Dictyoglomus thermophilum* for expression in the filamentous fungus *Trichoderma reesei*. FEMS Microbiol Lett 190:13–19

Tokuoka M, Tanaka M, Ono K, Takagi S, Shintani T, Gomi K (2008) Codon optimization increases steady-state mRNA levels in *Aspergillus oryzae* heterologous gene expression. Appl Environ Microbiol 74:6538–6546

Tsuboi H, Koda A, Toda T, Minetoki T, Hirotsune M, Machida M (2005) Improvement of the *Aspergillus oryzae* enolase promoter (P-enoA) by the introduction of cis-element repeats. Biosci Biotechnol Biochem 69:206–208

Tsuchiya K, Nagashima T, Yamamoto Y, Gomi K, Kitamoto K, Kumagai C, Tamura G (1994) High level secretion of calf chymosin using a glucoamylase-prochymosin fusion gene in *Aspergillus oryzae*. Biosci Biotechnol Biochem 58:895–899

Valkonen M, Ward M, Wang H, Penttilä M, Saloheimo M (2003) Improvement of foreign-protein production in *Aspergillus niger* var. *awamori* by constitutive induction of the unfolded-protein response. Appl Environ Microbiol 69:6979–6986

van den Hombergh JPTW, Sollewijn Gelpke MD, van de Vondervoort PJI, Buxton FP, Visser J (1997) Disruption of three acid proteases in *Aspergillus niger*: effects on protease spectrum, intracellular proteolysis, and degradation of target proteins. Eur J Biochem 247:605–613 van Gemeren IA, Punt PJ, Drint-Kuyvenhoven A, Broekhuijsen MP, van't Hoog A, Beijersbergen A, Verrips CT, van den Hondel CA (1997) The ER chaperone encoding *bipA* gene of black aspergilli is induced by heat shock and unfolded proteins. Gene (Amst) 198:43–52

Ward OP (2012) Production of recombinant proteins by filamentous fungi. Biotechnol Adv 30:1119–1139

Ward M, Wilson LJ, Kodama KH, Rey MW, Berka RM (1990) Improved production of chymosin in *Aspergillus* by expression as a glucoamylase-chymosin fusion. Bio/Technology 8:435–440

Wiebe MG, Karandikar A, Robson GD, Trinci AP, Candia JL, Trappe S, Wallis G, Rinas U, Derkx PM, Madrid SM, Sisniega H, Faus I, Montijn R, van den Hondel CA, Punt PJ (2001) Production of tissue plasminogen activator (t-PA) in *Aspergillus niger*. Biotechnol Bioeng 76:164–174

Yabuki M, Fukui S (1970) Presence of binding site for α-amylase and of masking protein for this site on mycelial cell wall of *Aspergillus oryzae*. J Bacteriol 104:138–144

Yamanishi M, Ito Y, Kintake R, Imamura C, Katahira S, Ikeuchi A, Moriya H, Matsuyama T (2013) A genome-wide activity assessment of terminator regions in *Saccharomyces cerevisiae* provides a "Terminatome" toolbox. ACS Synth Biol 2:337–347

Yoon J, Kimura S, Maruyama J, Kitamoto K (2009) Construction of quintuple protease gene disruptant for heterologous protein production in *Aspergillus oryzae*. Appl Microbiol Biotechnol 82:691–701

Yoon J, Aishan T, Maruyama J, Kitamoto K (2010) Enhanced production and secretion of heterologous proteins by the filamentous fungus *Aspergillus oryzae* via disruption of vacuolar protein sorting receptor gene *Aovps10*. Appl Environ Microbiol 76:5718–5727

Yoon J, Maruyama J, Kitamoto K (2011) Disruption of ten protease genes in the filamentous fungus *Aspergillus oryzae* highly improves production of heterologous proteins. Appl Microbiol Biotechnol 89:747–759

Chapter 15
Overproduction of L-Glutamate in *Corynebacterium glutamicum*

Hisashi Yasueda

Abstract L-Glutamic acid (L-Glu) is an industrially important amino acid for the production of the flavor enhancer, monosodium-L-glutamate (MSG), which expresses a savory taste known as *umami*. The fundamental fermentation process for MSG production was established in 1956 and is based on the L-glutamate-producing bacterium *Corynebacterium glutamicum*, which was originally isolated from a soil sample. Since the discovery of this epoch-making production method, extensive metabolic engineering studies designed to improve L-glutamate production have been conducted using the genomic information of *C. glutamicum*. This chapter focuses on recent studies of the molecular mechanism of L-glutamate secretion and metabolic pathway design for efficient overproduction of L-glutamate from glucose in *C. glutamicum* strains. The mechanism of L-glutamate overproduction is unique, and it has been demonstrated that the product of the *NCgl1221* gene, which is homologous to mechano-sensitive channels, plays a crucial role in L-glutamate secretion. It has also been shown that the specific activity of the 2-oxoglutarate dehydrogenase complex decreases during L-glutamate overproduction and that the enzymatic activity is controlled by a novel regulatory mechanism that involves OdhI and serine/threonine protein kinases. On the other hand, a new metabolic pathway was designed to enhance L-glutamate production by bypassing the CO_2-releasing pyruvate dehydrogenase reaction in the glycolytic pathway by introducing heterologous phosphoketolase. This attempt improved the conversion yield of L-Glu and reduced the CO_2 emission level during the fermentation process of L-glutamate overproduction by *C. glutamicum*.

Keywords *Corynebacterium glutamicum* • Fermentation • L-Glutamate • Metabolic pathway design • Secretion mechanism

H. Yasueda (✉)
Institute for Innovation, Ajinomoto Co., Inc.,
1-1 Suzuki-cho, Kawasaki-ku, Kawasaki 210-8681, Japan
e-mail: hisashi_yasueda@ajinomoto.com

H. Anazawa and S. Shimizu (eds.), *Microbial Production: From Genome Design to Cell Engineering*, DOI 10.1007/978-4-431-54607-8_15, © Springer Japan 2014

15.1 Introduction

15.1.1 *Discovery of the "Umami" Taste by Monosodium L-Glutamate*

Soup stock *dashi* extracted from broth of the kelp-like seaweed (*kombu*) has long been an indispensable part of Japanese cuisine because of its flavor-enhancing properties. In 1908, Kikunae Ikeda, a professor at the Tokyo Imperial University and chemist (Fig. 15.1a), succeeded in identifying the unique taste component in *kombu* to be a salt of L-glutamic acid (L-Glu) (Fig. 15.1b). He coined the term *umami* in Japanese to describe this taste (Yamaguchi and Ninomiya 2000). After the discovery of this key factor expressing the *umami* taste, Saburousuke Suzuki, a well-known entrepreneur in the chemical and pharmaceutical industry, collaborated with Ikeda to produce and commercialize a new seasoning containing monosodium L-glutamate (MSG). This seasoning was named *AJI-NO-MOTO*®, meaning "quintessence of flavor" (Sano 2009; Nakamura 2011).

15.1.2 *A Brief History of the Manufacturing Method of L-Glutamate*

In December 1908, the first attempt was made to produce L-glutamate on an industrial scale. The early manufacturing method was an acid hydrolysis and extraction process using wheat gluten as the source of L-glutamate. Wheat gluten was used because among industrially available raw materials, it has the highest content of L-glutamine, which is converted to L-Glu after protein hydrolysis with hydrochloric acid. L-Glu hydrochloride was isolated from the acid hydrolysate and then L-glutamate was purified as MSG. The first commercial MSG was successfully produced in 1909 and sold to general consumers in Japan. After the hydrolysis method for the commercial production of MSG was established, many technical improvements in the extraction process were continuously achieved using soybean as an alternate protein source (Sano 2009).

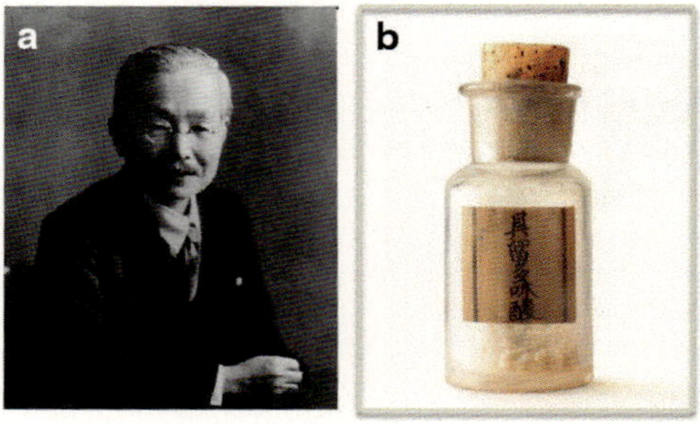

Fig. 15.1 Discovery of the *umami* taste by sodium L-glutamate. **(a)** Dr. Kikunae Ikeda (1864–1936) of the Tokyo Imperial University. **(b)** The first bottle of L-glutamate extracted from the kelp-like seaweed *kombu*. (Reproduced with permission from Ajinomoto Co., Inc.)

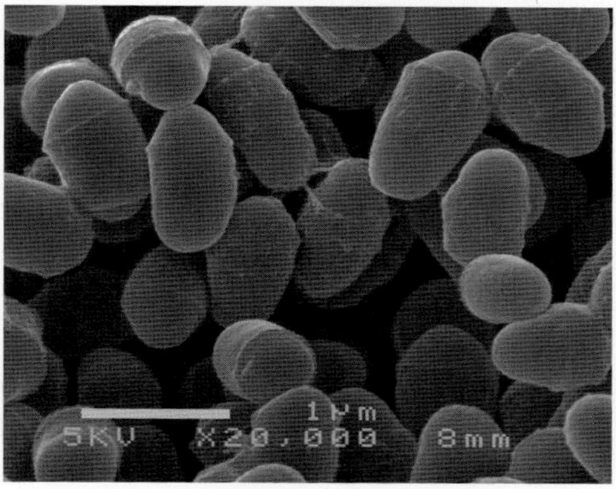

Fig. 15.2 Electron micrograph of *Corynebacterium glutamicum* 2256 ATCC13869. *Bar* 1 μm

As the demand for MSG increased, a new mass production process for L-Glu was required to ensure an adequate supply of raw materials, optimal use of by-products, and proper management of environmental issues associated with MSG manufacturing. In 1956, Kyowa Hakko Kogyo Co., Ltd. succeeded in developing the first industrial technology for fermentation of L-glutamate using *Corynebacterium glutamicum*, which was reported by Kinoshita et al. (1957). Since this report of an epoch-making method for L-glutamate production, many bacteria useful in production of MSG have been isolated, identified, and used for fermentative production, including *Brevibacterium lactofermentum* and *Brevibacterium flavum* (Yamada and Komagata 1972). These L-glutamate-producing bacteria are all coryneform bacteria (Fig. 15.2), which are rod-shaped, nonsporulating, gram-positive, nonpathogenic, and nonmotile and require biotin as a cofactor in carboxylation and transcarboxylation reactions for cell growth.

Besides L-glutamate, *C. glutamicum* has the ability to produce large amounts of other amino acids, including L-lysine, which is one of the principal amino acids used as an animal feed additive. The complete genomic DNA sequence of a representative wild-type strain of *C. glutamicum*, ATCC13032, was determined by some research teams independently (Kalinowski et al. 2003; Ikeda and Nakagawa 2003). The *C. glutamicum* genome is a single circular chromosome of 3.3 Mbp; the number of predicted open reading frames (ORFs) is 2,900–3,100. Using this genomic information, whole-genome DNA microarrays for *C. glutamicum* were developed for analysis of comprehensive gene expression networks under L-glutamate-overproducing conditions (Wendisch 2003; Kataoka et al. 2006). In addition, comprehensive analyses of the *C. glutamicum* proteome, metabolome, and fluxome have been performed to elucidate the molecular mechanism of L-glutamate overproduction and to improve the efficiency of L-glutamate production by metabolic engineering of *C. glutamicum* as a cell factory (Schluesener et al. 2005; Strelkov et al. 2004; Shirai et al. 2005).

15.2 Mechanism of Overproduction and Secretion of L-Glutamate

15.2.1 Induction of L-Glutamate Overproduction in C. glutamicum

Wild-type *C. glutamicum*, without strain breeding, secretes and accumulates more than several tens of grams per liter (g/l) of L-glutamate under appropriate culture conditions. Although biotin is required for cell growth, L-glutamate overproduction is induced under biotin-limiting condition. The secretion of L-glutamate into the culture medium can also be triggered by other special treatments or growth conditions, including the addition of selected detergents, such as Tween 40 or Tween 60, and penicillin (Kimura 2005; Shimizu and Hirasawa 2007). These treatments cause damage to cell-surface structures, including the cytoplasmic membrane and peptidoglycan layer of the cell wall. Biotin is a cofactor for the enzymes involved in fatty acid biosynthesis; the detergents and β-lactam antibiotics affect fatty acid biosynthesis and peptidoglycan formation, respectively. Therefore, it had been thought that L-glutamate passively leaked through the cell membrane as a result of the increased permeability of the membrane and accumulated in the medium. This idea was historically called the "leakage model" for L-glutamate production using *C. glutamicum*. However, because the secretion of L-glutamate occurs despite the unchanged permeability of the cell membrane to other ions and other amino acids or carboxylic acids, this working hypothesis was not validated.

DtsR1 is the carboxyltransferase subunit of a biotin-containing acetyl-CoA carboxylase involved in fatty acid synthesis in *C. glutamicum*. The *dtsR1*-disrupted mutant exhibited auxotrophy for oleic acid and its ester (Tween 80). In addition, Kimura et al. (1997) indicated that disruption of the *dtsR1* causes constitutive overproduction of L-glutamate, even in the presence of excess biotin, and is accompanied by reduction of the activity of the 2-oxoglutarate dehydrogenase complex (ODHC). ODHC, which catalyzes the conversion of 2-oxoglutarate to succinyl-CoA, is a branch-point enzyme complex between the tricarboxylic acid (TCA) cycle and L-glutamate biosynthesis (Figs. 15.3, 15.4a). The reduction of ODHC activity during secretion of L-glutamate in response to some inducible treatments, including biotin limitation was also reported (Kawahara et al. 1997). Therefore, a decrease in ODHC enzymatic activity could shift the metabolic flow from succinyl-CoA, in the TCA cycle, to L-Glu. These studies suggested that the target of biotin limitation is the acyl-CoA carboxylase complex, including DtsR1, and the inducible treatment seems to depress fatty acid biosynthesis, which could affect a physiological property of the cell membrane in *C. glutamicum*. In addition, a functional relationship between fatty acid biosynthesis and ODHC activity on L-glutamate overproduction was speculated (Fig. 15.3).

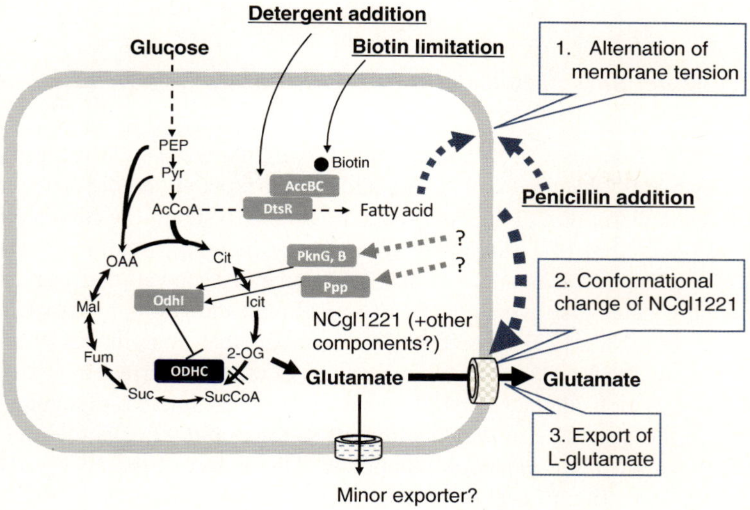

Fig. 15.3 Model for the induction of L-glutamate production in *C. glutamicum*. Treatments triggering L-glutamate overproduction alter membrane tension. The change in tension induces a structural transformation in NCgl1221, which enables it to secrete L-glutamate. OdhI is phosphorylated and dephosphorylated by PknG, B, and Ppp, respectively. The nonphosphorylated form of OdhI inhibits ODHC activity, which shifts the metabolic flow of 2-oxoglutarate (2-OG) from succinyl-CoA to L-glutamate

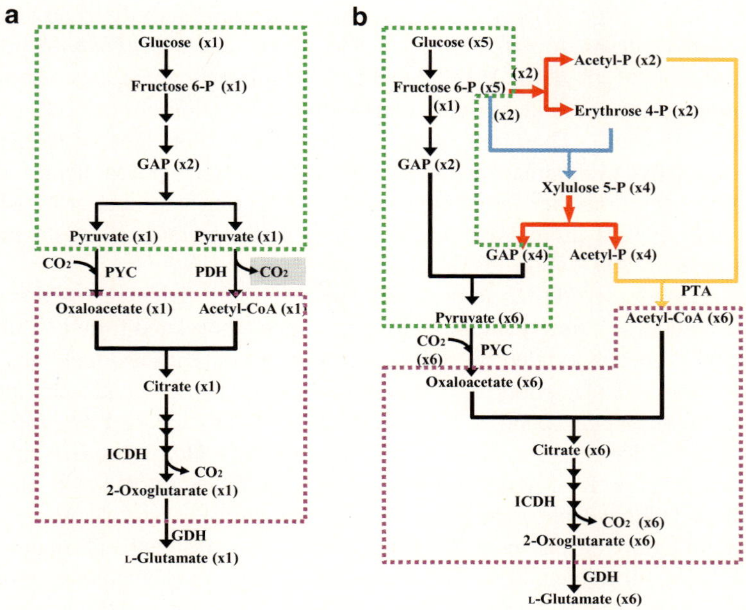

Fig. 15.4 Biosynthetic pathway from glucose to L-glutamate. (**a**) Metabolic pathway of wild-type *C. glutamicum*. (**b**) New pathway (indicated by *red lines*) designed to bypass CO_2 emission by introduction of PKT (Chinen et al. 2007). Under ideal conditions, this pathway produces 6 mol L-glutamate from 5 mol glucose without the loss of carbon through CO_2 release. The reactions of the pentose phosphate pathway and PTA are indicated by *blue and yellow lines*, respectively

15.2.2 Molecular Mechanism of L-Glutamate Overproduction

ODHC generally consists of three subunits: 2-oxoglutarate dehydrogenase (E1o), dihydrolipoamide *S*-succinyltransferase, and dihydrolipoamide dehydrogenase. In *C. glutamicum*, disruption of the *odhA* encoding the E1o subunit results in overproduction and secretion of L-glutamate without any inducing treatments (Asakura et al. 2007). However, Nakamura et al. (2007) recently showed that L-glutamate-overproducing *odhA* disruptants also carry additional mutations in the NCgl1221 gene locus. The NCgl1221 gene encodes a protein that contains an N-terminal region similar to the mechano-sensitive channels of other bacteria. They also found that specific mutations in the NCgl1221 gene lead to constitutive L-glutamate secretion, even in the absence of *odhA* disruption, and also render cells tolerant to the toxic effects of an L-Glu analogue compound. Disruption of the NCgl1221 gene eliminated the secretion of L-glutamate into the medium, causing an increase in the intracellular concentration of the L-glutamate pool under biotin-limiting conditions. In contrast, amplification of the wild-type NCgl1221 gene increased L-glutamate production in response to induction (Nakamura et al. 2007). The location of the NCgl1221 protein in the cell was analyzed by green fluorescent protein (GFP) tagging and confocal microscopy. The results suggest that NCgl1221 is a membrane protein with four transmembrane segments; it is localized to the cytoplasmic membrane and has its C-terminus in the cytoplasm (Yao et al. 2009). In addition, studies have shown that NCgl1221 has mechano-sensitive channel activity when expressed in *Escherichia coli* spheroplasts (Börngen et al. 2010) and *Bacillus subtilis* provacuoles (Hashimoto et al. 2010, 2012). Mechano-sensitive channels sense changes in cell membrane tension and mediate adaptation to alterations in osmotic pressure (Levina et al. 1999). Therefore, these results suggest that the NCgl1221 gene encodes a putative L-glutamate exporter. It is also proposed that treatments inducing L-glutamate secretion alter membrane tension, triggering a structural transformation of the NCgl1221 protein that enables it to export L-glutamate into the medium (Nakamura et al. 2007) (Fig. 15.3).

A novel mode of ODHC activity regulation has also been reported recently. ODHC activity was strongly inhibited by the nonphosphorylated form of OdhI protein, which is phosphorylated by serine/threonine protein kinases, including PknG and PknB, and is dephosphorylated by a phosphoserine/threonine protein phosphatase Ppp (Niebisch et al. 2006; Schultz et al. 2007). OdhI, encoded by *cg1630*, is a 15-kDa protein identified as an in vivo substrate of PknG. OdhA, the E1o subunit of ODHC, interacts preferably with nonphosphorylated OdhI. Boulahya et al. (2010) reported that OdhI is dephosphorylated during L-glutamate overproduction triggered by biotin limitation. Furthermore, Kim et al. (2011) investigated the phosphorylation status of OdhI during penicillin- and Tween 40-triggered L-glutamate overproduction and suggested that levels of nonphosphorylated OdhI substantially increase after such triggering treatments. On the other hand, deletion of *odhI* strongly impaired the overproduction of L-glutamate in *C. glutamicum* induced by treatments including biotin limitation, Tween 40 addition, and penicillin addition (Schultz et al. 2007).

These results suggest that OdhI plays an important role in the efficient production of L-glutamate in *C. glutamicum* under the inducing conditions through an interaction with ODHC. During L-glutamate overproduction, nonphosphorylated OdhI is abundant in the cell, resulting in reduction of ODHC activity and higher metabolic flux of 2-oxoglutarate toward L-glutamate. In addition, it was presumed that the phosphorylation status of OdhI is controlled by at least three proteins, in response to specific stimuli that arise from inducing treatments, through a yet-unknown signaling pathway. The present working model of the mechanism of L-glutamate overproduction by *C. glutamicum* is shown in Fig. 15.3. In this manner, the mechanism of L-glutamate overproduction in coryneform bacteria is unique and interesting.

15.3 Metabolic Engineering for L-Glutamate Overproduction

15.3.1 Metabolic Pathway Design for Efficient Production

As mentioned, MSG is used in the dietary life of humans and is produced on a large scale by fermentation. Therefore, it would be of great commercial value to improve the efficiency of the biological conversion of glucose to L-glutamate by metabolic engineering of the producing strain. In *C. glutamicum*, L-glutamate is originally biosynthesized from glucose through the glycolytic pathway, the oxidative branch of the TCA cycle, and the action of glutamate dehydrogenase (GDH) (Fig. 15.4a). During these reactions, one mole of CO_2 per mole of L-Glu biosynthesized is fixed to pyruvate by pyruvate carboxylase (PYC), whereas two moles of CO_2 are released by pyruvate dehydrogenase (PDH) and isocitrate dehydrogenase (ICDH). The overall reaction for the biosynthesis of L-Glu from glucose through this pathway is summarized in the following reaction formula:

$$\text{Glucose} + NH_3 + 3NAD^+ \rightarrow \text{L-Glu} + 3NADH + 3H^+ + CO_2$$

The maximum production yield of L-Glu from glucose through the original metabolic pathway is calculated to be 81.3 % by weight (or 100 %mol of L-Glu produced/mol of glucose consumed).

To improve the production yield, a novel pathway for the biosynthesis of L-glutamate was designed by inserting a phosphoketolase (PKT) derived from bifidobacteria into *C. glutamicum* (Chinen et al. 2007). This pathway created a metabolic route bypassing the PDH reaction, which releases CO_2 (Fig. 15.4b). In conjunction with phosphoacetyltransferase (PTA) and enzymes of the pentose phosphate pathway, PKT can supply acetyl-CoA from fructose-6-phosphate (F6P) without producing CO_2. By employing PKT, 6 mol L-Glu is produced from only 5 mol glucose, as described by the following formula:

$$5\text{Glucose} + 6NH_3 + 6NAD^+ \rightarrow 6\text{L-Glu} + 6NADH + 6H^+$$

Therefore, the incorporation of the active PKT pathway into the original pathway for the biosynthesis of L-Glu was predicted to increase the theoretical maximum production yield to 98.0 % by weight (or 120 %mol/mol).

PKTs (EC 4.1.2.9, EC 4.1.2.22) are thiamine diphosphate-dependent enzymes found in the pentose phosphate pathway of lactic acid bacteria and the D-fructose-6-phosphate (F6P) shunt of bifidobacteria (Heath et al. 1958; Goldberg and Racker 1962). PKTs catalyze the following reactions: F6P + Pi→ acetyl-phosphate + erythrose-4-phosphate; xylulose-5-phosphate (X5P) + Pi→acetyl-phosphate + glyceraldehyde-3-phosphate (GAP). Recently, the crystal structures of PKTs from bifidobacteria have been determined (Takahashi et al. 2010; Suzuki et al. 2010), and the catalytic reaction mechanisms have been analyzed according to structure–function relationships.

The *xfp*, encoding PKT of *Bifidobacterium animalis* JCM1190, was cloned using a polymerase chain reaction (PCR) method based on sequence information from known PKTs. It comprised an ORF encoding 825 amino-acid residues and was inserted into an expression vector pVK9 to generate pPS2-xfp. The plasmids pPS2-xfp and pVK9 were then each introduced into the L-glutamate-overproducing *C. glutamicum* strain CGS01, in which *odhA* has been disrupted, to assess the effect of PKT on L-glutamate productivity. In vitro and in vivo experiments confirmed that PKT expressed inside a heterologous *C. glutamicum* cell exhibits enzymatic activity under the physiological conditions (Chinen et al. 2007).

15.3.2 Enhancement in L-Glutamate Production by Installing the PKT Pathway

CGS01 strains harboring pVK9 or pPS2-xfp were initially cultivated in flasks, and the production yields of L-glutamate from these two strains were compared. The production yield of L-glutamate in the strains containing *xfp* (55.8 % by weight, on average; Fig. 15.5a) was higher than that in the control strains (48.5 % by weight on the average); however, the cell mass yield of the *xfp*-containing strains was low (Fig. 15.5a). In subsequent experiments, the cell masses were standardized by adding penicillin G at various cultivation time points. Because this treatment arrests cell growth without significant metabolic disturbance, the L-glutamate productivities of both producers can be compared with more precision. As a result, it was shown that the L-glutamate production yields of the strains expressing PKT were consistently higher than those of the control strains, even in cultures with the same final cell masses (by cell densities, indicated as the OD_{620} values of the cultures) (Fig. 15.5b). This finding indicates that PKT can improve L-glutamate production in *C. glutamicum*.

In the next series of experiments, the distribution of carbon during L-glutamate fermentation from glucose in the engineered *C. glutamicum* was investigated under controlled conditions using a jar fermenter (Fig. 15.6). The control strain (CGS01/pVK9), which produced L-glutamate in an average yield of 60.8 % by weight, generated about 5.51 g CO_2 and formed 1.78 g dried cell weight as biomass. The main

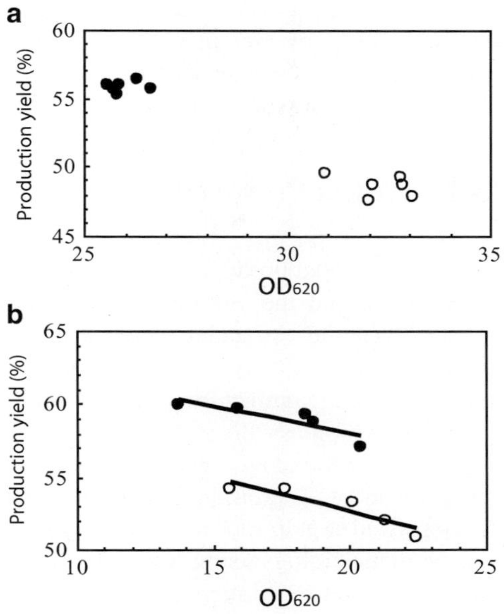

Fig. 15.5 Evaluation of L-Glu productivity in flask cultivation (Chinen et al. 2007). **(a)** In ordinary flask cultivation. **(b)** Cell growth was suppressed. L-Glu production yields (%) and final cell densities (OD_{620}) are plotted for two types of L-glutamate producers: the control strain CGS01/pVK9 (*open circles*) and the *xfp*-introducing strain CGS01/pPS2-xfp (*closed circles*)

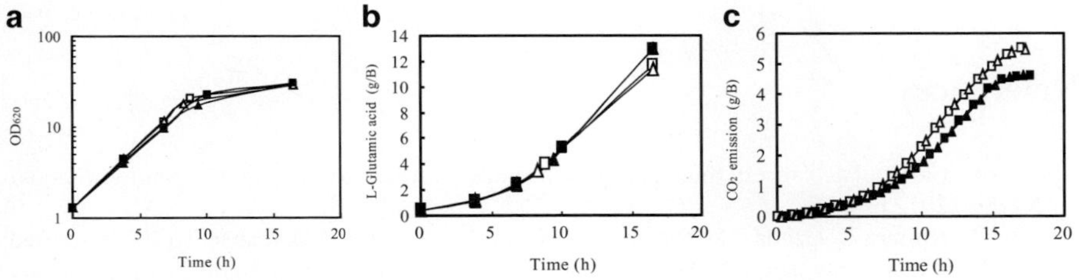

Fig. 15.6 Production of L-glutamate by metabolically engineered *C. glutamicum* strains in a jar fermenter. **(a)** Profiles of cell growth. **(b)** L-Glutamate accumulation and CO_2 emission levels (**c**). Cultivation temperature was maintained at 31.5 °C and growth rate was measured by determining the OD_{620} of the culture broth. Initial amount of glucose was 18.0 g. Data are from two independent experiments. Penicillin was added to the medium at an OD_{620} of about 20 to suppress additional cell growth. *Open squares* and *triangles*, a control strain; *closed squares* and *triangles*, CGS01/pPS2-xfp

byproduct of the fermentation was 2-oxoglutarate (0.52 g), along with small quantities of other metabolites such as malate and L-alanine. On the other hand, the PKT-expressing strain CGS01, which consumed the same amount of glucose in this batch fermentation, yielded 69.4 % L-glutamate by weight, while the amount of CO_2 gas released was reduced to about 84 % of that released by the control strain. No major differences were observed in the composition or amount of by-products released into the culture medium by the two strains. Therefore, the change in the carbon distribution from glucose suggests that decrease in the amount of carbon generated as CO_2 gas is mainly responsible for the increased amount of L-Glu produced.

In this experiment, improvement in the production yield of L-glutamate from glucose by 9 % by weight was performed with a significant increase (10 %) in productivity (Chinen et al. 2007).

15.4 Future Prospects

The current annual production of MSG in the world is estimated to be more than 2 million tons, and the volume of demand has continued to increase (Ajinomoto 2012). Given the estimated increase in the demand for MSG in future, the L-glutamate production system must continue evolving toward a more eco-friendly and sustainable manufacturing system. For example, use of carbon sources derived from unused biomass, the saving of energy and raw materials during the production processes, and the reduction of by-products and waste fluids are areas that can lead to improvement. Therefore, it is strongly expected that the producing strain is further improved to gain such abilities by application of systems metabolic engineering with synthetic biology using comprehensive omics technologies. It is also desired to develop innovative fermentation process technologies for maximizing the full potential of the engineered cell factory.

References

Ajinomoto (2012) Fact sheet: food products business. http://www.ajinomoto.com/jp/ir/ebook/food-oct2012.htm

Asakura Y, Kimura E, Usuda Y, Kawahara Y, Matsui K, Osumi T, Nakamatsu T (2007) Altered metabolic flux due to deletion of *odhA* causes L-glutamate overproduction in *Corynebacterium glutamicum*. Appl Environ Microbiol 73:1308–1319

Börngen K, Battle AR, Möker N, Morbach S, Marin K, Martinac B, Krämer R (2010) The properties and contribution of the *Corynebacterium glutamicum* MscS variant to fine-tuning of osmotic adaptation. Biochim Biophys Acta 1798:2141–2149

Boulahya K-A, Guedon E, Delaunay S, Schultz C, Boudrant J, Bott M, Goergen J-L (2010) OdhI dephosphorylation kinetics during different glutamate production processes involving *Corynebacterium glutamicum*. Appl Microbiol Biotechnol 87:1867–1874

Chinen A, Kozlov YI, Hara Y, Izui H, Yasueda H (2007) Innovative metabolic pathway design for efficient L-glutamate production by suppressing CO_2 emission. J Biosci Bioeng 103:262–269

Goldberg ML, Racker E (1962) Formation and isolation of a glycolaldehyde-phosphoketolase intermediate. J Biol Chem 237:3841–3842

Hashimoto K, Nakamura K, Kuroda T, Yabe I, Nakamatsu T, Kawasaki H (2010) The protein encoded by *NCgl1221* in *Corynebacterium glutamicum* functions as a mechanosensitive channel. Biosci Biotechnol Biochem 74:2546–2549

Hashimoto K, Murata J, Konishi T, Yabe I, Nakamatsu T, Kawasaki H (2012) Glutamate is excreted across the cytoplasmic membrane through the NCgl1221 channel of *Corynebacterium glutamicum* by passive diffusion. Biosci Biotechnol Biochem 76:1422–1424

Heath EC, Hurwitz J, Horecker BL, Ginsburg A (1958) Pentose fermentation by *Lactobacillus plantarum*. I. The cleavage of xylulose 5-phosphate by a phosphoketolase. J Biol Chem 231:1009–1029

Ikeda M, Nakagawa S (2003) The *Corynebacterium glutamicum* genome: features and impacts on biotechnological processes. Appl Microbiol Biotechnol 62:99–109

Kalinowski J, Bathe B, Bartels D, Bischoff N, Bott M, Burkovski A, Dusch N, Eggeling L, Eikmanns BJ, Gaigalat L, Goesmann A, Hartmann M, Huthmacher K, Krämer R, Linke B, McHardy AC, Meyer F, Möckel B, Pfefferle W, Pühler A, Rey DA, Rückert C, Rupp O, Sahm H, Wendisch VF, Wiegräbe I, Tach A (2003) The complete *Corynebacterium glutamicum* ATCC13032 genome sequence and its impact on the production of L-aspartate-derived amino acids and vitamins. J Biotechnol 104:5–25

Kataoka M, Hashimoto K-I, Yoshida M, Nakamatsu T, Horinouchi S, Kawasaki H (2006) Gene expression of *Corynebacterium glutamicum* in response to the conditions inducing glutamate overproduction. Lett Appl Microbiol 42:471–476

Kawahara Y, Takahasi-Fuke K, Shimizu E, Nakamatsu T, Nakamori S (1997) Relationship between the glutamate production and the activity of 2-oxoglutarate dehydrogenase in *Brevibacterium lactofermentum*. Biosci Biotechnol Biochem 61:1109–1112

Kim J, Hirasawa T, Saito M, Furusawa C, Shimizu H (2011) Investigation of phosphorylation status of OdhI protein during penicillin- and Tween 40-triggered glutamate overproduction by *Corynebacterium glutamicum*. Appl Microbiol Biotechnol 91:143–151

Kimura E (2005) L-Glutamate production. In: Eggeling L, Bott M (eds) Handbook of *Corynebacterium glutamicum*. CRC/Taylor & Francis, Boca Raton, pp 439–464

Kimura E, Abe C, Kawahara Y, Nakamatsu T, Tokuda H (1997) A *dtsR1* gene-disrupted mutant of *Brevibacterium lactofermentum* requires fatty acids for growth and efficiently produces L-GLUTAMATE in the presence of an excess of biotin. Biochem Biophys Res Commun 234:157–161

Kinoshita S, Udaka S, Shimono M (1957) Studies on the amino acid fermentation. Part I. Production of L-glutamic acid by various microorganisms. J Gen Appl Microbiol 3:193–205

Levina N, Totemeyer S, Stokes NR, Louis P, Jones MA, Booth IR (1999) Protection of *Escherichia coli* cells against extreme turgor by activation of MscS and MscL mechanosensitive channels: identification of genes required for MscS activity. EMBO J 18:1730–1737

Nakamura E (2011) One hundred years since the discovery of the "umami" taste from seaweed broth by Kikunae Ikeda, who transcended his time. Chem Asian J 6:1659–1663

Nakamura J, Hirano S, Ito H, Wachi M (2007) Mutations of the *Corynebacterium glutamicum* NCgl1221 gene, encoding a mechanosensitive channel homolog, induce L-glutamic acid production. Appl Environ Microbiol 73:4491–4498

Niebisch A, Kabus A, Schultz C, Weil B, Bott M (2006) Corynebacterial protein kinase G controls 2-oxoglutarate dehydrogenase activity via the phosphorylation status of the OdhI protein. J Biol Chem 281:12300–12307

Sano C (2009) History of glutamate production. Am J Clin Nutr 90:728S–732S

Schluesener D, Fischer F, Kruip J, Rögner M, Poetsch A (2005) Mapping and membrane proteome of *Corynebacterium glutamicum*. Proteomics 5:1317–1330

Schultz C, Niebisch A, Gebel L, Bott M (2007) Glutamate production by *Corynebacterium glutamicum*: dependence on the oxoglutarate dehydrogenase inhibitor protein OdhI and protein kinase PknG. Appl Microbiol Biotechnol 76:691–700

Shimizu H, Hirasawa T (2007) Production of glutamate and glutamate-related amino acids: molecular mechanism analysis and metabolic engineering. In: Wendisch VF (ed) Amino acid biosynthesis: pathways, regulation and metabolic engineering. Springer, Berlin, pp 1–38

Shirai T, Nakato A, Izutani N, Nagahisa K, Shioya S, Kimura E, Kawarabayashi Y, Yamagishi A, Gojobori T, Shimizu H (2005) Comparative study of flux redistribution of metabolic pathway in glutamate production by two coryneform bacteria. Metab Eng 7:59–69

Strelkov S, von Elstermann M, Schomburg D (2004) Comprehensive analysis of metabolites in *Corynebacterium glutamicum* by gas chromatography/mass spectrometry. Biol Chem 385:853–861

Suzuki R, Katayama T, Kim B-J, Wakagi T, Shoun H, Ashida H, Yamamoto K, Fushinobu S (2010) Crystal structures of phosphoketolase. Thiamine diphosphate-dependent dehydration mechanism. J Biol Chem 285:34279–34287

Takahashi K, Tagami U, Shimba N, Kashiwagi T, Ishikawa K, Suzuki E (2010) Crystal structure of *Bifidobacterium longum* phosphoketolase; key enzyme for glucose metabolism in *Bifidobacterium*. FEBS Lett 584:3855–3861

Wendisch VF (2003) Genome-wide expression analysis in *Corynebacterium glutamicum* using DNA microarrays. J Biotechnol 104:273–285

Yamada K, Komagata K (1972) Taxonomic studies on coryneform bacteria. IV. Morphological, cultural, biochemical, and physiological characteristics. J Gen Appl Microbiol 18:399–416

Yamaguchi S, Ninomiya K (2000) Umami and food palatability. J Nutr 130(4):921S–926S

Yao W, Deng X, Liu M, Zheng P, Sun Z, Zhang Y (2009) Expression and localization of the *Corynebacterium glutamicum* NCgl1221 protein encoding an L-glutamic acid exporter. Microbiol Res 164:680–687

Part V
Pharmaceuticals

Chapter 16
Microbial Hormones as a Master Switch for Secondary Metabolism in *Streptomyces*

Takeaki Tezuka and Yasuo Ohnishi

Abstract The genus *Streptomyces* can produce a wide variety of secondary metabolites, including antibiotics and other biologically active substances. These compounds are important starting scaffolds for the development of antibiotics, anthelminthic agents, anticancer agents, and immunosuppressants. Genome sequencing of some *Streptomyces* species has revealed that every *Streptomyces* strain has the potential to produce dozens of secondary metabolites, but only a fraction of these putative secondary metabolites are produced under standard fermentation conditions. Therefore, genetic approaches to engineering regulation are innovative ways to stimulate the expression of otherwise silent secondary metabolite gene clusters. In many *Streptomyces* species, γ-butyrolactone signaling molecules are used as a microbial hormone to induce secondary metabolite production. A-factor (2-isocapryloyl-3*R*-hydroxymethyl-γ-butyrolactone) is a representative γ-butyrolactone that triggers both morphological development and production of secondary metabolites, including streptomycin in *Streptomyces griseus*. Here, the A-factor regulatory cascade was reviewed with some results of our recent genome-wide analysis. Disruption of *arpA*, encoding the A-factor receptor protein ArpA, resulted in overproduction of streptomycin. Enhanced production of secondary metabolites by gene disruption of *arpA* homologues was also reported for other *Streptomyces* species. In addition, two other methods for the induction of secondary metabolite formation in *Streptomyces* are briefly described. New low molecular weight signaling molecules involved in the induction of secondary metabolite formation are also briefly described.

T. Tezuka • Y. Ohnishi (✉)
Department of Biotechnology, Graduate School of Agricultural and Life Sciences, The University of Tokyo 1-1-1 Yayoi, Bunkyo-ku, Tokyo 113-8657, Japan
e-mail: atezuka@mail.ecc.u-tokyo.ac.jp; ayasuo@mail.ecc.u-tokyo.ac.jp

H. Anazawa and S. Shimizu (eds.), *Microbial Production: From Genome Design to Cell Engineering*, DOI 10.1007/978-4-431-54607-8_16, © Springer Japan 2014

Keywords γ-Butyrolactone • Gene regulation • Microbial hormone • Secondary metabolism • *Streptomyces*

16.1 *Streptomyces* as a Producer of a Wide Variety of Secondary Metabolites

The genus *Streptomyces* comprises gram-positive, soil-dwelling, and filamentous bacteria. An important characteristic of *Streptomyces* is the ability to produce a wide variety of secondary metabolites, including antibiotics and other biologically active substances. *Streptomyces* has contributed greatly to fermentation industries all over the world. In 1944, Selman A. Waksman's team isolated streptomycin, the first antibiotic used to cure tuberculosis, from *Streptomyces griseus*. Because of further intensive screening for antibiotics, *Streptomyces* species produce more than two thirds of the clinically useful antibiotics of natural origin such as neomycin and chloramphenicol (Kieser et al. 2000), which makes this genus the largest antibiotic-producing prokaryote.

Secondary metabolites produced by *Streptomyces* and other actinomycetes are important starting scaffolds for the development of antibiotics, anthelminthic agents, anticancer agents, and immunosuppressants. Although the rate of isolation of novel natural products has been declining in recent years, it is now apparent with the advent of genome sequencing technology that *Streptomyces* and other actinomycetes have the potential to produce dozens of secondary metabolites (Bentley et al. 2002; Ikeda et al. 2003; Oliynyk et al. 2007; Udwary et al. 2007; Ohnishi et al. 2008). Under standard fermentation conditions, however, only a fraction of these putative secondary metabolites are produced in sufficient yields for detection. Awakening of cryptic genes for secondary metabolite formation is critical for discovering new secondary metabolites within a background of thousands of known compounds. Genetic approaches to engineering regulation are innovative ways to stimulate the expression of otherwise silent secondary metabolite gene clusters. Therefore, elucidating the regulatory mechanisms of secondary metabolite production is important not only for the biology of actinomycetes but also for the practical use of these prokaryotes in the production of useful secondary metabolites.

Another characteristic of the *Streptomyces* genus is the complex morphological differentiation resembling that of filamentous fungi. Substrate hyphae formed from a germinating spore branch frequently on agar medium and grow by cell-wall extension at the hyphal tips. Aerial hyphae emerge by reuse of materials assimilated into the substrate hyphae, which mostly lyse and die (Miguélez et al. 1999). After apical growth of aerial hyphae, septa are formed at regular intervals along the hyphae to form unigenomic spores, leading to spore chains consisting of many tens of spores. The complex morphogenesis has made this genus a model prokaryote for studying multicellular differentiation.

16.2 γ-Butyrolactone Signaling Molecules as a Microbial Hormone

The pioneer work of A.S. Khokhlov and his colleagues on A-factor (2-isocapryloyl-3*R*-hydroxymethyl-γ-butyrolactone) revealed that it induces both sporulation and streptomycin production in *S. griseus* (Khokhlov et al. 1967). Subsequent studies have indicated that A-factor and its derivatives are autoregulators that switch on morphological differentiation or secondary metabolism, or both (Horinouchi 2007; Yamada and Nihira 1998; Takano 2006). These low molecular weight compounds can be termed microbial hormones that function at concentrations as low as 10^{-9} M. A-factor is synthesized from dihydroxyacetone phosphate (DHAP), an intermediate of the glycolysis pathway, and β-ketoacyl-acyl carrier protein (ACP). AfsA is the key enzyme in the biosynthesis of A-factor, which catalyzes the condensation of DHAP and a β-keto acid derivative (Kato et al. 2007). Because the transcription of *afsA* is almost constant throughout growth, A-factor is accumulated in a growth-dependent manner, reaching a maximum of about 100 nM at or near the midpoint of the exponential phase. When the concentration of A-factor reaches a critical level, the expression of a variety of genes with various functions that are required for secondary metabolism and morphological development is activated.

Chemical signaling molecules with a γ-butyrolactone structure have been found in various *Streptomyces* species. For example, SCB1 controls the production of actinorhodin, undecylprodigiosin, and a polyketide compound in *S. coelicolor* A3(2) (Takano et al. 2000, 2005), and virginiae butanolides control virginiamycin production in *S. virginiae* (Okamoto et al. 1995) (Fig. 16.1a). These autoregulators, including A-factor, can move freely within the filamentous hyphae and spread into adjacent hyphae. Diffusible γ-butyrolactone-type compounds cause rapid sporulation or secondary metabolite formation, or both, in the entire population.

16.3 The A-Factor Regulatory Cascade in *S. griseus*

The key steps in the A-factor regulatory cascade, including AfsA, the key enzyme in A-factor biosynthesis; ArpA, the A-factor receptor protein; and AdpA, a global transcriptional activator, have been elucidated. When the concentration of A-factor reaches a critical level at or near the midpoint of the exponential growth phase, it binds ArpA, which is bound to and represses the promoter of *adpA*, and dissociates it from the promoter, leading to the expression of *adpA* (Ohnishi et al. 1999). ArpA belongs to the TetR family, and when A-factor binds to the ligand-binding pocket of ArpA it induces the relocation of a helix that links the ligand-binding pocket with the DNA-binding domain. As a result, ArpA dissociates from the DNA (Onaka et al. 1995; Onaka and Horinouchi 1997; Natsume et al. 2004). *adpA* is the sole target of ArpA (Kato et al. 2004), and the global transcriptional regulator AdpA

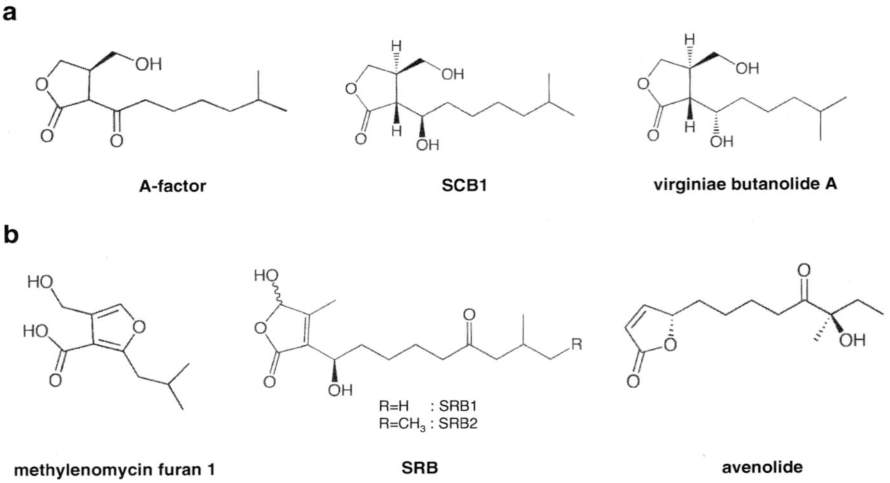

Fig. 16.1 Autoregulators found in *Streptomyces*. (**a**) γ-Butyrolactone signaling molecules in *Streptomyces* species. Chemical signaling molecules with a γ-butyrolactone ring control morphological or physiological differentiation, or both in *Streptomyces* species. A-factor switches on aerial mycelium formation and secondary metabolite production in *S. griseus*. SCB1 controls the production of actinorhodin, undecylprodigiosin, and a polyketide compound in *Streptomyces coelicolor* A3(2). Virginiae butanolides induce virginiamycin production in *Streptomyces virginiae*. (**b**) Autoregulators recently found in *Streptomyces*. Methylenomycin furans found in *S. coelicolor* A3(2) induce the methylenomycin production of the bacterium. Avenolide found in *S. avermitilis* induces the avermectin production of the bacterium. SRB1 and SRB2 found in *S. rochei* induce the lankacidin and lankamycin production of the bacterium

activates a variety of genes that are required for secondary metabolism and morphological development (Ohnishi et al. 2005) (Fig. 16.2). One of the AdpA targets is the *strR-aphD* operon within the streptomycin biosynthesis gene cluster. *strR* encodes a pathway-specific transcriptional activator for the streptomycin biosynthesis genes. Thus, the A-factor signal is transferred to the streptomycin biosynthesis genes through ArpA, AdpA, and StrR, causing the onset of streptomycin production (Ohnishi et al. 1999; Tomono et al. 2005b). *aphD* encodes streptomycin-6-phosphotransferase, which is a streptomycin-resistant determinant of the producer.

The linear chromosome of *S. griseus* consists of 8,545,929 bp, with an average GC content of 72.2 % and 7,138 predicted open reading frames, and contains 34 gene clusters or genes for the biosynthesis of known and unknown secondary metabolites (Ohnishi et al. 2008). The *S. griseus* transcriptomes were analyzed to reveal the genome-wide transcriptional regulation by A-factor. The transcriptomes of the wild type (wt) and *adpA*-deficient (Δ*adpA*) mutant strains extracted from cells in the early stationary phase in liquid culture were compared using DNA microarray. When genes with an increase or decrease in expression of more than twofold and a P value <0.05 (analyzed by the t test) were regarded as genes with significant variation in transcriptional level, more than 1,000 genes were transcriptionally upregulated or downregulated in the wt strain in comparison with the

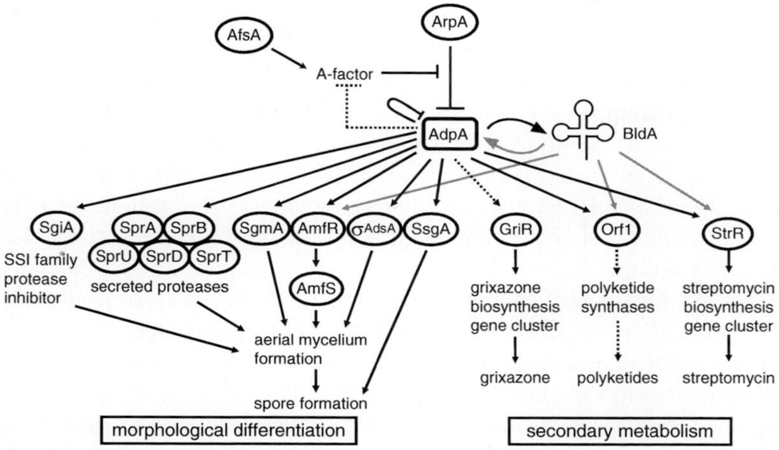

Fig. 16.2 The A-factor regulatory cascade. When the concentration of A-factor, produced by the action of the key enzyme AfsA in a growth-dependent manner, reaches a critical level, it binds ArpA, which binds and represses the promoter of *adpA* at an early growth stage, and dissociates ArpA from the DNA, inducing transcription of *adpA*. AdpA activates a variety of genes required for morphological differentiation and secondary metabolism. The targets of AdpA required for morphological differentiation include the following genes: *adsA*, encoding an extracytoplasmic function (ECF) sigma factor belonging to a subgroup of the primary σ^{70} family (Yamazaki et al. 2000); *amfR*, encoding a transcriptional regulatory protein that activates the *amf* operon (Ueda et al. 1993; Yamazaki et al. 2003b); genes encoding extracellular proteases, including a zinc-containing metalloendopeptidase (Kato et al. 2002), two trypsin-type proteases (Kato et al. 2005a), and three chymotrypsin-type proteases (Tomono et al. 2005a); a *Streptomyces* subtilisin inhibitor (SSI) gene (Hirano et al. 2006); *ssgA*, which is essential for spore septum formation (Yamazaki et al. 2003a); and *bldA*, the regulatory tRNA gene required for translation of TTA-containing genes (Higo et al. 2011). Because BldA is required for translation of TTA-containing *adpA*, a positive feedback loop composed of AdpA and BldA plays important roles in the initiation of morphological differentiation and secondary metabolism in *S. griseus*. To control its intracellular concentration at an appropriate level, AdpA represses its own transcription by binding the operator sites in the *adpA* promoter (Kato et al. 2005b). *Dotted lines* indicate indirect regulation or regulation that has not been confirmed yet to be direct. *Gray arrows* indicate regulation at the translation level

Δ*adpA* mutant. This result demonstrates that AdpA highly influences the transcription of many genes in *S. griseus*. The transcriptome analysis by DNA microarray also showed that at least 4 of the 34 clusters, in addition to the streptomycin biosynthesis gene cluster, were activated by AdpA. Thus, these 4 secondary metabolite gene clusters for unknown peptide and polyketide compounds are likely to be induced by A-factor. In addition, the effects of A-factor on global gene expression were determined by A-factor addition to the *afsA*-deficient mutant (Δ*afsA*) at the midpoint of the exponential growth phase (Hara et al. 2009). To follow the time-course of the gene expression triggered by A-factor, RNA samples were extracted from the Δ*afsA* cells at 5, 15, and 30 min, and at 1, 2, 4, 8, and 12 h, after A-factor addition, and compared with those obtained from cells in the absence of A-factor by DNA microarray analysis. When genes with a twofold or more increase or decrease in expression at least at one time point and a *P* value <0.05 (analyzed by the *t* test) were regarded as genes with significant variation in gene expression, 477 genes were suggested to be differentially transcribed after the addition of A-factor. Furthermore, the distribution of AdpA across the chromosome of *S. griseus*, determined by chromatin immunoprecipitation/chromatin affinity precipitation-seq (ChIP/ChAP-seq) analysis, indicated that AdpA bound to more than 1,200 sites in both liquid and solid cultures (Higo et al. 2012). The ChIP/ChAP-seq and DNA microarray analyses showed that AdpA directly controls more than 500 genes, indicating that the signal from A-factor is greatly amplified at the transcriptional regulatory step via AdpA as an amplifier.

As already mentioned, chemical signaling molecules with a γ-butyrolactone structure are widely distributed among *Streptomyces* species such as *S. bikiniensis*, *S. coelicolor* A3(2), *S. cyaneofuscatus*, *S. lavendulae*, *S. virginiae*, and *S. viridochromogenes* (Horinouchi and Beppu 1992; Horinouchi 2002). These species contain homologues of *afsA* and *arpA*, which implies that the mechanism of regulation of gene expression by these γ-butyrolactones may be the same as that of the ArpA–A-factor system in *S. griseus*. However, we have to pay attention to the differences between A-factor and other γ-butyrolactone systems. In contrast to A-factor, most γ-butyrolactones are only involved in secondary metabolism and are produced in a growth stage-dependent manner (usually just before the production of secondary metabolites) and not in a growth-dependent manner. Apparently, A-factor controls a higher stage of the regulatory hierarchy involved in secondary metabolism and morphological differentiation, without affecting vegetative growth, acting as a strict switch rather than a tuner.

16.4 ArpA Homologues as the Targets of Genetic Engineering

ArpA acts as a repressor of secondary metabolism during the early growth phase of *S. griseus*, suggesting that secondary metabolite production can be increased by disruption of the *arpA* gene. In fact, the *arpA*-deficient mutant began to produce streptomycin at an earlier stage of growth than the wt strain (Fig. 16.3), and the mutant produced approximately ten times more streptomycin than the wt strain (Miyake et al. 1990; Kato et al. 2004). Repression of secondary metabolism by the ArpA homologues was observed in other *Streptomyces* species. For example, an *sngR*-disruptant strain started natamycin production 6 h earlier and showed a 4.6-fold-higher production of natamycin than the wt strain in *S. natalensis* (Lee et al. 2005), and a *brp*-disruptant strain showed 150–300 % clavulanic acid and 120–220 % cephamycin C production compared with the parental strain in *S. clavuligerus* (Santamarta et al. 2005), indicating that SngR and Brp proteins exert a repressor role in antibiotic biosynthesis.

A gene encoding a γ-butyrolactone autoregulator receptor has been cloned from a non-*Streptomyces* genus of actinomycetes. In *Kitasatospora setae*, a *ksbA*-disruptant strain started producing bafilomycin 18 h earlier and showed a 2.4-fold-higher accumulation of bafilomycin than the wt strain (Choi et al. 2004). Although the

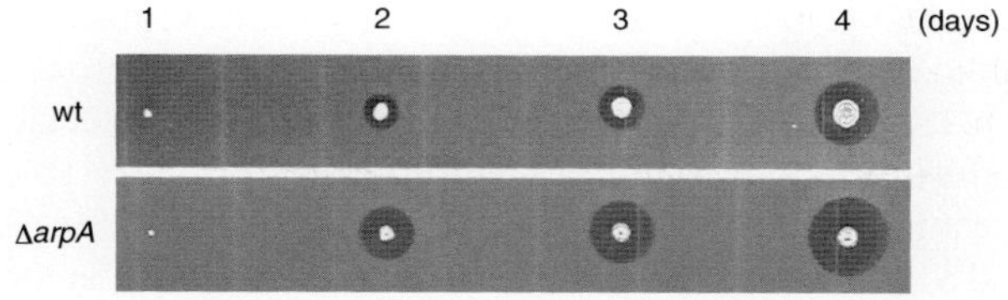

Fig. 16.3 Streptomycin production of an *arpA*-deficient (Δ*arpA*) mutant of *S. griseus*. Effects of the *arpA* gene disruption on streptomycin production were examined. The wild-type (wt) and Δ*arpA* mutant strains were inoculated on Bennett agar medium without glucose and grown at 28°C for the indicated periods. Then, *Bacillus subtilis* spores were overlaid and the plates were incubated at 37°C for 1 day. Streptomycin produced by the two strains was detected by the formation of a growth inhibition zone of *B. subtilis*. The Δ*arpA* mutant began to produce streptomycin at an earlier stage of growth than the wt strain, and produced approximately ten times more streptomycin than the wt strain

autoregulator receptors of the *Streptomyces* species have been well studied, little is known about closely related non-*Streptomyces* actinomycetes, which are also rich sources of medicinally important secondary metabolites. Thus, the fact that the γ-butyrolactone autoregulator receptor acts as a repressor of bafilomycin production in *K. setae* suggests that γ-butyrolactone receptors exist widely in the genera of non-*Streptomyces* actinomycetes and play important roles in controlling the production of secondary metabolites as in the *Streptomyces* species. Therefore, *arpA* homologues could be targets of genetic engineering for overproduction of secondary metabolites (or awakening of a cryptic secondary metabolite gene cluster) in non-*Streptomyces* actinomycetes as well as in *Streptomyces*.

16.5 Novel Low Molecular Weight Signaling Molecules and Other Methods for Induction of Secondary Metabolite Formation in *Streptomyces*

Because of their filamentous growth, actinomycetes have developed regulation by diffusible γ-butyrolactone hormones to communicate among the mycelia from a distance. Because any given strain contains its specific γ-butyrolactone and its receptor with strict ligand specificity, this regulatory system facilitates discrimination of signals from neighboring organisms, allowing the cell to recognize whether the neighbor is a member of the same species. Thus, this system enables entire hyphae to produce antibiotics at the same time, which is advantageous for survival in the natural habitats. Without this regulatory system, some hyphae would be killed by antibiotics produced by different hyphae of the same species.

In recent years, methylenomycin furans (2-alkyl-4-hydroxymethylfuran-3-carboxylic acids) were found to induce the production of methylenomycin in *S. coelicolor* A3(2) (Corre et al. 2008), and the butenolide-type compounds, SRB1 and SRB2, were isolated as signaling molecules that induce the production of lankacidin and lankamycin in *Streptomyces rochei* (Arakawa et al. 2012) (Fig. 16.1b). Another butenolide-type compound, avenolide [(4*S*,10*R*)-10-hydroxy-10-methyl-9-oxo-dodec-2-en-1,4-olide], was also purified as a signaling molecule that induces avermectin production in *S. avermitilis* (Kitani et al. 2011) (Fig. 16.1b). Homologues of the gene required for avenolide biosynthesis are present in *S. fradiae*, *S. ghanaensis*, and *S. griseoauranticus*, suggesting that this type of autoregulator may be a widespread class of *Streptomyces* signaling molecules involved in the regulation of antibiotic production. These novel compounds could form general antibiotic biosynthesis inducers in *Streptomyces* species with analogous functions to the γ-butyrolactone molecules. It is conceivable that *Streptomyces* spp. have employed various chemical substances as autoregulators for controlling their secondary metabolism during the long history of evolution, and the autoregulators must have concomitantly evolved with their specific receptors. Therefore, together with their specific receptors, other chemical substances may be revealed as novel autoregulators in future studies.

Onaka et al. (2011) developed a new method, called combined culture, which facilitates the screening of natural products from *Streptomyces* species. In combined culture, a mycolic acid-containing bacterium, such as *Tsukamurella pulmonis*, *Corynebacterium glutamicum*, *Rhodococcus erythropolis*, or *Mycobacterium smegmatis*, is added to a pure culture of a *Streptomyces* species to induce natural product biosynthesis. In fact, Onaka et al. (2011) isolated the novel antibiotic, alchivemycin A, from the coculture broth of *T. pulmonis* and *Streptomyces endus* S-522. Combined culture is an easy method for inducing the production of cryptic antibiotics and has been proven to be very useful for the screening of natural products.

Antibiotic production is dramatically activated by certain ribosomal mutations. For example, the streptomycin-resistant *Streptomyces lividans* strain, which has a point mutation (K88E) in the *rpsL* gene encoding the ribosomal protein S12, has been found to produce large amounts of actinorhodin under normal growth conditions. Streptomycin-resistant mutations also improved actinorhodin production in *S. coelicolor* A3(2) (Ochi 2007). The findings indicate that antibiotic production in *Streptomycetes* is improved or activated by introducing mutations into the translational machinery. A method for modulating ribosomes is to introduce mutations conferring resistance to drugs that target the ribosome, such as streptomycin, kanamycin, and chloramphenicol. This method is advantageous because drug-resistant mutants are easily selected on drug-containing plates. This technique is useful not only for strain improvement but also for searching for novel natural products, as the dormant genes are activated by this method.

16.6 Future Perspectives

The genome sequencing projects of *S. coelicolor* A3(2), *S. avermitilis*, and *S. griseus* revealed that the genomes of these organisms contain 25, 30, and 34 gene clusters or genes, respectively, for the biosynthesis of secondary metabolites (Bentley et al.

2002; Ikeda et al. 2003; Ohnishi et al. 2008), supporting the idea that *Streptomyces* is a rich source of natural products with clinical, agricultural, and biotechnological values. Because aerial hyphae emerge by programmed reuse of materials assimilated into the substrate mycelium such as DNA, proteins, and various storage compounds (Miguélez et al. 1999), it is conceivable that *Streptomyces* species produce a variety of antibiotics to suppress the growth of neighboring microorganisms, which could consume the materials from substrate mycelium for their own growth.

Next-generation sequencing (NGS) technology has revolutionized the sequencing of prokaryotic and eukaryotic genomes. With the dramatically reduced time and cost for sequencing, thousands of genome sequencing projects have been finished or are in progress. An important observation that emerged from early *Streptomyces* genome sequencing projects was that each strain contains genes that encode 20 or more potential secondary metabolites. Analysis of other microbial genomes originating from myxobacteria, cyanobacteria, and filamentous fungi showed the presence of a comparable number of pathways for putative secondary metabolites (Bode and Müller 2006; Ramaswamy et al. 2006; Galagan et al. 2003), suggesting that numerous gene clusters or genes for secondary metabolite biosynthesis will be sequenced by the NGS technology in the near future. In addition, the number of putative secondary metabolite biosynthesis gene clusters is greater than the number of currently known metabolites. These numerous gene clusters are a treasure trove for uncovering novel bioactive compounds, and many new metabolites must await discovery.

As it has become apparent that most of the secondary metabolite gene clusters are not expressed during standard fermentation, awakening of these sleeping genes by various approaches is critical to obtain enough material for structure elucidation and biological testing. In the genus *Streptomyces*, the widely distributed *arpA* homologues described in this chapter are promising targets of genetic manipulation for strain improvement. The investigation of regulation such as the A-factor regulatory cascade is available for the isolation of pharmacologically and biologically active natural products. Development of new antibiotics is still required because various bacteria, including staphylococci, enterococci, streptococci, and *Salmonella*, are exhibiting multidrug resistance. Although the empirical strain improvement approach used in the industry undoubtedly remains a powerful method to achieve higher titers, manipulating specific regulatory mechanisms can expedite and complement these efforts. Our rapidly growing understanding of complex regulatory systems of secondary metabolite formation will lead to new, more powerful strategies for novel drug discovery in the coming years.

References

Arakawa K, Tsuda N, Taniguchi A, Kinashi H (2012) The butenolide signaling molecules SRB1 and SRB2 induce lankacidin and lankamycin production in *Streptomyces rochei*. ChemBioChem 13:1447–1457

Bentley SD, Chater KF, Cerdeño-Tárraga AM et al (2002) Complete genome sequence of the model actinomycete *Streptomyces coelicolor* A3(2). Nature (Lond) 417:141–147

Bode HB, Müller R (2006) Analysis of myxobacterial secondary metabolism goes molecular. J Ind Microbiol Biotechnol 33:577–588

Choi SU, Lee CK, Hwang YI, Kinoshita H, Nihira T (2004) Cloning and functional analysis by gene disruption of a gene encoding a gamma-butyrolactone autoregulator receptor from *Kitasatospora setae*. J Bacteriol 186:3423–3430

Corre C, Song L, O'Rourke S, Chater KF, Challis GL (2008) 2-Alkyl-4-hydroxymethylfuran-3-carboxylic acids, antibiotic production inducers discovered by *Streptomyces coelicolor* genome mining. Proc Natl Acad Sci USA 105:17510–17515

Galagan JE, Calvo SE, Borkovich KA et al (2003) The genome sequence of the filamentous fungus *Neurospora crassa*. Nature (Lond) 422:859–868

Hara H, Ohnishi Y, Horinouchi S (2009) DNA microarray analysis of global gene regulation by A-factor in *Streptomyces griseus*. Microbiology 155:2197–2210

Higo A, Horinouchi S, Ohnishi Y (2011) Strict regulation of morphological differentiation and secondary metabolism by a positive feedback loop between two global regulators AdpA and BldA in *Streptomyces griseus*. Mol Microbiol 81:1607–1622

Higo A, Hara H, Horinouchi S, Ohnishi Y (2012) Genome-wide distribution of AdpA, a global regulator for secondary metabolism and morphological differentiation in *Streptomyces*, revealed the extent and complexity of the AdpA regulatory network. DNA Res 19:259–273

Hirano S, Kato JY, Ohnishi Y, Horinouchi S (2006) Control of the *Streptomyces* subtilisin inhibitor gene by AdpA in the A-factor regulatory cascade in *Streptomyces griseus*. J Bacteriol 188:6207–6216

Horinouchi S (2002) A microbial hormone, A-factor, as a master switch for morphological differentiation and secondary metabolism in *Streptomyces griseus*. Front Biosci 7: d2045–d2057

Horinouchi S (2007) Mining and polishing of the treasure trove in the bacterial genus *Streptomyces*. Biosci Biotechnol Biochem 71:283–299

Horinouchi S, Beppu T (1992) Autoregulatory factors and communication in actinomycetes. Annu Rev Microbiol 46:377–398

Ikeda H, Ishikawa J, Hanamoto A et al (2003) Complete genome sequence of and comparative analysis of the industrial microorganism *Streptomyces avermitilis*. Nat Biotechnol 21: 526–531

Kato JY, Suzuki A, Yamazaki H, Ohnishi Y, Horinouchi S (2002) Control by A-factor of a metalloendopeptidase gene involved in aerial mycelium formation in *Streptomyces griseus*. J Bacteriol 184:6016–6025

Kato JY, Miyahisa I, Mashiko M, Ohnishi Y, Horinouchi S (2004) A single target is sufficient to account for the biological effects of the A-factor receptor protein of *Streptomyces griseus*. J Bacteriol 186:2206–2211

Kato JY, Chi WJ, Ohnishi Y, Hong SK, Horinouchi S (2005a) Transcriptional control by A-factor of two trypsin genes in *Streptomyces griseus*. J Bacteriol 187:286–295

Kato JY, Ohnishi Y, Horinouchi S (2005b) Autorepression of AdpA of the AraC/XylS family, a key transcriptional activator in the A-factor regulatory cascade in *Streptomyces griseus*. J Mol Biol 350:12–26

Kato JY, Funa N, Watanabe H, Ohnishi Y, Horinouchi S (2007) Biosynthesis of gamma-butyrolactone autoregulators that switch on secondary metabolism and morphological development in *Streptomyces*. Proc Natl Acad Sci USA 104:2378–2383

Khokhlov AS, Tovarova II, Borisova LN et al (1967) The A-factor, responsible for streptomycin biosynthesis by mutant strains of *Actinomyces streptomycini*. Dokl Akad Nauk SSSR 177:232–235

Kieser T, Bibb MJ, Buttner MJ, Chater KF, Hopwood DA (2000) Practical *Streptomyces* genetics, 2nd edn. John Innes Foundation, Norwich

Kitani S, Miyamoto KT, Takamatsu S et al (2011) Avenolide, a *Streptomyces* hormone controlling antibiotic production in *Streptomyces avermitilis*. Proc Natl Acad Sci USA 108:16410–16415

Lee KM, Lee CK, Choi SU et al (2005) Cloning and in vivo functional analysis by disruption of a gene encoding the gamma-butyrolactone autoregulator receptor from *Streptomyces natalensis*. Arch Microbiol 184:249–257

Miguélez EM, Hardisson C, Manzanal MB (1999) Hyphal death during colony development in *Streptomyces antibioticus*: morphological evidence for the existence of a process of cell deletion in a multicellular prokaryote. J Cell Biol 145:515–525

Miyake K, Kuzuyama T, Horinouchi S, Beppu T (1990) The A-factor-binding protein of *Streptomyces griseus* negatively controls streptomycin production and sporulation. J Bacteriol 172:3003–3008

Natsume R, Ohnishi Y, Senda T, Horinouchi S (2004) Crystal structure of a gamma-butyrolactone autoregulator receptor protein in *Streptomyces coelicolor* A3(2). J Mol Biol 336:409–419

Ochi K (2007) From microbial differentiation to ribosome engineering. Biosci Biotechnol Biochem 71:1373–1386

Ohnishi Y, Kameyama S, Onaka H, Horinouchi S (1999) The A-factor regulatory cascade leading to streptomycin biosynthesis in *Streptomyces griseus*: identification of a target gene of the A-factor receptor. Mol Microbiol 34:102–111

Ohnishi Y, Yamazaki H, Kato JY, Tomono A, Horinouchi S (2005) AdpA, a central transcriptional regulator in the A-factor regulatory cascade that leads to morphological development and secondary metabolism in *Streptomyces griseus*. Biosci Biotechnol Biochem 69:431–439

Ohnishi Y, Ishikawa J, Hara H et al (2008) Genome sequence of the streptomycin-producing microorganism *Streptomyces griseus* IFO 13350. J Bacteriol 190:4050–4060

Okamoto S, Nakamura K, Nihira T, Yamada Y (1995) Virginiae butanolide binding protein from *Streptomyces virginiae*. Evidence that VbrA is not the virginiae butanolide binding protein and reidentification of the true binding protein. J Biol Chem 270:12319–12326

Oliynyk M, Samborskyy M, Lester JB et al (2007) Complete genome sequence of the erythromycin-producing bacterium *Saccharopolyspora erythraea* NRRL23338. Nat Biotechnol 25:447–453

Onaka H, Horinouchi S (1997) DNA-binding activity of the A-factor receptor protein and its recognition DNA sequences. Mol Microbiol 24:991–1000

Onaka H, Ando N, Nihira T et al (1995) Cloning and characterization of the A-factor receptor gene from *Streptomyces griseus*. J Bacteriol 177:6083–6092

Onaka H, Mori Y, Igarashi Y, Furumai T (2011) Mycolic acid-containing bacteria induce natural-product biosynthesis in *Streptomyces* species. Appl Environ Microbiol 77:400–406

Ramaswamy AV, Flatt PM, Edwards DJ et al (2006) The secondary metabolites and biosynthetic gene clusters of marine cyanobacteria. Applications in biotechnology. Front Mar Biotechnol 175–224

Santamarta I, Pérez-Redondo R, Lorenzana LM, Martín JF, Liras P (2005) Different proteins bind to the butyrolactone receptor protein ARE sequence located upstream of the regulatory *ccaR* gene of *Streptomyces clavuligerus*. Mol Microbiol 56:824–835

Takano E (2006) Gamma-butyrolactones: *Streptomyces* signalling molecules regulating antibiotic production and differentiation. Curr Opin Microbiol 9:287–294

Takano E, Nihira T, Hara Y et al (2000) Purification and structural determination of SCB1, a gamma-butyrolactone that elicits antibiotic production in *Streptomyces coelicolor* A3(2). J Biol Chem 275:11010–11016

Takano E, Kinoshita H, Mersinias V et al (2005) A bacterial hormone (the SCB1) directly controls the expression of a pathway-specific regulatory gene in the cryptic type I polyketide biosynthetic gene cluster of *Streptomyces coelicolor*. Mol Microbiol 56:465–479

Tomono A, Tsai Y, Ohnishi Y, Horinouchi S (2005a) Three chymotrypsin genes are members of the AdpA regulon in the A-factor regulatory cascade in *Streptomyces griseus*. J Bacteriol 187:6341–6353

Tomono A, Tsai Y, Yamazaki H, Ohnishi Y, Horinouchi S (2005b) Transcriptional control by A-factor of strR, the pathway-specific transcriptional activator for streptomycin biosynthesis in *Streptomyces griseus*. J Bacteriol 187:5595–5604

Udwary DW, Zeigler L, Asolkar RN et al (2007) Genome sequencing reveals complex secondary metabolome in the marine actinomycete *Salinispora tropica*. Proc Natl Acad Sci USA 104:10376–10381

Ueda K, Miyake K, Horinouchi S, Beppu T (1993) A gene cluster involved in aerial mycelium formation in *Streptomyces griseus* encodes proteins similar to the response regulators of two-component regulatory systems and membrane translocators. J Bacteriol 175:2006–2016

Yamada Y, Nihira T (1998) Microbial hormones and microbial chemical ecology. In: Barton DHR, Nakanishi K (eds) Comprehensive natural products chemistry, vol 8. Elsevier, Heidelberg, pp 377–413

Yamazaki H, Ohnishi Y, Horinouchi S (2000) An A-factor-dependent extracytoplasmic function sigma factor (σ^{AdsA}) that is essential for morphological development in *Streptomyces griseus*. J Bacteriol 182:4596–4605

Yamazaki H, Ohnishi Y, Horinouchi S (2003a) Transcriptional switch-on of *ssgA* by A-factor, which is essential for spore septum formation in *Streptomyces griseus*. J Bacteriol 185:1273–1283

Yamazaki H, Takano Y, Ohnishi Y, Horinouchi S (2003b) *amfR*, an essential gene for aerial mycelium formation, is a member of the AdpA regulon in the A-factor regulatory cascade in *Streptomyces griseus*. Mol Microbiol 50:1173–1187

Chapter 17
Enzymatic Production of Designed Peptide

Kuniki Kino

Abstract Peptides are expected to be one of the most promising compounds that are beneficial for improving our quality of life. Research on functional peptides has been carried out in various fields, including food science, medicine, and cosmetics; new findings are frequently reported. Oligopeptides such as dipeptides or tripeptides also have unique physiological functions and physical properties that cannot be found in the constitutive amino acids. However, only a few dipeptides, such as L-aspartyl-L-phenylalanine methyl ester (Asp-Phe-OMe, aspartame), as an artificial sweetener and L-alanyl-L-glutamine (Ala-Gln), as a patient infusion, are commercially used, which can be attributed to the lack of an efficient process for production of these oligopeptides. Therefore, the development of an oligopeptide manufacturing process is important for addressing the growing needs of functional peptides. Recently, bacterial enzymes that produce various dipeptides, oligopeptides, or homopoly(oligo) amino acids have been found. L-Amino acid α-ligase (Lal, EC 6.3.2.28) belongs to the ATP-dependent (ADP-forming) carboxylate-amine/thiol ligase superfamily that catalyzes the condensation of unprotected amino acids and is applicable to fermentative production. In this group, ATP and Mg^{2+} are generally required for peptide synthesis, and aminoacyl phosphate is synthesized as the reaction intermediate. Various Lals have been newly identified by in silico searches using a BLAST program and by different approaches including purification of putative Lal from microorganisms producing peptide antibiotics as secondary metabolites. Furthermore, using only an adenylation domain (A-domain) of nonribosomal peptide synthetase (NRPS), various aminoacyl prolines, which are dipeptides containing a proline residue at the C-terminus, or various amide compounds can be synthesized from unprotected amino acids and proline without any additional process. This chapter reviews the current knowledge about these unique enzymes and novel enzymatic production methods of designed peptides.

K. Kino (✉)
Department of Applied Chemistry, Waseda University,
3-4-1 Ohkubo, Shinjuku-ku, Tokyo 169-8555, Japan
e-mail: kkino@waseda.jp

H. Anazawa and S. Shimizu (eds.), *Microbial Production: From Genome Design to Cell Engineering*, DOI 10.1007/978-4-431-54607-8_17, © Springer Japan 2014

Keywords Dipeptide • L-Amino acid ligase • Nonribosomal peptide synthetase • Oligopeptide • Peptide synthesis

17.1 Introduction

Peptides are expected to be beneficial compounds improving our quality of life. Oligopeptides such as dipeptides or tripeptides also have unique physiological functions and physical properties that cannot be found in the constitutive amino acids. However, information about the unique and interesting functions of these oligopeptides is not sufficient, one of the major reasons being the lack of cost-effective manufacturing processes. Therefore, establishing cost-effective and eco-friendly methods of producing oligopeptides is expected to boost the exploration and development of functional oligopeptides. In this chapter, current and newly developed technologies for searching novel enzymes catalyzing the condensation of amino acids and for application to the production of various oligopeptides are summarized. These oligopeptides contain non-proteinogenic amino acids as well as proteinogenic amino acids and their simple derivatives.

17.2 Function and Application of Peptides

Peptides exhibit a broad range of biological activities. In particular, a number of low molecular weight, bioactive peptides have been found from the protease hydrolysate of various food proteins such as bovine casein peptone, wheat gluten, soybean extract, and bonito extract. Dipeptides consisting of two amino acids also possess unique physiological functions and physical properties that the constituent individual amino acids do not exhibit. In the field of food science, Nippon Suisan Kaisha, Ltd. reported a saltiness-strengthening agent obtained by adding a dipeptide containing L-glutamic acid (Glu) such as L-glutamyl-L-threonine (Glu-Thr) (Shimono, M and Sugiyama, K., Japanese patent WO2009/113563, 2009). In addition, Kao Corporation reported a similar function of L-leucyl-L-serine (Leu-Ser), which also enhanced saltiness (Koike, M., Japanese patent JP 2012–165740, 2012). Reduction of the salt content in foods is particularly important for people with high blood pressure. Furthermore, Ohinata et al. recently reported that L-arginyl-L-phenylalanine (Arg-Phe) decreased blood pressure and food intake in rodents (Kagebayashi et al. 2012). Arg-Phe does not act as an angiotensin I-converting enzyme (ACE)-inhibitory peptide (Hernández-Ledesma et al. 2011), as is frequently reported, and it shows vasorelaxing activity. In the medical field, L-glutamyl-L-tryptophan (Glu-Trp), which is named oglufanide, was reported to show antiangiogenic properties and inhibition of tumor growth in in vivo preclinical models (Sgadari et al. 2011). It passed phase I and II trials, but the phase III randomized, double-blind, placebo-controlled trial failed to show significant antitumor activity. Nitta et al. reported that L-leucyl-L-isoleucine (Leu-Ile) induced brain-derived neurotrophic factor in cultured neuronal cells and may act as an antidepressant (Furukawa-Hibi et al. 2011).

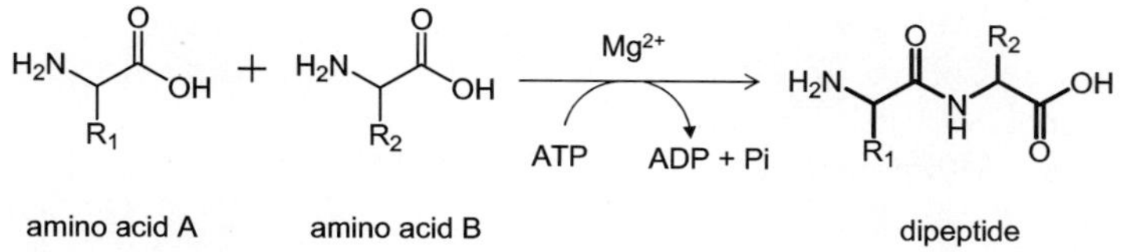

Fig. 17.1 Dipeptide synthesis by L-amino acid ligase (EC 6.3.2.28). L-Amino acid ligase (Lal) belongs to the ATP-dependent (ADP-forming) carboxylate-amine/thiol ligase superfamily that catalyzes the condensation of unprotected amino acids. In this group, ATP and Mg^{2+} are generally required for peptide synthesis, and aminoacyl phosphate is synthesized as the reaction intermediate

L-Histidyl-β-alanine (His-β-Ala), which is the reverse sequence of β-alanyl-L-HISTIDINE (β-Ala-His, carnosine), was reported to induce sedative and hypnotic effects (Tsuneyoshi et al. 2008). Carnosine has been well studied and has been shown to be useful as an antioxidant (Di Bernardini et al. 2011). The structural differences between β-Ala-His and His-β-Ala are small, but their bioactivity is quite different. In the cosmetic field, Kao Corporation reported the inhibitory effect of L-phenylalanyl-β-alanine (Phe-β-Ala) toward elastase, acting as a hair growth inhibitor (Tsuji, N. and Moriwaki, S., Japanese patent JP 2001–226232, 2001). These functional peptides just described here are newly discovered. Furthermore, dipeptides have also been reported to act as catalysts by themselves (Weber and Pizzarello 2006; Hernández and Juaristi 2011). These previous studies indicate that the development of a peptide manufacturing process is important for addressing the growing needs of these functional peptides.

17.3 L-Amino Acid α-Ligase

L-Amino acid ligase (Lal, EC 6.3.2.28) is a microbial enzyme that catalyzes dipeptide synthesis from unprotected amino acids by hydrolysis of ATP to ADP and phosphate. Lal belongs to the ATP-dependent (ADP-forming) carboxylate-amine/thiol ligase superfamily (Galperin and Koonin 1997) that contains glutathione synthetase and D-alanine-D-alanine ligase and, therefore, catalyzes ligation in an ATP-dependent manner through an aminoacyl-phosphate reaction intermediate (Fan et al. 1995). The first reported Lal, YwfE, has been identified from *Bacillus subtilis* 168 by in silico search (Tabata et al. 2005). This enzyme synthesized various dipeptides from amino-acid substrates in an ATP-dependent manner and did not catalyze oligopeptide synthesis (Fig. 17.1). This condensation process is achieved in one step, and degradation of reaction products does not occur. Because amino-acid substrates and ATP can be supplied to microbial cells, fermentative production of dipeptides has been developed (Tabata and Hashimoto 2007). Therefore, enzymatic peptide synthesis using Lal could be the very efficient. After that, various Lals that exhibit unique substrate specificity were newly obtained from microorganisms by in silico analysis using microbial genome information and various databases: Rsp1486a protein from *Ralstonia solanacearum* JCM10489 (Kino et al. 2008a), BL00235

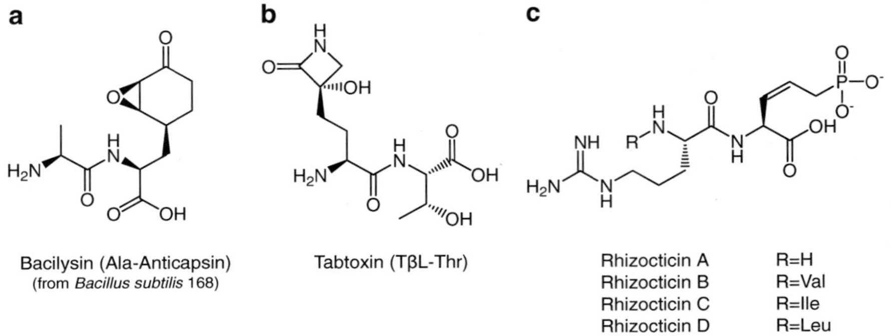

Fig. 17.2 Structure of bacilysin, tabtoxin, and rhizocticin derivatives. **(a)** Bacilysin is a dipeptide antibiotic produced by *Bacillus subtilis* 168 and composed of L-alanine (Ala) and anticapsin. **(b)** Tabtoxin is a dipeptidic phytotoxin produced by *Pseudomonas syringae* strains, composed of tabtoxinine-β-lactam (TβL) and L-threonine (Thr). **(c)** Rhizocticin is a peptide antibiotic produced by *Bacillus subtilis* NBRC3134. Four types have been reported. Rhizocticin A: L-ARGINYL-L-2-amino-5-phosphono-3-*cis*-pentenoic acid (Arg-APPA); rhizocticin B: L-valyl-L-arginyl-l-2-amino-5-phosphono-3-*cis*-pentenoic acid (Val-Arg-APPA); and rhizocticin C and D, which are the same as rhizocticin B but contain L-isoleucine (Ile) and L-leucine (Leu), respectively, in place of L-valine (Val)

protein from *Bacillus licheniformis* NBRC12200 (Kino et al. 2008b), the PSPPH_4299 protein from *Pseudomonas syringae* pv. *phaseolicola* 1448A (Arai and Kino 2008), and the plu1440 protein from *Photorhabdus luminescens* subsp. *laumondii* TT01 (Kino et al. 2010a).

17.3.1 Lals from Microorganisms Producing Peptidic Secondary Metabolites

In the course of searching new Lals by in silico analysis, new insight was provided into the biosynthesis of peptidic secondary metabolites, for example, bacilysin (Inaoka et al. 2003) and phaseolotoxin (Mitchell 1976). The *ywfE* gene is part of the biosynthetic gene cluster of a bacilysin peptide antibiotic that consists of L-alanine (Ala) and L-anticapsin. Thus, the physiological significance of the YwfE protein lies in its participation in peptide antibiotic biosynthesis (Fig. 17.2). Similarly, the PSPPH_4299 gene in the phaseolotoxin (peptide-phytotoxin) biosynthetic gene cluster from *P. syringae* pv. *phaseolicola* 1448A encodes a novel Lal, and the PSPPH_4299 protein synthesized various hetero-dipeptides containing basic amino acids in an ATP-dependent manner, and also synthesized alanyl-homoarginine, part of the phaseolotoxin scaffold (Arai and Kino 2008). In contrast, the role of the Rsp1486a and BL00235 proteins remain unclear.

Arai et al. (2013a) have recently focused on the putative Lal involved in the biosynthesis of dipeptidic phytotoxin, designated as tabtoxin (Fig. 17.2). This compound is composed of tabtoxinine-β-lactam (TβL) and L-threonine (Thr) (Stewart 1971);

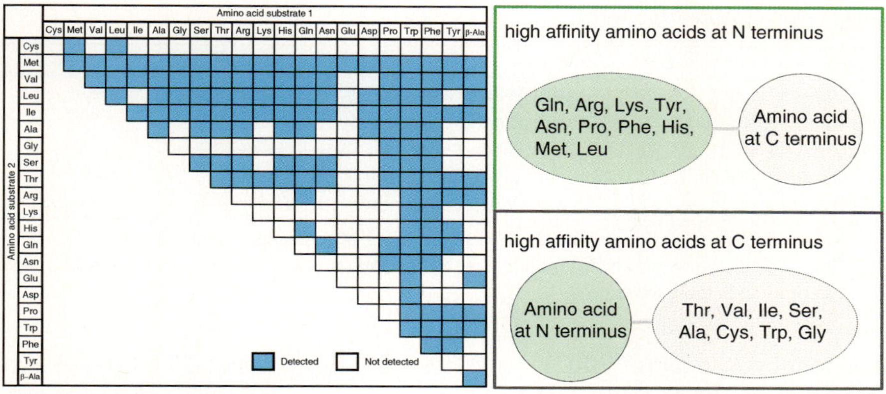

Fig. 17.3 Overview of substrate specificity of TabS. The reaction mixtures were analyzed by LC-ESI MS. A *filled square* indicates the formation of the corresponding dipeptide. The amino acids used as the substrates are shown in the vertical and horizontal columns and are assigned by *three-letter codes*

TβL is located at the N-terminus and Thr is at the C-terminus, resulting in TβL-Thr. TβL-Thr is hydrolyzed, and the resulting TβL irreversibly inhibits glutamine synthetase (EC 6.3.1.2), causing characteristic chlorosis in plants. The gene named *tabS* was cloned from *Pseudomonas syringae* NBRC14081 (ATCC27881) and was overexpressed in *Escherichia coli* cells. Surprisingly, this protein TabS was useful for the synthesis of various functional peptides with high selectivity because of distinct substrate specificity both at the N-terminus and C-terminus. Liquid chromatography-electrospray ionization-mass spectrometry (LC-ESI MS) analysis showed that TabS had broad substrate specificity, detecting *m/z* peaks corresponding to dipeptides in 136 combinations among 231 combinations of amino-acid substrates (Fig. 17.3). The substrate specificity of TabS is the broadest among known Lals. TabS synthesized Gln-Thr with 96 % yield. Gln-Thr can be converted to Glu-Thr using N-terminal amidase (Arai et al. 2013b), and the resulting peptide enhances saltiness in foods. Hence, Gln-Thr is usable as a precursor of Glu-Thr, which is an important peptide for people with high blood pressure, for reducing the salt levels in foods. In addition, L-proline (L-Pro) analogues, such as L-pipecolic acid, hydroxy-l-Pro, and L-azetidine-2-carboxylic acid, were found to be acceptable substrates at the N-terminus. Furthermore, TabS could selectively synthesize various functional peptides such as Arg-Phe, Leu-Ile, and Gln-Trp as a precursor of Glu-Trp and Leu-Ser (Table 17.1). These peptides were the main products, although the levels of by-products such as homopeptides and other heteropeptides were low or were not detected. Gln-Trp can be also converted to a Glu-Trp functional peptide by N-terminal amidase. TabS also synthesized a functional peptide containing β-Ala, for example, β-Ala-L-histidine (β-Ala-His), His-β-Ala, and Phe-β-Ala. All these peptides have beneficial effects for humans as already described.

Furthermore, to obtain a novel Lal, different approaches including purification of putative Lal from microorganisms producing peptide antibiotics as secondary metabolites were examined. Kino et al. (2009) focused on the rhizocticin peptide antibiotic

Table 17.1 Functional peptides synthesized by TabS

Peptide (reaction product)	Function	Product[a] (mM)	Yield[b] (%)
Arg-Phe	Antihypertensive effect	7.7±0.06	62
Leu-Ile	Antidepressive effect	9.6±0.09	77
Gln-Trp (precursor of Glu-Trp)	Glu-Trp shows an antiangiogenic activity	6.8±0.02	54
Leu-Ser	Enhancement of saltiness	10.5±0.18	83

[a]The reaction products were analyzed by HPLC: averages of three measurements
[b]Yield was calculated based on the amount of ATP added in the reaction mixture; hence, a 12.5 mM reaction product is the maximum

produced by *B. subtilis* NBRC3134 (ATCC6633) (Kugler et al. 1990) (Fig. 17.2). Depending on chemical structure, four types of rhizocticins have been reported: rhizocticin A: L-arginyl-L-2-amino-5-phosphono-3-*cis*-pentenoic acid (Arg-APPA); rhizocticin B: L-valyl-L-arginyl-L-2-amino-5-phosphono-3-cis-pentenoic acid (Val-Arg-APPA); and rhizocticin C and D, which are the same as rhizocticin B but contain L-isoleucine (Ile) and L-leucine (Leu), respectively, in place of L-valine (Val). Kino et al. hypothesized that the rhizocticins are synthesized by two Lals successively: one catalyzes Arg-APPA synthesis, and the other catalyzes ligation of Val, Leu, or Ile and Arg-APPA. By detecting L-arginine hydroxamate synthesis activity, the putative Lal named RizA was purified, and the gene was cloned. Further examination showed that RizA catalyzed Arg-Xaa dipeptide synthesis (Xaa, arbitrary amino acids) but not the synthesis of tripeptides. Hence, RizA might be involved in the biosynthesis of rhizocticin A, and other enzymes might be required for the synthesis of tripeptides such as rhizocticin B, C, and D. DNA sequence analysis of unknown regions around *rizA* was then carried out, and a novel gene, located about 9,000 bp upstream of *rizA*, that encodes a protein possessing an ATP-grasp motif was newly found. This gene was named *rizB*, and it was revealed that recombinant RizB synthesized heteropeptides such as Val-Arg-Xaa peptides. Furthermore, RizB synthesized homo-oligo amino acids of Val, Leu, Ile, and L-methionine (Met) (Kino et al. 2010b). After publication of the studies on RizA and RizB by Kino et al., Borisova et al. (2010) reported on the whole biosynthetic gene cluster and proposed the biosynthesis pathway; RizA and RizB were designated RhiM and RhiC by the authors, respectively. Then, the genes encoding the proteins homologous to RizB were cloned from various microorganisms, and it was found that these proteins also synthesized homo-oligo amino acids and showed distinct substrate specificities.

17.3.2 Lals for Oligopeptide Synthesis

The gene encoding *rizB* (*rhiC*) is located in the rhizocticin biosynthetic gene cluster (Fig. 17.4). This protein seemed to be involved in rhizocticin biosynthesis, catalyzing addition of Val, Leu, or Ile to Arg-APPA. LC-ESI MS analysis showed that RizB synthesized homo-oligo amino acids consisting of two to five amino acids

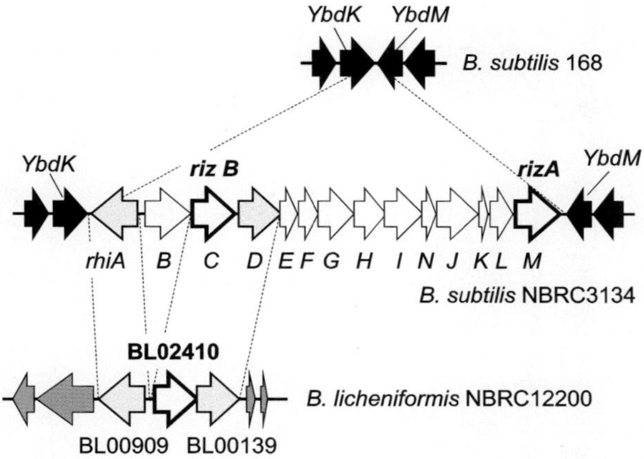

Fig. 17.4 Rhizocticin biosynthetic gene cluster and comparison with the genes from other *Bacillus* strains. The rhizocticin biosynthetic gene cluster is composed of 14 genes including 2 genes encoding Lal: *rizA* (*rhiM*) and *rizB* (*rhiC*). This gene cluster appears to be inserted between *ybdK* and *ybdM* of *B. subtilis* 168. The genome from *Bacillus licheniformis* NBRC 12200 has a region highly homologous to this cluster partially

Enzyme [Length (AA), Homology (%)]	Substrate	Product						
		2mer	3mer	4mer	5mer	6mer	7mer	8mer
RizB [408 AA, 100%]	Val	●	●	●	●			
	Leu	●	●	●	●			
	Ile	●	●	●				
	Met	●	●	●	●			
	Trp	●						
	Phe							
	Tyr							
BL02410 [405 AA, 62%]	Val	●	●	●	●			
	Leu	●	●	●	●			
	Ile	●	●	●				
	Met	●	●	●	●			
	Trp	●	●					
	Phe	●	●					
	Tyr							
Haur_2023 [408 AA, 39%]	Val	●	●	●				
	Leu	●	●	●				
	Ile	●	●					
	Met	●	●	●				
	Trp	●						
	Phe	●	●					
	Tyr							

Enzyme [Length (AA), Homology (%)]	Substrate	Product						
		2mer	3mer	4mer	5mer	6mer	7mer	8mer
spr0969 [400 AA, 33%]	Val	●	●	●	●○	●○		
	Leu	●	●	●	●○	●○		
	Ile	●	●	●○	○	○		
	Met	●	●	●	●	●		
	Trp	●						
	Phe	●						
	Tyr							
BAD_1200 [395 AA, 27%]	Val	●	●	●	●			
	Leu	●	●	●	●	●		
	Ile	●	●	●				
	Met	●	●	●	●	●		
	Trp	●	●	●	●○	○	○	○
	Phe	●	●	●○	●○	○		
	Tyr	●	●	●	●			
CV_0806 [392 AA, 23%]	Val	●	●	●				
	Leu	●	●	●				
	Ile	●						
	Met	●	●					
	Trp	●						
	Phe							
	Tyr							

Fig. 17.5 Summary of reaction products catalyzed by RizB and the proteins homologous to RizB. The reactions were conducted using a single amino acid as a substrate, and the reaction products were analyzed by LC-ESI MS. Homo-oligo amino acids in which their corresponding m/z peaks were detected are shown by a *black* or *white circle*. *Black* indicates detection in the supernatant of reaction mixture, and *white* indicates detection in the precipitate

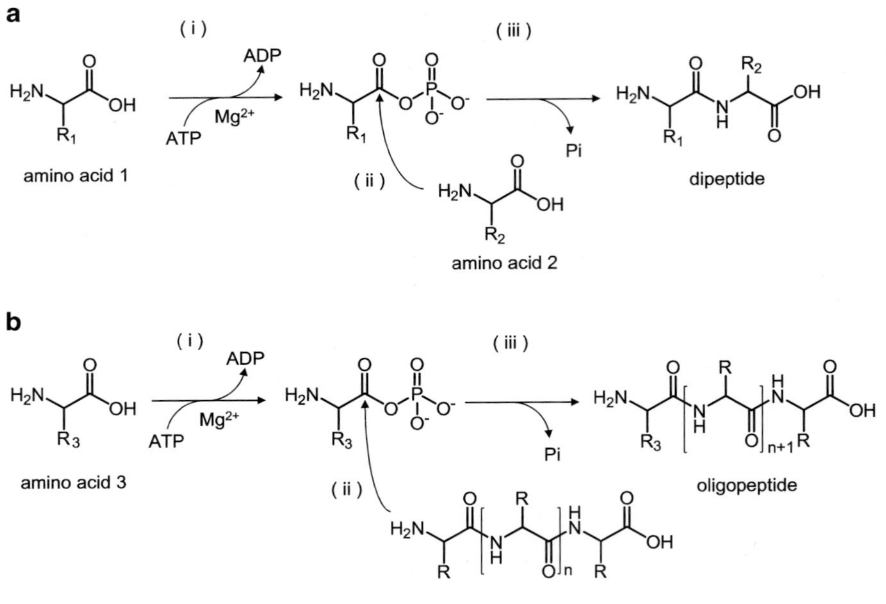

Fig. 17.6 Amide ligation mechanism in Lal. **(a)** Dipeptide synthesis: (*i*) an amino acid serves as an N-terminal substrate and undergoes initial phosphorylation at its carboxyl group, and an amino-acyl phosphate is synthesized as the reaction intermediate; (*ii*) the aminoacyl phosphate is then nucleophilically attacked by the other amino acid; and (*iii*) phosphate is released, resulting in the dipeptide. **(b)** Oligopeptide synthesis

when Val, Leu, Ile, and Met were used as a substrate (Fig. 17.5). In addition, high performance liquid chromatography (HPLC) analysis showed that RizB synthesized tripeptides as a main product when Val was used as a substrate.

RizB sheared low homology with known Lals catalyzing dipeptide synthesis, about 20 %, although significant homology is not found among dipeptide synthesizing Lals.

As already described, Lals belong to the ATP-dependent carboxylate-amine/thiol ligase superfamily. A mechanism of dipeptide synthesis by Lal has been proposed (Fig. 17.6): (1) an amino acid serves as an N-terminal substrate and undergoes initial phosphorylation at its carboxyl group, and an aminoacyl phosphate is synthesized as the reaction intermediate; (2) the aminoacyl phosphate is then nucleophilically attacked by the other amino acid; and (3) phosphate is released, resulting in the dipeptide. The synthesis of aminoacyl phosphate as a reaction intermediate is the common step among the enzymes belonging to the ATP-dependent carboxylate-amine/thiol ligase superfamily. In the case of RizB, this reaction mechanism is fundamentally the same as that of the Lals that catalyze only dipeptide synthesis. The reaction with C-protected amino acids showed that RizB ligated a single amino acid to a dipeptide at its N-terminus, resulting in a tripeptide. First, a dipeptide is synthesized from amino acids, as occurs in reactions catalyzed by known Lals; the resulting dipeptide then nucleophilically attaches to an aminoacyl phosphate of amino acid, forming a tripeptide. Further, the resulting tripeptide nucleophilically attaches to an aminoacyl phosphate of another amino acid, forming a tetrapeptide. As a result, the peptide is elongated at its N-terminus in a stepwise manner. Therefore, the prominent difference between RizB and known Lals that catalyze only dipeptide synthesis is that the former uses dipeptides, tripeptides, and pentapeptides as C-terminal substrates and the latter use C-terminal amino acid substrates only.

RizB has been reported as the first Lal catalyzing oligopeptide synthesis. The new Lal catalyzing oligopeptide synthesis was then searched using a BLAST search program. Arai and Kino (2010) reported five proteins homologous to RizB: BL02410 from *B. licheniformis* NBRC 12200, Haur_2023 from *Herpetosiphon aurantiacus* ATCC 23779, spr0969 from *Streptococcus pneumoniae* ATCC BAA-255, BAD_1200 from *Bifidobacterium adolescentis* JCM 1275, and CV_0806 from *Chromobacterium violaceum* NBRC 12614. These proteins commonly showed activity toward Val, Leu, Ile, and Met, such as RizB, but each showed distinct characteristics. In particular, spr0969 synthesized longer peptides than RizB, and BAD_1200 showed higher activity toward aromatic amino acids than toward the branched-chain ones (Fig. 17.5). Furthermore, in the reaction mixtures of the spr0969 protein, white precipitates were produced when Val and Ile were used as substrates, and in the case of Leu, no such precipitate was observed, but the reaction solution of Leu became a gel. These precipitates were dissolved in 50 % acetic acid and were analyzed by LC-ESI MS. In the case of Val, the 5-mer and 6-mer of Val were detected more clearly in the 50 % acetic acid dissolving the precipitate than in the supernatant of the reaction mixture, and therefore must have been the components of the white precipitate. In addition, LC-ESI MS analysis showed that the precipitate in the reaction mixture of Ile must have been the 4-mer, 5-mer, and 6-mer of Ile, and that the hydrogel in the reaction mixture of Leu must have been composed of the 5-mer and 6-mer of Leu. Precipitates were additionally produced in the reaction mixture of the BAD_1200 protein when Trp and Phe were used as substrates, although no precipitates were produced when branched-chain amino acids were used. LC-ESI MS analysis showed that the precipitates were composed of the 4-mer, 5-mer, and 6-mer of Phe and the 5-mer, 6-mer, 7-mer, and 8-mer of Trp. CV_0806 might use di- or tripeptide preferably to the monomer of amino acids as the C-terminal substrate. Amino-acid substrates with CV_0806 result in low di- and tripeptide synthesis, but large amounts of tripeptide and tetrapeptide were synthesized when the 2-mer or the 3-mer of Val was used as substrate with Val. The activities of BL02410 and Haur_2023 were similar to that of RizB.

The molecular mass was estimated by gel filtration. The results showed that the CV_0806 protein is a monomeric enzyme and that the others are dimeric enzymes.

The role of these homologous proteins in each microbial cell is not clear. However, RizB and BL02410 shows high homology on the amino acid sequence, about 62 %, and the two proteins that are located at upstream and downstream of BL02410 are also similar to those of RizB (Fig. 17.5). RhiA from *B. subtilis* NBRC 3134 and BL00909 from *B. licheniformis* NBRC 12200 code for a LysR-type transcriptional regulator and show 64 % homology. In addition, RhiD from *B. subtilis* NBRC 3134 and BL00139 from *B. licheniformis* NBRC 12200 code for a transporter and show 67 % homology. A horizontal gene transfer may have occurred, although it is not clear which is the donor.

17.4 Poly-α-Glutamic (α-PGA) Acid Synthetase, RimK

Ribosomal protein S6 (RPS6) is a part of the 30S subunit in *E. coli* K-12. RimK has been reported to catalyze the modification of RPS6 by adding up to four residues of Glu to the C-terminal Glu residue, generating RPS6-Glu, RPS6-Glu-Glu, RPS6-Glu-Glu-Glu, and RPS6-Glu-Glu-Glu-Glu. The role of these Glu oligopeptide tails with the α-amide linkage remains unclear. Kino et al. (2011) reported a novel catalytic activity of RimK. Without RPS6, this enzyme catalyzed the Glu polymerization reaction to produce α-PGA. At pH 9, the reaction product showed polydisperse polymer products consisting of 8–46 Glu residues. A higher pH resulted in the production of shorter α-PGA: 8–26 residues at pH 9.5 and 8–12 residues at pH 10. During the enzyme reaction, ATP was converted to ADP, suggesting that Glu is activated as an aminoacyl phosphate. In fact, RimK belongs to the ATP-dependent carboxylate-amine/thiol ligase superfamily. In addition, RimK was found to catalyze the phosphorylation of L-glutamyl-L-glutamic acid (Glu-Glu) during the polymerization reaction. This result suggested that peptide elongation seemed to have occurred at the C-terminus in a stepwise manner (Fig. 17.7), which is different from Lals elongating the peptides at the N-terminus. Kino et al. (2011) further reported that the enzyme reaction with Glu-Glu and 19 proteinogenic amino acids produced 15 kinds of Glu-Glu-Xaa (where Xaa represents proteinogenic amino acids used, except for Arg, L-lysine, His, and Pro). Thus, RimK has relaxed substrate specificity toward C-terminal substrates.

17.5 Aminoacyl Proline Synthesis Using the A-Domain of Nonribosomal Peptide Synthetases

Numerous peptidic secondary metabolites are produced by nonribosomal peptide synthetases (NRPSs) as well as the ribosomal system in nature. NRPSs have been found in various microorganisms such as fungi, streptomycetes, and other bacteria, and can accept nonnatural and non-proteinogenic amino-acid building blocks as substrates, thereby offering greater structural diversity. The enzymes are huge multifunctional proteins made up of a series of modules, each of which takes charge of adding one amino acid to a growing peptide. Each module contains at least three enzymatic units called domains. An adenylation domain (A-domain) recognizes the substrate amino acid and activates it as an aminoacyl-AMP accompanied by the hydrolysis of ATP to AMP and pyrophosphate. Subsequently, the activated amino acid is transferred to the 4′-phosphopantetheine moiety of the thiolation domain (T-domain) with the release of AMP. Then, the adjacent condensation domain (C-domain) catalyzes the formation of the peptide bond. Finally, the thioesterase domain (Te-domain) catalyzes the release of the product peptide from the enzyme protein. The order and number of modules of a NRPS system determine the sequence and length of the peptide product (Finking and Marahiel 2004).

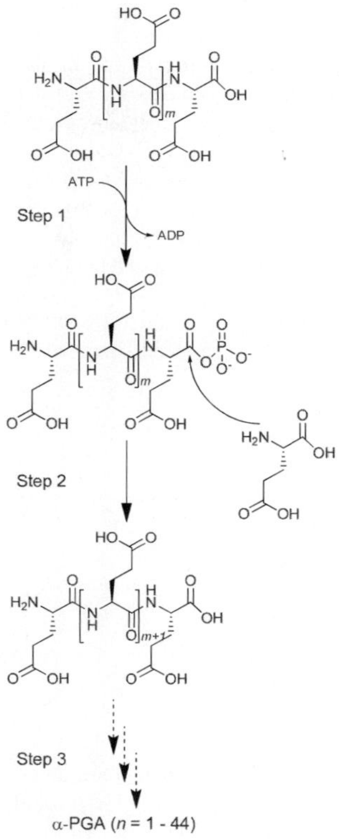

Fig. 17.7 Amide ligation mechanism for poly-α-glutamic acid (α-PGA) synthesis. This Glu polymerization reaction is catalyzed by RimK from *Escherichia coli*, which has been reported to catalyze the modification of RPS6 (ribosomal protein S6) by adding up to four residues of Glu to the C-terminal Glu residue. RimK was found to catalyze the phosphorylation of L-glutamyl-L-glutamic acid (Glu-Glu) during the polymerization reaction

Tyrocidine is an antibiotic produced by *Brevibacillus parabrevis* through a NRPS system that consists of three subunits, TycA, TycB, and TycC. TycA is a well-characterized NRPS that contains one module responsible for the integration of D-phenylalanine (D-Phe). TycA integrates the first D-Phe residue through activating and epimerizing L-Phe, followed by peptide bond formation by the condensation domain of the first module of TycB (TycB1). In the natural tyrocidine assembly line, D-Phe-L-Pro chain is transferred to the L-Phe-*S*-phosphopantetheinyl (Ppant) chain of the second module of TycB, and the further elongation step of the peptide proceeds. Recently, the substrate specificity of the adenylation domain (A-domain) of TycA was reported in detail, showing that TycA activates not only Phe but also aromatic and some kinds of aliphatic amino acids, although the activity was quite low (<1 % toward Phe). Hence, this system was considered to be capable to produce diverse aminoacyl prolines.

Kino et al. (unpublished data) have showed that various aminoacyl prolines such as Phe-Pro, Tyr-Pro, Trp-Pro, Leu-Pro, Val-Pro, and Met-Pro were synthesized using only TycA as well as the TycA/TycB1 system. Especially, when Trp, Leu, Val, and Met were used as substrate for TycA, corresponding aminoacyl prolines were respectively synthesized with high yield. Moreover, D-Pro as well as L-Pro was able

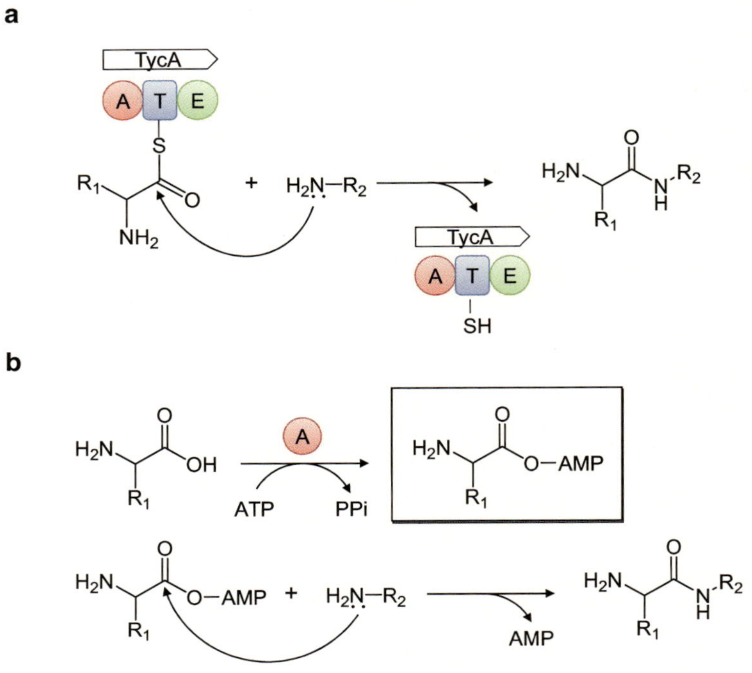

Fig. 17.8 Amino acid amide synthesis catalyzed by (**a**) TycA module and (**b**) a domain of TycA module from NRPS for tyrocidine synthesis. *A*, adenylation domain; *T*, thiolation domain; *E*, epimerization domain. Various compounds as nucleophiles were able to accept C-terminus-independent substrates such as *cis*-4-hydroxy-Pro, *trans*-4-hydroxy-Pro, pipecolinic acid, azetidine-2-carboxylic acid, and various amines including cyclic amines

to use a substrate as C-terminus, depending on the chirality, and aminoacyl proline was synthesized in the same amounts. It was considered that this peptide-bond formation proceeded because D-Pro or L-Pro nucleophilically attached to an aminoacyl phosphate, that is, an activated amino acid as N-terminus substrate, without condensation domain (Fig. 17.8). Various compounds as nucleophiles were able to accept substrates of C-terminus independently, for example, *cis*-4-hydroxy-Pro, *trans*-4-hydroxy-Pro, pipecolinic acid, azetidine-2-carboxylic acid, and various amines including cyclic amines. Furthermore, Kino et al. (unpublished data) found that that aminoacyl prolines were synthesized using only the A-domain of TycA as catalyst, and clarified that it was able to synthesize the amides as already described using another A-domain such as BacB1 and BacC4, which was the module of NRPS relating to bacitracin biosynthesis, although the substrate at the N-terminus corresponded to the amino acid. This novel method for amide bond formation using A domain from NRPSs was expected to be one of the effective production processes for diverse amide compounds.

References

Arai T, Kino K (2008) A novel L-amino acid ligase is encoded by a gene in the phaseolotoxin biosynthetic gene cluster from *Pseudomonas syringae* pv. *phaseolicola* 1448A. Biosci Biotechnol Biochem 72:3048–3050

Arai T, Kino K (2010) New L-amino acid ligases catalyzing oligopeptide synthesis from various microorganisms. Biosci Biotechnol Biochem 74:1572–1577

Arai T, Arimura Y, Ishikura S, Kino K (2013a) L-Amino acid ligase from Pseudomonas syringae producing tabtoxin can be used for the enzymatic synthesis of various functional peptides. Appl Environ Microbiol 79:5023–5029

Arai T, Noguchi A, Takano E, Kino K (2013b) Application of protein N-terminal amidase in enzymatic synthesis of dipeptides containing acidic amino acids specifically at the N-terminus. J Biosci Bioeng 115:382–387

Borisova SA, Circello BT, Zhang JK, van der Donk WA, Metcalf WW (2010) Biosynthesis of rhizocticins, antifungal phosphonate oligopeptides produced by *Bacillus subtilis* ATCC6633. Chem Biol 17:28–37

Di Bernardini R, Harnedy P, Bolton D, Kerry J, O'Neill E, Mullen AM, Hayes M (2011) Antioxidant and antimicrobial peptidic hydrolysates from muscle protein sources and by-products. Food Chem 124:1296–1307

Fan C, Moews PC, Shi Y, Walsh CT, Knox JR (1995) A common fold for peptide synthetases cleaving ATP to ADP: glutathione synthetase and D-alanine:D-alanine ligase of *Escherichia coli*. Proc Natl Acad Sci USA 92:1172–1176

Finking R, Marahiel MA (2004) Biosynthesis of nonribosomal peptides. Annu Rev Microbiol 58:453–487

Furukawa-Hibi Y, Nitta A, Ikeda T, Morishita K, Liu W, Ibi D, Alkam T, Nabeshima T, Yamada K (2011) The hydrophobic dipeptide Leu-Ile inhibits immobility induced by repeated forced swimming via the induction of BDNF. Behav Brain Res 220:271–280

Galperin MY, Koonin EV (1997) A diverse superfamily of enzymes with ATP-dependent carboxylate-amine/thiol ligase activity. Protein Sci 6:2639–2643

Hernández JG, Juaristi E (2011) Asymmetric aldol reaction organocatalyzed by (*S*)-proline-containing dipeptides: improved stereoinduction under solvent-free conditions. J Org Chem 76:1464–1467

Hernández-Ledesma B, del Mar CM, Recio I (2011) Antihypertensive peptides: production, bioavailability and incorporation into foods. Adv Colloid Interface Sci 165:23–35

Inaoka T, Takahashi K, Ohnishi-Kameyama M, Yoshida M, Ochi K (2003) J Biol Chem 278:2169–2176

Kagebayashi T, Kontani N, Yamada Y, Mizushige T, Arai T, Kino K, Ohinata K (2012) Novel CCK-dependent vasorelaxing dipeptide, Arg-Phe, decreases blood pressure and food intake in rodents. Mol Nutr Food Res 56:1456–1463

Kino K, Nakazawa Y, Yagasaki M (2008a) Dipeptide synthesis by L-amino acid ligase from *Ralstonia solanacearum*. Biochem Biophys Res Commun 371:536–540

Kino K, Noguchi A, Nakazawa Y, Yagasaki M (2008b) A novel L-amino acid ligase from *Bacillus licheniformis*. J Biosci Bioeng 106:313–315

Kino K, Kotanaka Y, Arai T, Yagasaki M (2009) A novel L-amino acid ligase from *Bacillus subtilis* NBRC3134, a microorganism producing peptide-antibiotic rhizocticin. Biosci Biotechnol Biochem 73:901–907

Kino K, Noguchi A, Arai T, Yagasaki M (2010a) Identification and characterization of a novel L-amino acid ligase from *Photorhabdus luminescens* subsp. *laumondii* TT01. J Biosci Bioeng 110:39–41

Kino K, Arai T, Tateiwa D (2010b) A novel L-amino acid ligase from *Bacillus subtilis* NBRC3134 catalyzed oligopeptide synthesis. Biosci Biotechnol Biochem 74:129–134

Kino K, Arai T, Arimura Y (2011) Poly-α-glutamic acid synthesis using a novel catalytic activity of RimK from *Escherichia coli* K-12. Appl Environ Microbiol 77:2019–2025

Kugler M, Loeffler W, Rapp C, Kern A, Jung G (1990) Rhizocticin A, an antifungal phosphono-oligopeptide of *Bacillus subtilis* ATCC 6633: biological properties. Arch Microbiol 153:276–281

Mitchell RE (1976) Isolation and structure of a chlorosis-inducing toxin of *Pseudomonas phaseolicola*. Phytochemistry 15:1941–1947

Sgadari C, Bacigalupo I, Barillari G, Ensoli B (2011) Pharmacological management of Kaposi's sarcoma. Expert Opin Pharmacother 12:1669–1690

Stewart WW (1971) Isolation and proof of structure of wildfire toxin. Nature (Lond) 229:174–178

Tabata K, Hashimoto S (2007) Fermentative production of L-alanyl-L-glutamine by a metabolically engineered *Escherichia coli* strain expressing L-amino acid alpha-ligase. Appl Environ Microbiol 73:6378–6385

Tabata K, Ikeda H, Hashimoto S (2005) *ywf*E in *Bacillus subtilis* codes for a novel enzyme, L-amino acid ligase. J Bacteriol 187:5195–5202

Tsuneyoshi Y, Yamane H, Tomonaga S, Morishita K, Denbow DM, Furuse M (2008) Reverse structure of carnosine-induced sedative and hypnotic effects in the chick under acute stress. Life Sci 82:1065–1069

Weber AL, Pizzarello S (2006) The peptide-catalyzed stereospecific synthesis of tetroses: a possible model for prebiotic molecular evolution. Proc Natl Acad Sci USA 103:12713–12717

Part VI
Functional Foods

Chapter 18
Microbial Production of Functional Polyunsaturated Fatty Acids and Their Derivatives

Jun Ogawa, Eiji Sakuradani, Shigenobu Kishino, Akinori Ando, Kenzo Yokozeki, and Sakayu Shimizu

Abstract Microorganisms are promising as producers of various nutritionally functional lipids such as polyunsaturated fatty acids (PUFAs). They further transform PUFAs into unique molecular species beyond common PUFAs. This chapter describes PUFA production through chemical mutant and molecular breeding of an oleaginous filamentous fungus *Mortierella alpina* 1S-4 and PUFA transformation by anaerobic bacteria existing in the gastrointestinal tract such as lactic acid bacteria. *M. alpina* 1S-4 and its mutants and transformants have led to the production of

J. Ogawa (✉) • E. Sakuradani
Division of Applied Life Sciences, Graduate School of Agriculture, Kyoto University, Kitashirakawa-oiwakecho, Sakyo-ku, Kyoto 606-8502, Japan
e-mail: ogawa@kais.kyoto-u.ac.jp

S. Kishino
Division of Applied Life Sciences, Graduate School of Agriculture, Kyoto University, Kitashirakawa-oiwakecho, Sakyo-ku, Kyoto 606-8502, Japan

Industrial Microbiology, Graduate School of Agriculture, Kyoto University, Kitashirakawa-oiwakecho, Sakyo-ku, Kyoto 606-8502, Japan

A. Ando
Division of Applied Life Sciences, Graduate School of Agriculture, Kyoto University, Kitashirakawa-oiwakecho, Sakyo-ku, Kyoto 606-8502, Japan

Research Unit for Physiological Chemistry, Kyoto University, Kitashirakawa-oiwakecho, Sakyo-ku, Kyoto 606-8502, Japan

K. Yokozeki
Industrial Microbiology, Graduate School of Agriculture, Kyoto University, Kitashirakawa-oiwakecho, Sakyo-ku, Kyoto 606-8502, Japan

S. Shimizu
Division of Applied Life Sciences, Graduate School of Agriculture, Kyoto University, Kitashirakawa-oiwakecho, Sakyo-ku, Kyoto 606-8502, Japan

Faculty of Bio-environmental Science, Kyoto Gakuen University, 1-1 Nanjyo, Sogabe, Kameoka, Kyoto 621-8555, Japan

H. Anazawa and S. Shimizu (eds.), *Microbial Production: From Genome Design to Cell Engineering*, DOI 10.1007/978-4-431-54607-8_18, © Springer Japan 2014

oils containing not only common n-6 and n-3 PUFAs but also rare PUFAs such as n-9 n-7, n-4, n-1, and nonmethylene interrupted PUFAs. Unique PUFA-transforming activities were found in anaerobic bacteria. They isomerized PUFAs, such as linoleic acid, arachidonic acid, and EPA to corresponding conjugated fatty acids and further transformed to partially saturated fatty acids with hydroxy fatty acids as intermediates through the so-called biohydrogenation pathway. The functions of these unique PUFAs have attracted much attention for improving our health and for developing new materials for functional foods and chemicals.

Keywords Arachidonic acid • Conjugated linoleic acid • *Lactobacillus* • *Mortierella* • Polyunsaturated fatty acid

Abbreviations

AA	Arachidonic acid (20:4n-6)
ALA	α-Linolenic acid (18:3n-3)
CLA	Conjugated linoleic acid
CLA1	9(*Z*),11(*E*)-octadecadienoic acid
CLA2	9(*E*),11(*E*)-octadecadienoic acid
DHA	4(*Z*),7(*Z*),10(*Z*),13(*Z*),16(*Z*),19(*Z*)-docosahexaenoic acid (22:6n-3)
DGLA	Dihomo-γ-linolenic acid (20:3n-6)
EPA	5(*Z*),8(*Z*),11(*Z*),14(*Z*),17(*Z*)-eicosapentaenoic acid (20:5n-3)
GLA	γ-Linolenic acid (18:3n-6)
MA	Mead acid (20:3n-9)
PUFA	Polyunsaturated fatty acids

18.1 General Introduction

Many attempts have been made to produce useful fatty acids or oils containing these acids by means of microbial processes. Because natural sources rich in polyunsaturated fatty acids (PUFAs) are limited to a few seed oils, which contain γ-linolenic acid (18:3n-6, GLA) and fish oils, which contain 5(*Z*),8(*Z*),11(*Z*),14(*Z*),17(*Z*)-eicosapentaenoic acid (20:5n-3, EPA) and 4(*Z*),7(*Z*),10(*Z*),13(*Z*),16(*Z*),19(*Z*)-docosahexaenoic acid (22:6n-3, DHA), recent investigations have focused on microorganisms as alternative sources of oils containing these PUFAs. The term "single cell oils" is used for unique oils produced by microorganisms (Cohen and Ratledge 2010). One of the most successful examples is the production of oils containing arachidonic acid (20:4n-6, AA) by an oleaginous filamentous fungus, *Mortierella alpina*.

Several *Mortierella* fungi have been found in Japan as potent producers of triacylglycerols containing AA, dihomo-γ-linolenic acid (20:3n-6, DGLA), and/or EPA (Yamada et al. 1987; Shinmen et al. 1989). A soil isolate strain, *M. alpina* 1S-4, produced AA on cultivation in a medium containing glucose and yeast extract. It also produced EPA when grown at low temperature (20 °C) or in a medium supplemented with α-linolenic acid (18:3n-3, ALA). When grown in the presence of sesame seed oil, the same fungus accumulated DGLA in place of AA because of the specific inhibition of Δ5-desaturase by sesamin and related lignan compounds present in the oil. Mutants that were considered to be defective (or to exhibit low activity) in Δ5, Δ6, Δ12, Δ9, and ω3-desaturase and others such as elongase activities were derived from *M. alpina* 1S-4 (Jareonkitmongkol et al. 1992a; Certik et al. 1998; Sakuradani et al. 2009; Sakuradani and Shimizu 2009). These mutants have led to the production of oils containing not only common n-6 and n-3 PUFAs but also rare n-3, n-7, n-4, n-1, and non-methylene-interrupted PUFAs.

This progress in PUFA production by microorganisms resulted in development of functional foods containing PUFAs and opened questions about how these PUFAs are metabolized by microorganisms in our gastrointestinal tract. The PUFA metabolism by anaerobic bacteria in mammalian gastrointestinal microorganisms were investigated and found to show unique PUFA-transforming activities (Ogawa et al. 2001, 2005; Kishino et al. 2009a). *Lactobacillus plantarum* isomerized linoleic acid to conjugated linoleic acid (CLA) effectively with hydroxyl fatty acids as intermediates through part reactions in biohydrogenation. *Clostridium bifermentans* saturated C20 PUFAs of AA and EPA into corresponding partially saturated fatty acids with conjugated isomers of C20 PUFAs as intermediates.

These PUFAs produced and transformed by microorganisms have a unique molecular structure beyond common PUFAs, and their functions have attracted much attention for improving our health and for developing materials for functional foods and chemicals.

18.2 Production of C20 PUFAs by *M. alpina* 1S-4

Mortierella alpina 1S-4 has the unique ability to synthesize a wide range of fatty acids, and it has several advantages not only as an industrial strain but also as a model for lipogenesis studies. The main biosynthetic pathways for PUFAs in *M. alpina* 1S-4 are shown in Fig. 18.1a. Chemical and molecular breeding of the strain improved the production of common PUFAs and further enabled the production of unique rare PUFAs (Sakuradani et al. 2009; Sakuradani and Shimizu 2009) (Fig. 18.1b, c).

18.2.1 n-6 PUFAs

The main product of *M. alpina* 1S-4, AA, is synthesized through the n-6 pathway, which involves Δ12- and Δ6-desaturases, elongase (EL2), and Δ5-desaturase (Fig. 18.1a). Depending on the conditions, the total amount of AA varies between 3 and 20 g/l (30–70 % of total cellular fatty acids), with 70–90 % of the AA produced being present in triacylglycerols (Higashiyama et al. 1998, 2002).

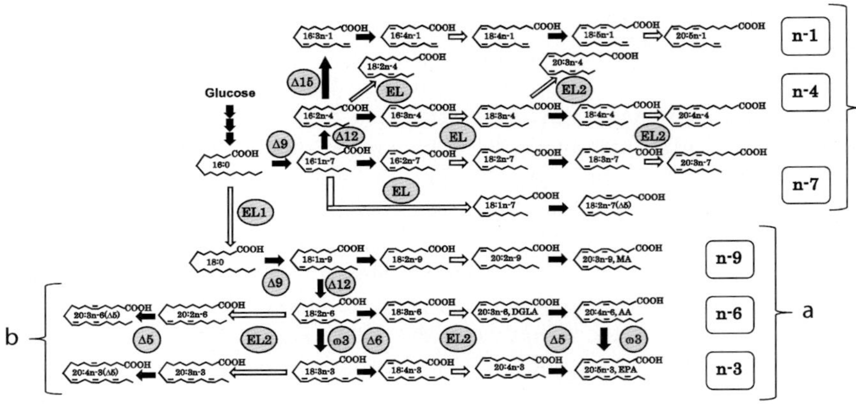

Fig. 18.1 Pathways for the biosynthesis of polyunsaturated fatty acids (PUFAs) in *Mortierella alpina* 1S-4 and its mutants. *AA* arachidonic acid, *ΔN* ΔN desaturase, *DGLA* dihomo-γ-linolenic acid, *EL*, *EL1*, and *EL2* fatty acid elongase isozymes, *EPA* eicosapentaenoic acid, *MA* Mead acid, *ω3* ω3 desaturase. *Bold* and *open arrows* represent desaturation and elongation reactions, respectively

On addition of 3 % sesame seed oil to a glucose yeast extract medium, *M. alpina* 1S-4 accumulated DGLA, a precursor of AA biosynthesis. The production of DGLA by *M. alpina* 1S-4 reached 1.7 g/l, whereas the production of AA amounted to only 0.7 g/l. The effective factors responsible for this phenomenon, Δ5-desaturase inhibitors, were isolated from a sesame seed extract or sesame oil and identified as lignan compounds, that is, (+)-sesamin, (+)-episesamin, (+)-sesaminol, and (+)-episesaminol (Shimizu et al. 1991). Several chemical mutants defective in Δ5-desaturation, which is characterized by a high DGLA level and a reduced AA level, were obtained (Jareonkitmongkol et al. 1992b, 1993a). One of the mutants, mutant S14, produced 2.4 g/l DGLA in the culture broth (43.3 % of total fatty acids; AA content, <1 %) on growth at 28 °C for 7 days in a 5-l jar fermentor (Jareonkitmongkol et al. 1993a).

18.2.2 n-3 PUFAs

Mortierella alpina 1S-4 produces EPA (approximately 10 % of total fatty acids) below the growth temperature of 20 °C through the n-3 pathway and direct ω3-desaturation of AA (Fig. 18.1a). The strain produced levels of EPA of more than 0.3 g/l at 12 °C. The strain exhibits higher EPA production on the addition of α-linolenic acid (18:3n-3)-containing oils, such as linseed oil, to the medium (1.88 g/l culture broth; 66.6 mg/g dry mycelia, from 3 % linseed oil at 12 °C) (Shimizu et al. 1989).

By using a Δ12-desaturase-defective mutant, Mut48, derived from *M. alpina* 1S-4, an EPA-rich oil with a low level of AA was obtained through converting exogenous ALA (1 g/l culture broth; 64 mg/g dry mycelia), accounting for 20 % of the total mycelial fatty acids at 20 °C for 10 days) (Jareonkitmongkol et al. 1993b).

Overexpression of the ω3-desaturase gene in the wild strain and S14 (Δ5-desaturation-defective mutant) led to higher production of EPA (0.8 g/l, 30 %) and 20:4n-3 (1.8 g/l, 35 %), which usually comprise about 10 % of total fatty acids in the wild strain and S14 cultivated at low temperature (<20 °C), respectively (Ando et al. 2009).

18.2.3 n-9 PUFAs

Δ12-Desaturase-defective mutants accumulate high levels of n-9 PUFAs, such as Mead acid (20:3n-9, MA), through the n-9 pathway (Jareonkitmongkol et al. 1992c) (Fig. 18.1a). One of these mutants, JT-180, yields a large amount of MA (2.6 g/l, 49 % in oil) on commercial production as a result of its enhanced Δ5- and Δ6-desaturase activities, not including n-6 and n-3 PUFAs (Sakuradani et al. 2002). Double mutants defective in both Δ12- and Δ5-desaturase activities accumulate eicosadienoic acid (20:2n-9) as a final product of n-9 PUFAs in large quantities (Kamada et al. 1999).

Gene disruption in *M. alpina* 1S-4 was performed by a RNAi method with double-stranded RNA (Takeno et al. 2005). The Δ12-desaturase gene-silenced strains accumulated octadecadienoic acid (18:2n-9), 20:2n-9, and MA-like Δ12-desaturation defective mutants.

18.2.4 Non-Methylene-Interrupted PUFAs

Δ6-Desaturase-defective mutants accumulate linoleic acid (18:2n-6) as the main fatty acid in the mycelial oil (up to 32 %) (Jareonkitmongkol et al. 1993c). These mutants are characterized by the accumulation of eicosadienoic acid (20:2n-6) and non-methylene-interrupted n-6 eicosatrienoic acid (20:3n-6Δ5) synthesized from linoleic acid through two subsequent reactions without Δ6 desaturation, that is, elongation and Δ5-desaturation (Fig. 18.1b). 20:4n-3Δ5 was detected when the mutant was grown at a temperature lower than 24 °C or in a culture medium supplemented with either ALA or 20:3n-3 (Fig. 18.1b).

18.2.5 n-7, n-4, and n-1 PUFAs

The fatty acid profile of elongase (EL1 for the conversion of palmitic acid, 16:0 to 18:0)-defective mutants, is characterized by high levels of 16:0 and palmitoleic acid (16:1n-7), with small amounts of various kinds of n-7 and n-4 PUFAs (shown in Fig. 18.1c), which are not detected in the wild strain (Sakuradani 2010; Sakuradani et al. 2010a, b). The total content of these PUFAs in the oil reaches about 30 %.

These fatty acids are thought to be derived from the 16:1n-7 and accumulated through the n-7 and n-4 pathways via Δ6-desaturation of 16:1n-7 to 16:2n-7 and Δ12-desaturation of 16:1n-7 to 16:2n-4, respectively. Therefore, 16:1n-7 corresponds to 18:1n-9, and the n-7 and n-4 pathways to the n-9 and n-6 pathways, respectively, in the wild strain (Fig. 18.1c). In a similar manner, n-1 PUFAs can be produced through the n-1 pathways via Δ15 desaturation of 16:2n-4 to 16:3n-1 catalyzed by ω3-desaturase (Fig. 18.1c). The same n-1 PUFAs were produced from n-1 hexadecenoic acid (16:1n-1) added to the medium. The transformant of EL1-defective mutant M1 obtained on RNAi of the Δ12-desaturase gene accumulated n-7 PUFAs and decreased n-4 PUFAs. This finding indicates that n-4 PUFAs are biosynthesized from n-7 PUFAs via Δ12-desaturation.

As just described, chemical and molecular breeding of *M. alpina* strains enabled the production of various PUFAs beyond common PUFAs and facilitated the elucidation of the PUFA biosynthetic pathway of the strain.

18.3 PUFA Transformation by Anaerobic Bacteria in the Gastrointestinal Tract

As described here, it has been possible to make microorganisms produce most of major PUFAs, and some of them, such as AA, have been used as functional food components. The next stage might be analysis of the PUFA metabolisms in our gastrointestinal tract by anaerobic microorganisms to evaluate the effects of metabolites on the health and apply the reactions found in these anaerobic bacteria to the transformation of these PUFAs into rare functional fatty acids (Ogawa et al. 2005; Kishino et al. 2009a, 2013).

18.3.1 Biohydrogenation of PUFAs by Anaerobic Bacteria

Fatty acid saturation metabolism, so-called biohydrogenation, is considered to be a detoxifying metabolism of anaerobic bacteria to transform toxic free PUFAs to less toxic free saturated fatty acids. Biohydrogenation was intensively studied in ruminant animals to create healthier food products with a low saturated and high n-3 fatty acid content and with potentially health-promoting CLA. One earlier study reported that *Butyrivibrio fibrisolvens* transformed linoleic acid to 9(*Z*),11(*E*)-octadecadienoic acid (CLA1) and further to *trans*-vaccenic acid (Kepler and Tove 1967). Recently, the detailed biohydrogenation pathway was revealed in *L. plantarum* AKU 1009a (Kishino et al. 2009b, 2013). The fatty acids recognized as the substrates by washed cells of *L. plantarum* AKU1009a had the common structure of a C18 fatty acid with a 9(*Z*),12(*Z*) diene system. The 9(*Z*),12(*Z*) diene system is converted to the 9(*Z*),11(*E*) and 9(*E*),11(*E*) diene systems and further saturated to the 10(*E*) monoene system by washed cells of *L. plantarum* AKU 1009a. These biohydrogenation pathways involve many potential reactions for fatty acid transformation, as described next.

18.3.2 Metabolic Basis of PUFA Biohydrogenation in Lactic Acid Bacteria

The biohydrogenation pathway of *L. plantarum* consists of multiple reactions (Kishino et al. 2011a, b, 2013). The first reaction of the biohydrogenation is hydration of the carbon–carbon double bond at the $\Delta 9$ position to generate 10-hydroxy fatty acid catalyzed by a hydratase, CLA-HY. The second reaction is dehydrogenation of the hydroxy group at C10 to generate 10-oxo fatty acid catalyzed by a dehydrogenase, CLA-DH. The third reaction is isomerization of the carbon–carbon double bond at $\Delta 12$ to generate the conjugated enone structure, 10-oxo-*trans*-11-fatty acid, catalyzed by a isomerase, CLA-DC. The fourth reaction is hydrogenation of the carbon–carbon double bond at the $\Delta 11$ position to generate the carbon–carbon single bond catalyzed by a enone reductase, CLA-ER. The fifth reaction is hydrogenation of the oxo group at $\Delta 10$ to generate 10-hydroxy fatty acid catalyzed by CLA-DH. The last reaction is dehydration of a hydroxy group at $\Delta 10$ to generate *cis*-9 and *trans*-10 monoenoic fatty acids catalyzed by CLA-HY. Through a branched pathway of this saturation metabolism, conjugated fatty acids are generated by the combined actions of three enzymes, CLA-HY, CLA-DH, and CLA-DC (Fig. 18.2a, c). The branched pathway starts from hydrogenation of the oxo group at $\Delta 10$ in 10-oxo-trans-11-fatty acid to generate 10-hydroxy-*trans*-11-fatty acid catalyzed by CLA-DH. The last reaction is dehydration of the hydroxy group at $\Delta 10$ in 10-hydroxy-*trans*-11-fatty acid to generate *cis*-9, *trans*-11 and *trans*-9, *trans*-11 conjugated fatty acids catalyzed by CLA-HY (Fig. 18.2a, b). C18 fatty acids with $\Delta 9$ and $\Delta 12$ diene system such as ALA, GLA, and stearidonic acid undergo the same transformations in *L. plantarum* AKU 1009a, indicating that the corresponding

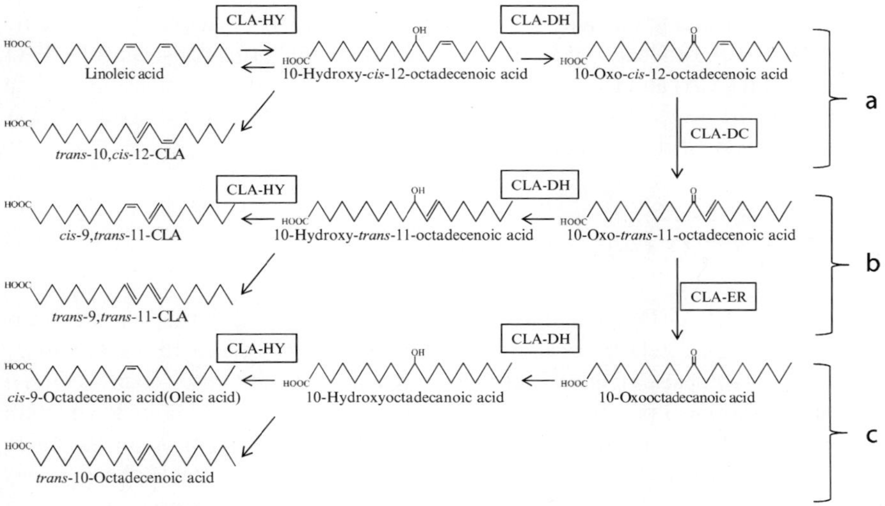

Fig. 18.2 Proposed biohydrogenation pathway of linoleic acid in *Lactobacillus plantarum*

intermediates, such as hydroxy, oxo, conjugated, and partially saturated fatty acids, are produced by the combined action of these enzymes (Kishino et al. 2009b).

18.3.3 Conjugated Fatty Acid Production

Conjugated fatty acids such as CLA have attracted much attention as biologically beneficial functional lipids. *L. plantarum* AKU 1009a was selected as a potential strain for CLA production from linoleic acid (Kishino et al. 2002a, 2003a). Washed cells of *L. plantarum* produced 40 mg/ml CLA (33 % molar yield) from 12 % (w/v) linoleic acid in 108 h. The resulting CLA, mainly in a free fatty acid form, comprised a mixture of CLA1 (38 % of total CLA) and CLA2 (62 % of total CLA), and accounted for 50 % of the total fatty acids obtained (Fig. 18.3a). In a similar manner, lactic acid bacteria transformed ALA and GLA into the corresponding conjugated trienoic acids (Kishino et al. 2003b, 2010): those produced from ALA were identified as 9(*Z*),11(*E*),15(*Z*)-octadecatrienoic acid and 9(*E*),11(*E*),15(*Z*)-octadecatrienoic acid, and those from GLA as 6(*Z*),9(*Z*),11(*E*)-octadecatrienoic acid and 6(*Z*),9(*E*),11(*E*)-octadecatrienoic acid (Fig. 18.3b). The conjugated fatty acid production process was further improved by using *Escherichia coli* transformant coexpressing the hydratase, dehydrogenase, and isomerase (Kishino et al. 2011b).

Based on the finding that hydroxy fatty acids are intermediates in CLA production, the transformation of hydroxy fatty acids by lactic acid bacteria was investigated. Lactic acid bacteria transformed ricinoleic acid (12-hydroxy-9(*Z*)-octadecenoic acid) into CLA (a mixture of CLA1 and CLA2) (Kishino et al. 2002b). *L. plantarum* JCM 1551 was selected as a potential catalyst for CLA production from ricinoleic acid (Ando et al. 2003). With the free acid form of ricinoleic acid as the substrate and washed cells of *L. plantarum* as the catalyst, 2.4 mg/ml CLA was produced from 3.4 mg/ml ricinoleic acid in 90 h, the molar yield as to ricinoleic acid being 71 %. The CLA produced, mainly in a free fatty acid form, consisted of CLA1 (21 % of total CLA) and CLA2 (79 % of total CLA). Ricinoleic acid is abundant in castor oil. In the presence of lipase, castor oil became an effective substrate for CLA production by lactic acid bacteria (Ando et al. 2004).

18.3.4 Hydroxy Fatty Acid Production

The hydratase CLA-HY, catalyzing the first step of fatty acid saturation, was overexpressed in *E. coli* and the *E. coli* transformant was used as the catalyst for hydroxy fatty acid synthesis. Under optimized reaction conditions, about 30 g/l 10-hydroxy-18:0 was produced from oleic acid with more than 90 % yield and with strict stereospecificity for the *S*-isomer. The ability of hydration of double bonds in fatty acid was further screened in lactic acid bacteria, and it was found that *L. brevis* produced a variety of hydroxyl fatty acids including a dihydroxy fatty acid (10-hydroxy, 13-hydroxy, and 10,13-dihydroxy fatty acids) from C18 fatty acids with a 9(Z),12(Z) nonconjugated diene system (Takeuch et al. 2013).

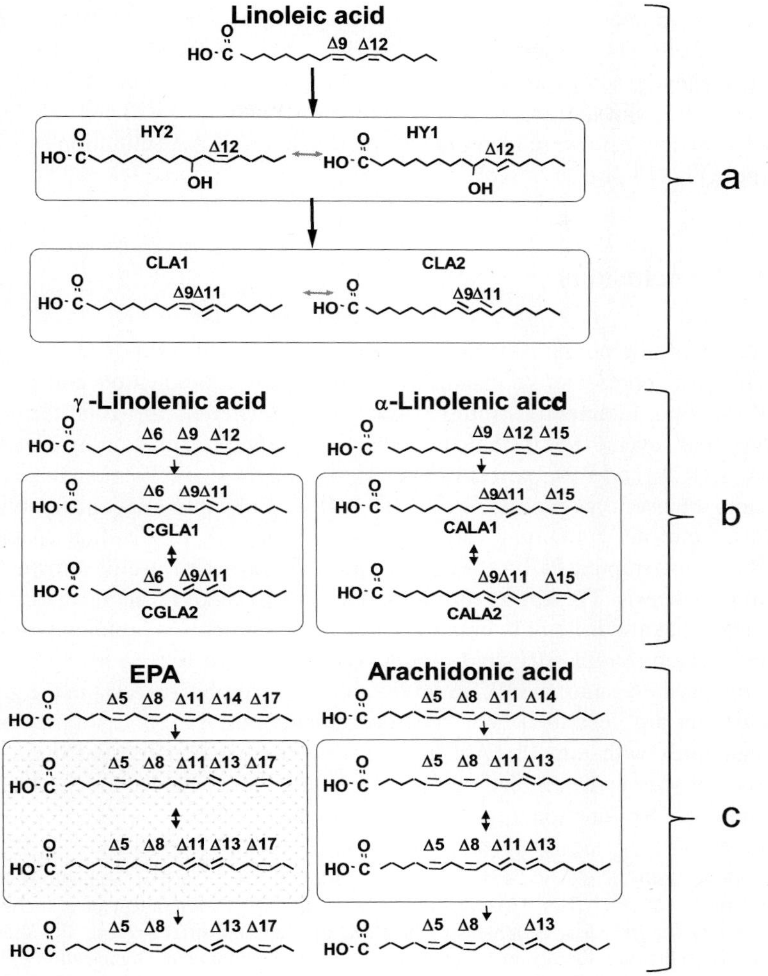

Fig. 18.3 Pathways for the conjugated fatty acids and partially saturated PUFA synthesis in anaerobic bacteria

18.3.5 Partially Saturated PUFA Production via Conjugated PUFA

An anaerobic bacterium, *Clostridium bifermentans,* saturated C20 PUFAs of AA and EPA into 5(*Z*),8(*Z*),13(*E*)-20:3 and 5(*Z*),8(*Z*),13(*E*),17(*Z*)-20:4, respectively (Fig. 18.3c) (Ogawa et al. 2005). Similar reactions were observed with C18 and C20 fatty acids with a ω-6, ω-9 nonconjugated (*Z*),(*Z*)-diene system such as ALA, GLA, and DGLA. The ω-6, ω-9 nonconjugated (*Z*),(*Z*)-diene systems were transformed into ω-7(*E*)-monoene systems. The conjugated isomers of EPA, 5(*Z*),8(*Z*),11(*Z*),13(*E*),17(*Z*)-eicosapentaenoic acid and 5(*Z*),8(*Z*),11(*E*),13(*E*),17(*Z*)-eicosapentaenoic acid, were identified as EPA saturation intermediates (Fig. 18.3c). Similarly, 5(*Z*),8(*Z*),11(*Z*),13(*E*)-eicosatetraenoic acid and 5(*Z*),8(*Z*),11(*E*),13(*E*)-eicosatetraenoic acid were identified as intermediates of AA saturation by *C. bifermentans* (Fig. 18.3c).

18.4 Conclusions

The ability of microorganisms to produce unique PUFAs is introduced in this chapter with two representative examples of fermentative production and enzymatic transformation. Filamentous fungi of the genus *Mortierella* are good fermentative producers of several useful single cell oils, especially those containing PUFAs such as AA, DGLA, and EPA. A soil isolate, *M. alpina* 1S-4, and its mutants and transformants are useful as producers of these PUFAs and exhibit expanded abilities to produce new oils containing rare PUFAs such as n-9, n-7, n-4, n-1, and non-methylene-interrupted PUFAs. Furthermore, they provided useful information on fungal lipogenesis. The ability of microorganisms to catalyze the formation of rare fatty acids that are difficult to obtain from plants or animals is promising. As examples of microbial rare fatty acid production, conjugated fatty acid, hydroxy fatty acid, and partially saturated PUFA production by anaerobic bacteria in the gastrointestinal tract are described here. Microorganisms may be good sources of these unusual lipids with rare PUFAs. Investigations of the functions of these unique PUFAs will open up a new methodology for improving our health and for developing new materials for functional foods and chemicals.

Acknowledgments This work was partially supported by the Industrial Technology Research Grant Program in 2007 (No. 07A08005) and the Project for Development of a Technological Infrastructure for Industrial Bioprocesses on R&D of New Industrial Science and Technology Frontiers from the New Energy and Industrial Technology Development Organization (NEDO) of Japan, Grants-in-Aid for Scientific Research (No. 19780056, No. 16688004, and No. 18208009), COE for Microbial-Process Development Pioneering Future Production Systems, and Advanced Low Carbon Technology Research and Development Program (ALCA) from the Ministry of Education, Culture, Sports, Science and Technology of Japan, and by the Bio-Oriented Technology Research Advancement Institution of Japan.

References

Ando A, Ogawa J, Kishino S et al (2003) CLA production from ricinoleic acid by lactic acid bacteria. J Am Oil Chem Soc 80:889–894

Ando A, Ogawa J, Kishino S et al (2004) Conjugated linoleic acid production from castor oil by *Lactobacillus plantarum* JCM 1551. Enzyme Microb Technol 35:40–45

Ando A, Sumida Y, Negoro H et al (2009) Establishment of *Agrobacterium tumefaciens*-mediated transformation of an oleaginous fungus, *Mortierella alpina* 1S-4, and its application for eicosapentaenoic acid producer breeding. Appl Environ Microbiol 75:5529–5535

Certik M, Sakuradani E, Shimizu S (1998) Desaturase-defective fungal mutants: useful tools for the regulation and overproduction of polyunsaturated fatty acids. Trends Biotechnol 16:500–505

Cohen Z, Ratledge C (eds) (2010) Single cell oils, microbial and algal oils, 2nd edn. AOCS, Urbana

Higashiyama K, Yaguchi T, Akimoto K et al (1998) Enhancement of arachidonic acid production by *Mortierella alpina*. J Am Oil Chem Soc 75:1501–1505

Higashiyama K, Fujikawa S, Park E et al (2002) Production of arachidonic acid by *Mortierella* fungi. Biotechnol Bioprocess Eng 7:252–262

Jareonkitmongkol S, Kawashima H, Shimizu S et al (1992a) Production of 5,8,11-*cis*-eicosatrienoic acid by a Δ12-desaturase-defective mutant of *Mortierella alpina* 1S-4. J Am Oil Chem Soc 69:939–944

Jareonkitmongkol S, Kawashima H, Shirasaka N et al (1992b) Production of dihomo-γ-linolenic acid by a Δ5-desaturase-defective mutant of *Mortierella alpina* 1S-4. Appl Environ Microbiol 58:2196–2200

Jareonkitmongkol S, Shimizu S, Yamada H (1992c) Fatty acid desaturation-defective mutants of an arachidonic-acid-producing fungus, *Mortierella alpina* 1S-4. J Gen Microbiol 138:997–1002

Jareonkitmongkol S, Sakuradani E, Shimizu S (1993a) A novel Δ5-desaturase-defective mutant of *Mortierella alpina* 1S-4 and its dihomo-γ-linolenic acid productivity. Appl Environ Microbiol 59:4300–4304

Jareonkitmongkol S, Shimizu S, Yamada H (1993b) Occurrence of two nonmethylene-interrupted fatty acids in a Δ6-desaturase-defective mutant of the fungus *Mortierella alpina* 1S-4. Biochim Biophys Acta 1167:137–141

Jareonkitmongkol S, Shimizu S, Yamada H (1993c) Production of an eicosapentaenoic acid-containing oil by a Δ12 desaturase-defective mutant of *Mortierella alpina* 1S-4. J Am Oil Chem Soc 70:119–123

Kamada N, Kawashima H, Sakuradani E et al (1999) Production of 8,11-*cis*-eicosadienoic acid by a Δ5 and Δ12 desaturase-defective mutant derived from the arachidonic acid producing fungus *Mortierella alpina* 1S-4. J Am Oil Chem Soc 76:1269–1274

Kepler CR, Tove SB (1967) Biohydrogenation of unsaturated fatty acids. III. Purification and properties of a linoleate Δ12-*cis*, Δ11-*trans*-isomerase from *Butyrivibrio fibrisolvens*. J Biol Chem 242:5686–5692

Kishino S, Ogawa J, Ando A et al (2002a) Ricinoleic acid and castor oil as substrates for conjugated linoleic acid production by washed cells of *Lactobacillus plantarum*. Biosci Biotechnol Biochem 66:2283–2286

Kishino S, Ogawa J, Omura Y et al (2002b) Conjugated linoleic acid production from linoleic acid by lactic acid bacteria. J Am Oil Chem Soc 79:159–163

Kishino S, Ogawa J, Ando A et al (2003a) Structural analysis of conjugated linoleic acid produced by *Lactobacillus plantarum*, and factors affecting isomer production. Biosci Biotechnol Biochem 67:179–182

Kishino S, Ogawa J, Ando A et al (2003b) Conjugated α-linolenic acid production from α-linolenic acid by *Lactobacillus plantarum* AKU1009a. Eur J Lipid Sci Technol 105:572–577

Kishino S, Ogawa J, Yokozeki K (2009a) Metabolic diversity in biohydrogenation of polyunsaturated fatty acids by lactic acid bacteria involving conjugated fatty acid production. Appl Microbiol Biotechnol 84:87–97

Kishino S, Ogawa J, Yokozeki K et al (2009b) Microbial production of conjugated fatty acids. Lipid Technol 21:177–181

Kishino S, Ogawa J, Ando A et al (2010) Microbial production of conjugated γ-linolenic acid from γ-linolenic acid by *Lactobacillus plantarum* AKU 1009a. J Appl Microbiol 108: 2012–2018

Kishino S, Ogawa J, Yokozeki K et al (2011a) Linoleic acid isomerase in *Lactobacillus plantarum* AKU1009a proved to be a multi-component enzyme system requiring oxidoreduction cofactors. Biosci Biotechnol Biochem 75:318–322

Kishino S, Park SB, Takeuchi M et al (2011b) Novel multi-component enzyme machinery in lactic acid bacteria catalyzing C=C double bond migration useful for conjugated fatty acid synthesis. Biochem Biophys Res Commun 416:188–193

Kishino S, Takeuchi M, Park SB et al (2013) Polyunsaturated fatty acid saturation by gut lactic acid bacteria affecting host lipid composition. Proc Natl Acad Sci USA 110:17808–17813

Ogawa J, Matsumura K, Kishino S et al (2001) Conjugated linoleic acid (CLA) accumulation via 10-hydroxy-12-octadecaenoic acid during microaerobic transformation of linoleic acid by *Lactobacillus acidophilus*. Appl Environ Microbiol 67:1246–1252

Ogawa J, Kishino S, Ando A et al (2005) Production of conjugated fatty acids by lactic acid bacteria. J Biosci Bioeng 100:355–364

Sakuradani E (2010) Advances in the production of various polyunsaturated fatty acids through oleaginous fungus *Mortierella alpina* breeding. Biosci Biotechnol Biochem 74:908–917

Sakuradani E, Shimizu S (2009) Single cell oil production by *Mortierella alpina*. J Biotechnol 144:31–36

Sakuradani E, Kamada N, Hirano Y et al (2002) Production of 5,8,11-*cis*-eicosatrienoic acid by a Δ5 and Δ6 desaturation activity-enhanced mutant derived from a Δ12 desaturation activity-defective mutant of *Mortierella alpina* 1S-4. Appl Microbiol Biotechnol 60:281–287

Sakuradani E, Ando A, Ogawa J et al (2009) Improved production of various polyunsaturated fatty acids through filamentous fungus *Mortierella alpina* breeding. Appl Microbiol Biotechnol 84:1–10

Sakuradani E, Ando A, Ogawa J et al (2010a) Arachidonic acid-producing *Mortierella alpina*: creation of mutants, isolation of the related enzyme genes, and molecular breeding. In: Cohen Z, Ratledge C (eds) Single cell oils, microbial and algal oils, 2nd edn. AOCS, Urbana, pp 29–49

Sakuradani E, Ogawa J, Kishino S et al (2010b) Oils, microbial production. In: Flickinger MC (ed) Encyclopedia of industrial biotechnology: bioprocess, bioseparation, and cell technology. Wiley, New York, pp 3693–3706

Shimizu S, Kawashima H, Akimoto K et al (1989) Conversion of linseed oil to an eicosapentaenoic acid-containing oil by *Mortierella alpina* 1S-4 at low temperature. Appl Microbiol Biotechnol 32:1–4

Shimizu S, Akimoto K, Shinmen Y et al (1991) Sesamin is a potent and specific inhibitor of Δ5 desaturase in polyunsaturated fatty acid biosynthesis. Lipids 26:512–516

Shinmen Y, Shimizu S, Akimoto K et al (1989) Production of arachidonic acid by *Mortierella* fungi: selection of a potent producer and optimization of culture conditions for large-scale production. Appl Microbiol Biotechnol 31:11–16

Takeno S, Sakuradani E, Tomi A et al (2005) Improvement of the fatty acid composition of an oil-producing filamentous fungus, *Mortierella alpina* 1S-4, through RNA interference with Δ12-desaturase gene expression. Appl Environ Microbiol 71:5124–5128

Takeuch M, Kishino S, Tanabe K et al (2013) Hydroxy fatty acid production by *Pediococcus* sp. Eur J Lipid Sci Technol. doi:10.1002/ejlt.201200414

Yamada H, Shimizu S, Shinmen Y (1987) Production of arachidonic acid by *Mortierella elongata* 1S-5. Agric Biol Chem 51:785–790

Chapter 19
Enzymatic Production of Oligosaccharides

Takashi Kuroiwa

Abstract Enzymatic production methods of physiologically active oligosaccharides using bioreactors are summarized. First, types of enzyme bioreactors used for oligosaccharide production are introduced, and their features are discussed in relationship to reaction mechanisms of oligosaccharide formation and the basis of operational characteristics of each type of bioreactor. Then, an example of research on development of bioreactors for producing functional oligosaccharides that is conducted by the author's research group is demonstrated. Preparation of stable immobilized enzyme is a key point to develop a stable bioreactor system because generally enzymes are very unstable in free form. However, immobilization of enzymes often affects not only catalytic activity but also selectivity of produced oligosaccharides. Here, an index for evaluating the effect of operational conditions on the yield of target oligosaccharides is introduced. Physiologically active chitosan oligosaccharides could be produced continuously with high selectivity and good stability by choosing adequate bioreactors and by optimizing their operation conditions.

Keywords Bioreactor • Chitosan • Continuous production • Immobilized enzyme • Oligosaccharide

19.1 Introduction

Oligosaccharides consist of two to ten residues of monosaccharides, which are bonded via glycoside linkages. Generally they are produced by hydrolysis of polysaccharides or by synthesis from mono- or disaccharides via condensation or

T. Kuroiwa (✉)
Faculty of Engineering, Tokyo City University,
1-28-1 Tamazutsumi, Setagaya-ku, Tokyo 158-8557, Japan
e-mail: tkuroiwa@tcu.ac.jp

H. Anazawa and S. Shimizu (eds.), *Microbial Production: From Genome Design to Cell Engineering*, DOI 10.1007/978-4-431-54607-8_19,

transglycosylation reactions. To date, various kinds of oligosaccharides have been found. Their physiological activities are dependent upon their composition of constituent sugars and degree of polymerization. For food application of oligosaccharides, it is important to produce them not only in large amounts at low cost but also as safe food materials. Thus, enzymatic methods for producing oligosaccharides have been widely applied because they can act under moderate conditions and have high specificity for various saccharine materials without use of harmful chemicals.

A reactor that is operated using biocatalysts including enzymes and microorganisms is called a "bioreactor." In this chapter, general features of oligosaccharide production using enzyme bioreactors are summarized and recent studies on bioreactors for producing oligosaccharides conducted by the author and coworkers are introduced.

19.2 Enzyme Bioreactors for Oligosaccharide Production

19.2.1 Type of Enzyme Bioreactors

Selection of the type of bioreactor is important for developing a production process using enzymatic reactions. In this section, effect of bioreactor type on enzymatic production of oligosaccharides is summarized. Typical enzyme bioreactors are shown schematically in Fig. 19.1.

19.2.1.1 Packed Bed Bioreactors

A packed bed reactor (PBR) consists of enzymes immobilized on insoluble supports such as resins and gels packed into a column-type reaction vessel. Reaction products can be obtained at the outlet of the column by feeding a substrate solution to the column (Fig. 19.1a). Enzymes are trapped in the reactor by immobilization and can be used continuously and repetitively. For PBRs, productivity per unit volume of reactor can be improved compared to that of stirred tank reactors. Furthermore, the loss of enzymatic activity from to shear stress during reaction operation is not significant. PBR cannot be utilized for heterogeneous substrate solutions containing solid compound. It is difficult to feed highly viscous solutions because of the large pressure drop during operation.

19.2.1.2 Stirred Tank Bioreactors

Stirred tank bioreactors (Fig. 19.1b) consist of a tank vessel with stirrers such as turbines and propellers. This type of reactor has a very simple structure and is adaptable for both small-scale batch reactions and large-scale continuous operation. The composition of reaction mixture is uniform throughout the reactor. The stirred tank reactor can be used for heterogeneous solutions including particle suspensions

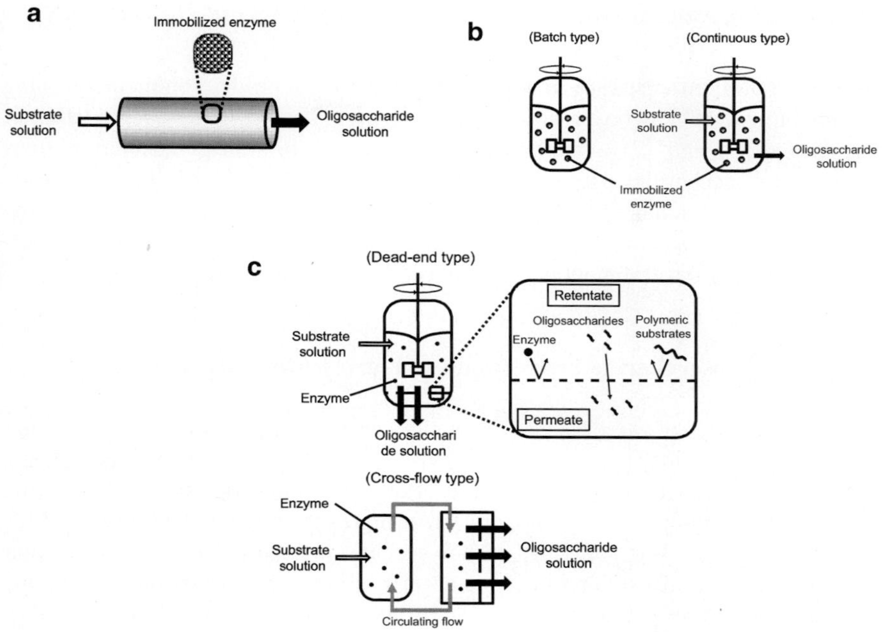

Fig. 19.1 Typical enzyme bioreactors: packed bed bioreactor (**a**), stirred tank bioreactor (**b**), and membrane bioreactor (**c**)

because the reactor contents are fully mixed constantly. The composition of outlet flow of the continuous stirred tank reactor (CSTR) is constant and the same as that in the reactor. The CSTR is an ideal-flow reactor that is opposite to the PBR from the aspect of liquid mixing in the reactor. Mechanical damage of biocatalysts caused by agitation of the reaction mixture often decrease the performance of the reactor.

19.2.1.3 Membrane Bioreactors

A membrane bioreactor (MBR) contains an enzyme and a membrane with the appropriate pore size and physicochemical properties, such as microfiltration (MF), ultrafiltration (UF), and nanofiltration (NF) membranes (Fig. 19.1c). The substrate is continuously fed into the reaction mixture while the products are continuously withdrawn in the permeate. Enzyme molecules are retained in the reaction mixture and can be used continuously without any physical or chemical immobilization treatment. The size of the product molecules in the permeate can be controlled by proper selection of the pore size of the membrane. MBRs are classified as a dead-end flow type or a cross-flow type by flow direction of the reaction mixture to the membrane. In the case of polymeric substrates or a particle suspension system, MBRs tend have decreased productivity as the result of membrane fouling.

19.2.2 *Oligosaccharide Production Using Enzyme Bioreactors*

Based on the oligosaccharide formation mechanism, the production methods using enzyme bioreactors can be classified into two types: (1) a method by degradation of polysaccharides and (2) a method by condensation or transglycosylation from mono- and disaccharides. Here, the characteristics of both types of oligosaccharide production are summarized. Choice of enzyme immobilization methods and the type of reactors based on the reaction mechanism and properties of used materials are important for development of efficient production processes.

19.2.2.1 Oligosaccharide Production by Hydrolysis of Polysaccharides

The characteristics of oligosaccharide production by hydrolysis of high molecular weight polysaccharides are as follows: (i) viscosity of substrate solution is high at a low concentration; (ii) reaction mixture is often a heterogeneous system containing solid or fibrous compounds; (iii) operation at high substrate concentration is difficult because of the low solubility of polysaccharides and high viscosity of their solutions; and (iv) diffusion problems and steric hindrance for immobilized enzyme systems are more serious for polymeric substrates. Production processes for malto-oligosaccharides from starches, pectate oligosaccharides from pectic acids, inulo-oligosaccharides from inulins, and chitosan oligosaccharides from chitosans are included in this type.

Figure 19.2 shows typical methods for immobilizing enzymes. Because enzyme molecules are immobilized inside solid particles and the substrate diffusion near enzymes is quite restricted, crosslinking and entrapment methods for enzyme immobilization are rarely applied for polysaccharide hydrolysis. Therefore, the support-binding method is frequently used for oligosaccharide production by hydrolysis of polysaccharides.

In the case of Mechaelis–Menten type enzyme reactions in which the reaction rate increases with the increase of substrate concentration, PBRs are more efficient and are often applied for oligosaccharide production because of their less axial mixing property. In addition, because the axial mixing in bioreactors causes decrease of the yield of oligosaccharides, intermediates of polysaccharide hydrolysis, PBRs, are more suitable for production of oligosaccharides with a higher degree of polymerization. However, PBRs cannot be applied for highly viscous or heterogeneous substrate solutions because of large pressure drop in the reactor or severe column clogging by solid components. In such cases, the CSTR-type reactors are suitable.

In enzymatic hydrolysis of polysaccharides, enzymes and polymeric substrates are polymeric compounds whereas the oligosaccharides produced are low molecular weight compounds. Therefore, by using MBRs with a UF membrane with adequate molecular weight cutoff, production of oligosaccharides and their separation from the enzymes and unreacted polysaccharides can be achieved at the same time. However, it is problematic that the productivity of oligosaccharides tends to decrease by membrane fouling. Use of viscous and heterogeneous solutions is limited in the MBR system, as well as in PBRs.

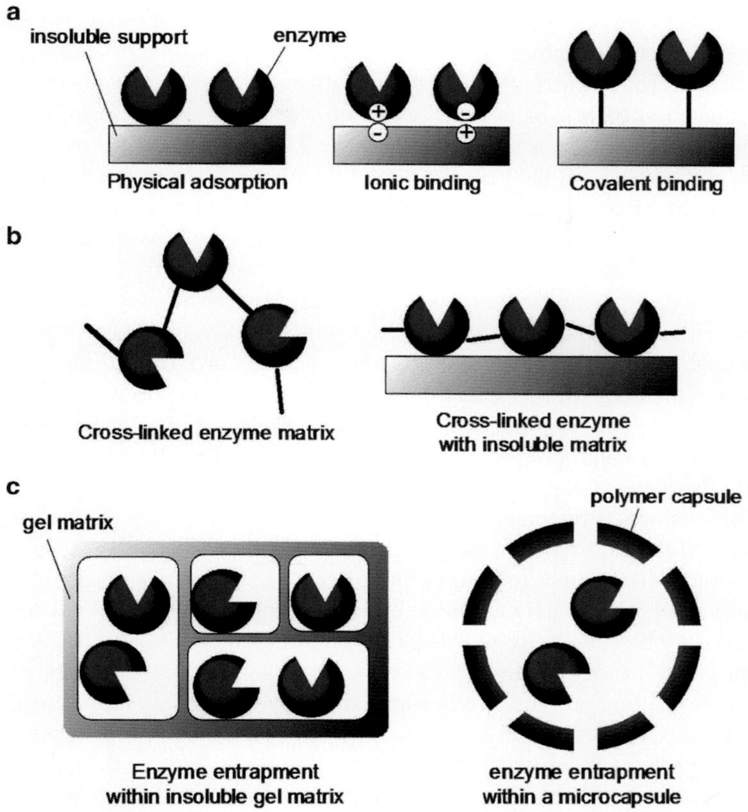

Fig. 19.2 Schematic illustration of typical methods for enzyme immobilization: support binding method (**a**), cross-linking method (**b**), and entrapping method (**c**)

When immobilized enzymes are used for oligosaccharide production by polysaccharide hydrolysis, physical factors such as diffusional limitations of substrates significantly affects the distribution of degree of polymerization of the produced oligosaccharides. Especially, when the target oligosaccharides are reaction intermediates, the yield of the target product is affected by mass transfer properties, and thus it is necessary to adequately design the operational conditions of bioreactors.

19.2.2.2 Oligosaccharide Production by Condensation and Transglycosylation

Many glycosyl hydrolases can also catalyze condensation reactions, the reverse reactions of hydrolysis. Also, some glycosyl hydrolases show strong transglycosylation activity. Therefore, oligosaccharide synthesis from small sugars such as mono- and disaccharides can be achieved by condensation and transglycosylation reactions using such enzymes. The features of these reactions are as follows: (1) the reactions are conducted at high substrate concentrations, and thus the effect of substrate or product inhibition becomes more significant; and (2) diffusion of not only substrates but also products affects the performance of the immobilized enzymes. Many kinds of oligosaccharides such as fructo-oligosaccharides, galacto-oligosaccharides, gluco-oligosaccharides, isomalto-oligosaccharides, and manno-oligosaccharides are synthesized enzymatically from corresponding small sugars.

As methods for enzyme immobilization, the support-binding method is often used as well as polysaccharide hydrolysis. A wider variety of support materials such as particles, sheets, films, and membranes are used for enzyme immobilization. Entrapment methods using polymer gels and crosslinking methods are also applied. In the case of synthetic production of oligosaccharides, lower operational limitations from substrate properties compared with the case of polysaccharide hydrolysis diversify the types of support materials and immobilization methods.

In the production methods using condensation or transglycosylation, the PBRs are frequently used. PBRs are more efficient, compared to other types of reactors, for reactions with product inhibition. In contrast, in reactions with significant substrate inhibition, the CSTRs are more efficient compared to PBRs. The MBRs are also used sometimes because of lower risk of membrane fouling by small molecule substrates than that of polymeric substrates.

In the production process of oligosaccharides by condensation or transglycosylation reactions, as well as polysaccharide hydrolysis, use of immobilized enzymes affects the produced oligosaccharide composition because of mass transfer limitation. Furthermore, mass transfer problems also affect the apparent stability of immobilized enzymes (Mozaffar et al. 1986).

19.3 Production of Physiologically Active Oligosaccharides from Marine Biomass Using Enzyme Bioreactors

Production of useful materials from renewable biomass resources has attracted much attention as a technology for achieving sustainable material circulation. One of the biomass resources expected to be utilized effectively are non-food polysaccharides such as celluloses, chitinous compounds, and pectins. Chitins and chitosans (Fig. 19.3) are obtained from marine biomass such as the shells of crabs and shrimps, and their extensive use would be desirable. Especially, chitosans, deacetylated chitins, have primary amino groups in their structure and possess various unique properties that are not found in other polysaccharides, and it is expected these substances can be used as a medical supply, as a functional food, and as a biologically active substance.

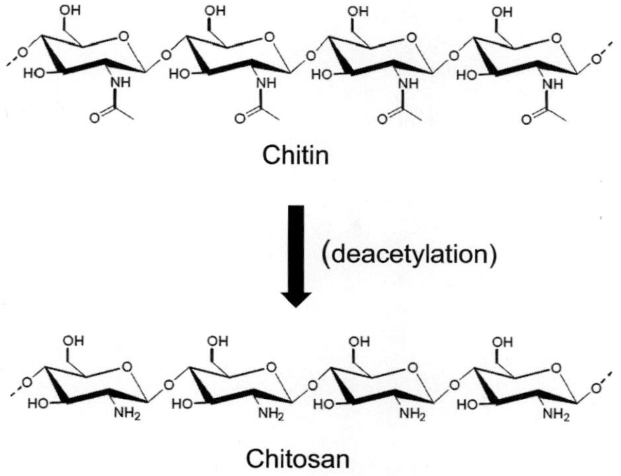

Fig. 19.3 Chemical structure of chitin and chitosan

Chitosan oligosaccharides, which are obtained by partial hydrolysis of chitosans, exhibit many beneficial physiological activities similar to those of chitosans. Thus, recently they have attracted attention as water-soluble functional materials. In this section, production methods of chitosan oligosaccharides using immobilized enzyme bioreactors developed by the author's research group are introduced.

19.3.1 Chitosan and Chitosan Oligosaccharides

Chitosan oligosaccharides consist of two to ten D-glucosamine units that are bound via β-1,4-glycoside linkages. They are not digestible by humans and exhibit not only anticariogenicy similar to other oligosaccharides but also various physiological activities such as antimicrobial, antitumor, and immune-enhancing activities (for review of their physiological activities, see Kuroiwa et al. 2005a; Kim and Rajapakse 2005; Xia et al. 2011). Their functional properties are greatly dependent upon their molecular weight; higher oligosaccharides (pentamers and larger oligosaccharides) are especially active. Therefore, to utilize chitosan oligosaccharides as functional food materials and pharmaceuticals, products containing a large fraction of higher oligosaccharides are desirable.

19.3.2 Production Method of Chitosan Oligosaccharides

Chitosan oligosaccharides are obtained from partial hydrolysis of chitosan. There are two typical methods for producing chitosan oligosaccharides: chemical hydrolysis using acids and enzymatic hydrolysis using chitosanolytic enzymes. In the chemical method, chitosans are hydrolyzed typically by hydrochloric acid of concentration

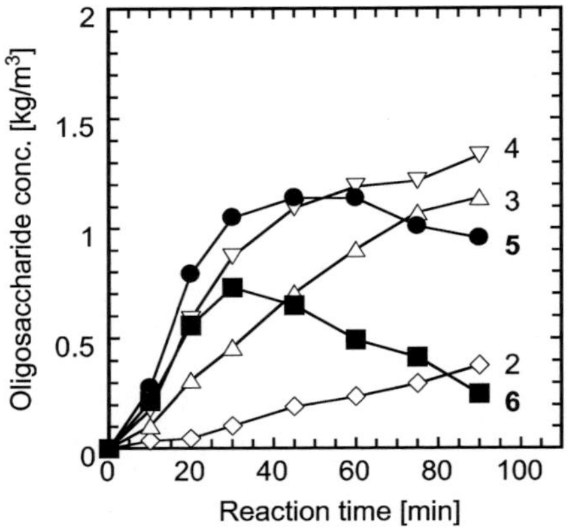

Fig. 19.4 Time-course of chitosan oligosaccharide concentrations in batch hydrolysis of chitosan by free chitosanase. Amount of the enzyme used was 63 U in 100 ml chitosan solution (5 kg/m^3), and reaction temperature was 35 °C. *Number with each symbol* represents the degree of polymerization of chitosan oligosaccharides (2 = dimers, 3 = trimers,…)

higher than 6 mol/l at 50–80 °C. This method is widely used for industrial processes of chitosan hydrolysis and has the advantage of low cost. However, the reaction selectivity is poor, and a large amount of monomeric D-glucosamine is produced. Therefore, the yield of higher oligosaccharides is quite low. In the enzymatic method, on the other hand, there is little formation of monomeric D-glucosamine and the yield of higher oligosaccharides can be improved. Furthermore, enzymatic reactions can proceed at near atmospheric pressure, moderate temperatures, and weakly acidic conditions. Figure 19.4 shows a typical time-course of batch hydrolysis of chitosan by the free chitosanase (from *Bacillus pumilus*). As shown in Fig. 19.4, oligosaccharides from dimers to hexamers were produced. The maximum yield of the target higher oligosaccharides (pentamers and hexamers) using this chitosanase was 40–50 % to the used chitosan. This yield was higher than the values reported for conventional acid hydrolysis (~15 %). However, the total concentration of pentamers and hexamers decreased after reaching the maximum value as a result of enzymatic hydrolysis on these intermediates. From this result, it was concluded that to produce higher chitosan oligosaccharides efficiently the reaction should be stopped when their total concentration reached the maximum value. To stop the reaction immediately, therefore, the enzymes should be used in an immobilized form because immobilized enzymes can be separated easily from the reaction mixture.

19.3.3 Production of Chitosan Oligosaccharides Using Enzyme Bioreactors

19.3.3.1 Immobilization and Stabilization of Chitosanase by the Multipoint Attachment Method

Based on these results, immobilization of the chitosanase was studied. By examination of various immobilization methods and support materials, it was found that the chitosanase could be immobilized stably by the multipoint attachment method using

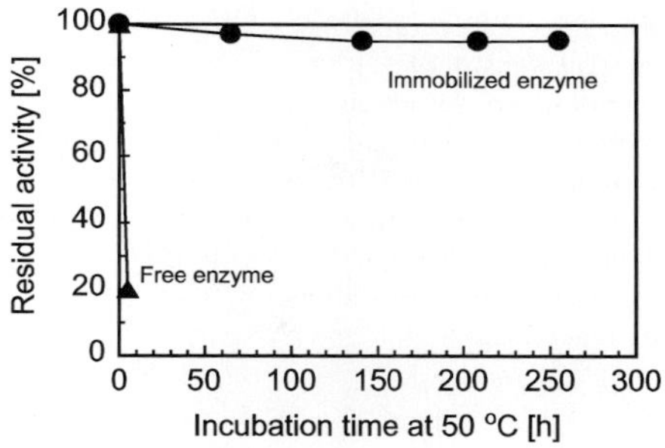

Fig. 19.5 Thermal stability of free and immobilized chitosanases. (Reproduced from Ichikawa et al. 2002, with permission from Elsevier)

chemically activated agar gel supports (Ichikawa et al. 2002). Figure 19.5 shows the comparison of stabilities of free and immobilized chitosanases. The immobilized enzyme was dramatically stabilized compared to free enzyme. In this immobilization method, the thermal stability of immobilized enzymes was significantly affected by the surface density of aldehyde groups, which formed covalent bonding between enzyme molecules and the support, on the activated agar gels used for immobilization. This result indicates formation of multipoint covalent bonding between enzyme molecules and the support material (Ichikawa et al. 2002; Kuroiwa et al. 2005b).

19.3.3.2 Continuous Production of Chitosan Oligosaccharides Using Bioreactors

The composition of chitosan oligosaccharides produced in batch and continuous chitosan hydrolysis reactions strongly depends on the experimental conditions during enzyme immobilization and hydrolysis reactions (Kuroiwa et al. 2002, 2003). As an example, the effect of surface enzyme density on the yield of pentamers and hexamers in continuous hydrolysis of chitosan using the PBR is shown in Fig. 19.6. When the surface enzyme density was 1.01×10^3 U/m^2, the maximum yield was similar to that of the free enzyme system. As the surface enzyme density increased, the maximum yield decreased. It is also found that the size of support particles, reaction temperature, substrate concentration, and flow rate of reaction mixture also affect the maximum yield of the target oligosaccharides even if the surface enzyme density was fixed (Kuroiwa et al. 2002, 2003). These factors are related, by the Damköhler number (*Da*), to the rates of reaction and mass transfer near the surface of the supports (Bailey and Ollis 1986):

$$Da = \frac{V_{\max}}{\left(k_{\mathrm{F}} C_0\right)} = \frac{(maximum\ reaction\ rate)}{(maximum\ mass\ transfer\ rate)},$$

where V_{max} is the maximum reaction rate per unit surface area [kg/(m^2 s)], k_F is the film mass transfer coefficient (m/s), and C_0 is the initial substrate concentration (kg/m^3). The *Da* values under various conditions were calculated and plotted for the maximum

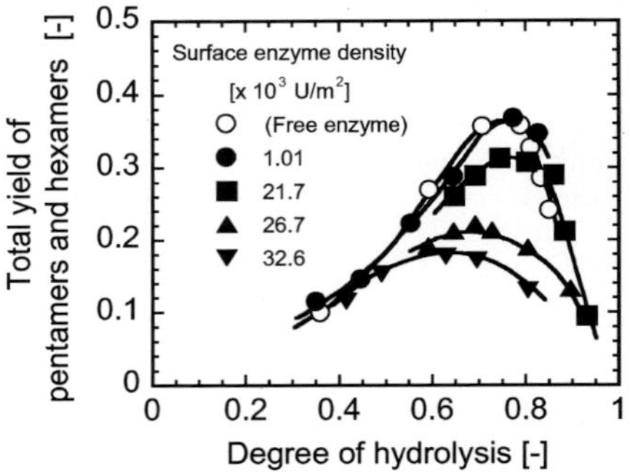

Fig. 19.6 Effect of surface enzyme density of immobilized chitosanase on yield of pentamers and hexamers based on initial substrate concentration in continuous hydrolysis of chitosan using a packed bed bioreactor

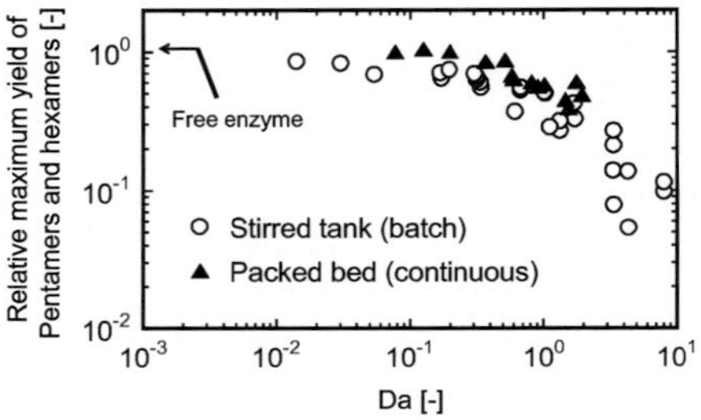

Fig. 19.7 Correlation of maximum yield of pentamers and hexamers of chitosan oligosaccharides with Damköhler number (*Da*). (Reproduced from Kuroiwa et al. 2003, with permission from Wiley)

yield of pentamers and hexamers normalized by their maximum yield in the free enzyme system (Fig. 19.7) (Kuroiwa et al. 2003). In Fig. 19.7, the results of batch experiments are also plotted for the *Da* values. A higher yield was obtained under the condition that gave a smaller *Da*. The relationships between *Da* and the maximum yield in batch and continuous experiments were similar in the same range of *Da*, indicating *Da* is a useful index to obtain a high yield of pentamers and hexamers and that an optimum condition for the packed bed reactor operation can be predicted from the data of preliminary batch experiments. The continuous production of pentamers and hexamers of chitosan oligosaccharides was carried out using a packed bed reactor under the optimum condition determined based on *Da*. Under the optimal condition, the continuous production of the target oligosaccharides was achieved stably for a month without significant decrease in the yield of pentamers and hexamers (>35 %). The estimated half-life of the reactor was about 200 days, determined from the change in reducing sugar concentration at the outlet of the reactor.

In addition, a MBR equipped with a UF membrane and immobilized chitosanase was developed to achieve selective and stable production of the target pentamers and hexamers (Kuroiwa et al. 2009a). Under the optimized conditions, pentamers and hexamers of chitosan oligosaccharides were steadily produced at the yield of 46 % for a month (Fig. 19.8). Besides these reactors, a stirred tank reactor equipped with multidisk impellers directly immobilizing chitosanases, which was applicable for a highly concentrated substrate solution having high viscosity, was developed for production of chitosan oligosaccharides at high concentration (Ming et al. 2006a, b). Furthermore, chitosanases immobilized on magnetic nanoparticles were

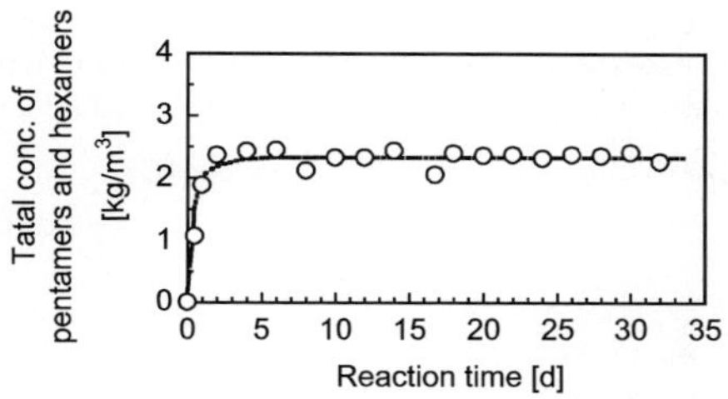

Fig. 19.8 Continuous production of pentamers and hexamers of chitosan oligosaccharides using a membrane bioreactor with immobilized chitosanase. Chitosan concentration in the feed solution was 5 kg/m^3. *d*, days. (Reproduced from Kuroiwa et al. 2009a, with permission from Elsevier)

also developed to improve dispersibility and ease in recovery of enzymes (Kuroiwa et al. 2008).

Generally, obtained chitosan oligosaccharides are a mixture of oligosaccharides with different degrees of polymerization and the content of useful oligosaccharides is less than 50 %. Therefore, the author and coworkers developed a purification method using nanofiltration (Kuroiwa et al. 2009b): the content of pentamers and hexamers in the total oligosaccharide products could be increased up to approximately 80 % by the nanofiltration treatment.

19.4 Concluding Remarks

In this chapter, the enzymatic production method of functional oligosaccharides using bioreactors has been summarized and the production method of oligosaccharides from a marine biomass (chitosan) has been demonstrated.

Various oligosaccharide-producing enzymes have been found so far, and their reaction properties have been investigated. To produce oligosaccharides efficiently, it is obviously important to reveal the reaction mechanism of those enzymes and to improve their activity and reaction selectivity. Also, to fully utilize their ability for producing oligosaccharides, it is necessary to design adequate bioreactors and to optimize the operational conditions based on the reaction characteristics of used enzymes.

Recently, production of functional oligosaccharides from various biomasses has attracted much attention from the point of view of effective utilization of natural and sustainable resources. In addition, utilization of oligosaccharides as value-added food materials would be extended with progress of research on the physiological activities of various oligosaccharides. Under these circumstances, understanding the catalytic properties of useful enzymes, improving their activity, and developing efficient production processes would all become increasingly important. Collaboration of researchers and engineers in different fields, such as biochemists, food researchers, and chemical engineers, is necessary for making progress in the development of novel technologies.

Acknowledgments This chapter is written based on work supervised by Emeritus Professor Sukekuni Mukataka, Professor Seigo Sato, and Professor Sosaku Ichikawa of University of Tsukuba, Japan. The author is highly grateful for their excellent suggestions and useful discussions.

References

Bailey JE, Ollis DF (1986) Biochemical engineering fundamentals, 2nd edn. McGraw-Hill, New York

Ichikawa S, Takano K, Kuroiwa T, Hiruta O, Sato S, Mukataka S (2002) Immobilization and stabilization of chitosanase by multipoint attachment to agar gel support. J Biosci Bioeng 93:201–206

Kim SK, Rajapakse N (2005) Enzymatic production and biological activities of chitosan oligosaccharides (COS): a review. Carbohydr Polym 62:357–368

Kuroiwa T, Ichikawa S, Hiruta O, Sato S, Mukataka S (2002) Factors affecting the composition of oligosaccharides produced in chitosan hydrolysis using immobilized chitosanases. Biotechnol Prog 18:969–974

Kuroiwa T, Ichikawa S, Sato S, Mukataka S (2003) Improvement of the yield of physiologically active oligosaccharides in continuous hydrolysis of chitosan using immobilized chitosanases. Biotechnol Bioeng 84:121–127

Kuroiwa T, Ichikawa S, Sato S, Mukataka S (2005a) Physiological activities and efficient production processes of chitosan oligosaccharides. Nippon Shokuhin Kagaku Kogaku Kaishi 52:285–296 (in Japanese)

Kuroiwa T, Shoda H, Ichikawa S, Sato S, Mukataka S (2005b) Immobilization and stabilization of pullulanase from *Klebsiella pneumoniae* by the multipoint attachment method using activated agar gel supports. Process Biochem 40:2637–2642

Kuroiwa T, Noguchi Y, Nakajima M, Sato S, Mukataka S, Ichikawa S (2008) Production of chitosan oligosaccharides using chitosanases immobilized on amylose-coated magnetic nanoparticles. Process Biochem 43:62–69

Kuroiwa T, Izuta H, Nabetani H, Nakajima M, Sato S, Mukataka S, Ichikawa S (2009a) Selective and stable production of physiologically active chitosan oligosaccharides using an enzymatic membrane bioreactor. Process Biochem 44:283–287

Kuroiwa T, Izuta H, Nabetani H, Nakajima M, Sato S, Mukataka S, Ichikawa S (2009b) Purification of physiologically active chitosan oligosaccharides by means of nanofiltration membrane. Membrane 34:336–341

Ming M, Kuroiwa T, Ichikawa S, Sato S, Mukataka S (2006a) Production of chitosan oligosaccharides by chitosanase directly immobilized on an agar gel-coated multidisk impeller. Biochem Eng J 28:289–294

Ming M, Kuroiwa T, Ichikawa S, Sato S, Mukataka S (2006b) Production of chitosan oligosaccharides at high concentration by immobilized chitosanase. Food Sci Technol Res 12:85–90

Mozaffar Z, Nakanishi K, Matsuno R (1986) Mechanism for reversible inactivation of immobilized β-galactosidase from *Bacillus circulans* during continuous production of galacto-oligosaccharides. Appl Microbiol Biotechnol 25:229–231

Xia W, Liu P, Zhang J, Chen J (2011) Biological activities of chitosan and chitooligosaccharides. Food Hydrocoll 25:170–179

Part VII
Cosmetics

Chapter 20
Cosmetic Ingredients Fermented by Lactic Acid Bacteria

Naoki Izawa and Toshiro Sone

Abstract There is an interesting relationship between the skin and fermentation of lactic acid bacteria (LAB) or bifidobacteria. Supernatants of these bacteria contain lactate and amino acids, which contribute to the hydration of the skin. Many cosmetic ingredients have been developed using LAB and bifidobacteria. In this chapter, four cosmetic ingredients that are being developed are introduced. Skim milk fermented by *Streptococcus thermophilus* (SE) has skin hydration, antioxidative, and pH control effects. Moreover, the cell protective effect of this ingredient has been proven in recent research. *Aloe vera* fermented by *Lactobacillus plantarum*, which was selected from 119 strains of LAB (AE), possesses fourfold greater skin hydration effect than nonfermented *A. vera* juice. Soybean milk fermented by *Bifidobacterium breve* has the potential to enhance hyaluronic acid production in three-dimensional culture of human cells. *S. thermophilus* YIT 2084 was proven able to produce hyaluronic acid. Although hyaluronic acid is a conventional cosmetic ingredient, it has the added value of being safe owing to its production using *S. thermophilus*, which is generally recognized as safe. It is believed that the technology introduced here will be useful for the development of next-generation cosmetic ingredients.

Keywords Cosmetic ingredients • Fermentation • Lactic acid bacteria

20.1 Introduction

Probiotics are defined as "viable bacteria that exhibit beneficial effects for health based on an improvement in the balance of intestinal bacterial flora" (Fuller 1989) or "living microorganisms which when administered in adequate amounts confer a

N. Izawa (✉) • T. Sone
Yakult Central Institute for Microbiological Research,
1796 Yaho, Kunitachi-shi, Tokyo 186-8650, Japan
e-mail: naoki-izawa@yakult.co.jp

H. Anazawa and S. Shimizu (eds.), *Microbial Production: From Genome Design to Cell Engineering*, DOI 10.1007/978-4-431-54607-8_20, © Springer Japan 2014

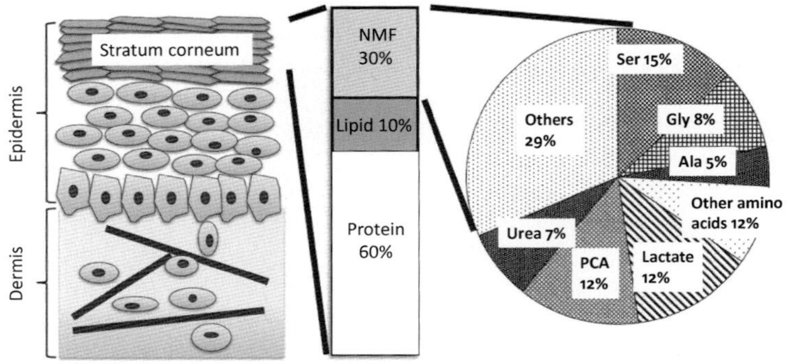

Fig. 20.1 Composition of skin and natural moisturizing factor. *NMF* natural moisturizing factor, *PCA* pyrrolidone carboxylic acid. (Modified from Izawa 2012)

benefit on the host" (Guarner and Schaafsma 1998). Some lactic acid bacteria (LAB) and bifidobacteria, such as *Lactobacillus casei* strain Shirota and *Bifidobacterium breve* strain Yakult, are known as probiotics (Nomoto 2005). LAB have long been used as a starter culture for traditional fermented foods such as cheese, yogurt, and pickles to enhance the taste and flavor of these foods. In recent studies, their preventive effect against infection has been revealed, for example, decreases in incidences of infection (Gleeson et al. 2011) and acute diarrhea in children (Sur et al. 2011). Moreover, LAB are widely used as a functional food, and the products and supernatant of LAB fermentation are also useful as cosmetic ingredients. In March 2012, 79 LAB-fermented cosmetic ingredients were introduced in Japan. *Streptococcus thermophilus*, *Lactobacillus* spp., and *Lactococcus* spp. are mainly used for fermentation. Various substrates are used as media such as skim milk, soybeans, fruit, and plants.

In this chapter, the relationship between the skin and cosmetic ingredients produced by LAB fermentation, examples of developing cosmetic ingredients by LAB fermentation, and hyaluronic acid (HA) production by novel LAB are introduced.

20.1.1 Skin and Lactic Acid Bacteria Products

Figure 20.1 shows the structure of the skin and the composition of the stratum corneum (SC) and natural moisturizing factor (NMF). The skin is composed of two major layers, the dermis and epidermis. The epidermis is further divided into two main layers, the viable epidermis and the SC, which is the topmost layer. The most important function of the SC is to act as a barrier, preventing dehydration from the loss of body water. The SC contains 30 % NMF, composed of 40 % amino acids, such as serine, glycine, and alanine, and 12 % lactate, which can hold water in the stratum corneum (Spier and Pascher 1956).

On the other hand, LAB are commonly defined as being gram-positive, nonsporulating, catalase-negative, aerotolerant, acid-tolerant, nutritionally fastidious, strictly fermentative organisms that lack cytochromes and produce lactic acid as the major end product of carbohydrate metabolism (Axelsson 1998). LAB have cell wall-bound proteinase that initiates the transformation of extracellular proteins into oligopeptides (Savijoki et al. 2006). The protease activity and catabolic production of proteins and peptides enables LAB to produce amino acids in broth (Lee et al. 2001). Many amino acids are contained in various LAB-fermented foods such as cheese (Visser 1993), sausage (Hierro et al. 1999), and Japanese sake (Iwano et al. 2004).

Thus, a LAB-fermented ingredient contains large amounts of lactic acid and amino acids, which comprise NMF. Therefore, these ingredients exert hydration effect when they are applied on the skin. Furthermore, a certain combination of a substrate and a LAB strain could provide other useful functional effects on the skin.

20.2 Example of Ingredients

In this section, two cosmetic ingredients produced by LAB and one produced by *Bifidobacterium* fermentation that have hydration and other useful effects on the skin are introduced.

20.2.1 Skim Milk Fermented by S. thermophilus

The supernatant of skim milk fermented by *S. thermophilus* YIT 2001 (formerly named ST-1), which is hereafter referred to as SE, has a skin hydration effect (Yamada et al. 1982) because it contains lactic acid and amino acids similarly to NMF (Hiraki 1975). Proline-rich peptides in the SE show an antioxidative effect (Chiba 2007). SE can also potentially maintain the pH at weakly acidic (4.5–6.5) (Hiraki 1975). A weakly acidic skin pH is important for maintaining healthy human skin by retaining the resident bacterial flora (Lambers et al. 2006).

The supernatant of skim milk fermented by *S. thermophilus* YIT 2084, hereafter referred to as SE2, has antioxidant and pH-maintaining effects similarly to SE. It also has a greater skin hydration effect than SE because it contains HA, which exerts an extremely strong hydration effect in the presence of other moisturizers. It is considered that there is a synergistic action among HA, lactic acid, and amino acids as moisturizers in the SE2 supernatant (Izawa et al. 2012). Furthermore, SE2 protects skin cells from oxidative stress. This effect was considered different from the antioxidative effect because oxidative stress was applied after SE2 was removed from the skin cells. HA is recognized by receptors such as CD44 and receptor for hyaluronan-mediated motility (RHAMM) and it might trigger various cell signals including those for wound healing (Weindl et al. 2004). Besides triggering cell signals, HA might adhere to cells, forming a barrier against oxidative stress.

20.2.2 Aloe vera *Fermented by* Lactobacillus plantarum

Aloe vera has antibacterial, antiviral, antimycotic, wound-healing, and antiinflammatory effects (Eshun and He 2004). To enhance the skin hydration effect of cosmetic ingredients produced by LAB fermentation, 119 strains of LAB including *Lactobacillus*, *Lactococcus*, and *Leuconostoc* spp. were screened. The supernatant of *A. vera* fermented by *L. plantarum* YIT 0102, which was isolated from sauerkraut, shows the greatest (400 %) skin hydration effect. *L. plantarum* can convert malic acid to lactic acid via the malo-lactic fermentation pathway, and the combination of fructose and lactic acid increased the water content in the skin (Sone et al. 2005). On the basis of the results of amino-acid analysis, the amounts of serine, glycine, and alanine, which are the major components of amino acids in NMF, were increased. The cell envelope proteinase varies among the LAB strains (Kunji et al. 1996). The proteolytic system of *L. plantarum* YIT 0102 breaks down proteins in *A. vera* into these amino acids selectively.

20.2.3 Soybean Milk Fermented by Bifidobacterium breve

Although *Bifidobacterium breve* is not defined as a LAB, the supernatant of soybean fermented by *B. breve* strain Yakult is introduced in this section because it has some common denominators to LAB. For example, it is a well-known probiotic, is commonly used in preparing traditional foods, and produces organic acids. BE also has an intriguing effect on the skin.

Soybean contains isoflavones as inactive glycosides, such as genistin and daidzin. The aglycon of these isoflavones, such as genistein and daidzein, has the potential to enhance the production of HA in skin cell (keratinocyte) culture (Miyazaki et al. 2002). HA, which contributes to skin hydration and elasticity (Miyazaki et al. 2004), is produced in human keratinocytes and fibroblasts. An image of a histochemically stained human skin cell culture treated with BE is shown in Fig. 20.2. BE has the potential to enhance HA production in three-dimensional human skin cell culture because *B. breve* strain Yakult has the potential to convert isoflavone glycoside to aglycon via β-glucosidase (Miyazaki et al. 2003 and Ishikawa et al. 2003).

20.3 HA Production by Novel LAB

HA is a linear, high molecular weight polysaccharide composed of alternating *N*-acetylglucosamine (GlcNAc) and glucuronic acid (GlcA) moieties (Laurent and Fraser 1992). HA is widely used in the cosmetic, medical, and food industries, for example, as skin moisturizers and in orthopedic surgery, rheumatism treatment, drug delivery, and wound healing (Weindl et al. 2004; Kogan et al. 2007). HA is obtained from rooster combs or by microbial fermentation using certain attenuated strains of pathogenic Lancefield group A and C streptococci. The pathogenic

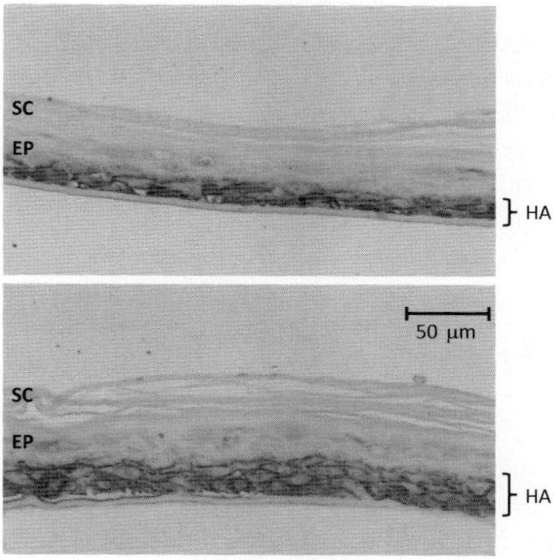

Fig. 20.2 Images of histochemically stained three-dimensional skin culture system treated with BE (The supernatant of soybean fermented by *Bifidobacterium breve*). *SC* stratum corneum, *EP* epidermis, *HA* hyaluronic acid. (Modified from Miyazaki et al. 2003)

Lancefield group A and C streptococci are not ideal sources of HA because the risks of mutation of the bacterial strains and possible coproduction of toxins and pyrogens are difficult to eliminate completely. Thus, at present, alternative sources of HA production are being explored.

20.3.1 Streptococcus thermophilus

Streptococcus thermophilus is widely used for the manufacture of yogurt and cheese. It has two major characteristics: first, it is generally recognized as safe (GRAS) and, second, it can produce various extracellular polysaccharides (Vaningelgem et al. 2004).

To obtain HA produced by fermentation by GRAS microorganisms, 46 strains of *S. thermophilus* were screened and *S. thermophilus* YIT 2084, which can produce 8 mg/l HA in skim milk, was selected (Izawa et al. 2009).

20.3.2 Fermentation Conditions

To increase HA yield by *S. thermophilus* YIT 2084, fermentation conditions such as pH, temperature, aeration rate, agitation speed, and supplementation with soybean peptides were investigated using a 2-l jar fermenter. Aeration rate and agitation speed did not affect HA production. pH is the strongest factor affecting HA production. About 100 mg/l HA was obtained at a pH of 6.8. This production rate was maintained at 33–40 °C. Moreover, supplementation with a high molecular weight soybean peptide increased production to 200 mg/l (Izawa et al. 2010).

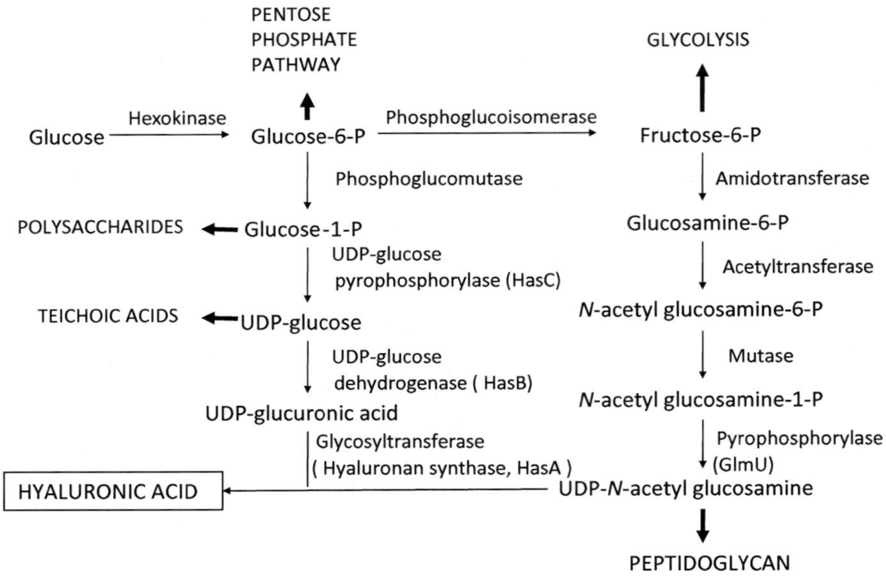

Fig. 20.3 Biosynthetic pathway of hyaluronic acid production in *Streptococcus* spp. (Modified from Chong et al. 2005)

20.3.3 Construction of Recombinant Strains

HA production by recombinant GRAS microorganisms was attempted. The proposed biosynthetic pathway of pathogenic streptococci for HA is presented in Fig. 20.3 (Chong et al. 2005). HA is synthesized from two nucleotide sugars, namely, UDP-glucuronic acid (UDP-GlcA) and UDP-*N*-acetyl glucosamine (UDP-GlcNAc), as catalyzed by HA synthase (HasA or glycosyltransferase). The precursors of HA, UDP-GlcA, and UDP-GlcNAc were synthesized by UDP-glucose dehydrogenase (HasB) from UDP-glucose and by pyrophosphorylase (GlmU) from *N*-acetyl-glucosamine-1-phosphate, respectively. It is considered that *S. thermophilus* YIT 2084 produces HA via almost the same pathway except for the uptake of lactose and its degradation into the glucose and galactose moiety by β-galactosidase in the first step. Widner et al. (2005) reported that multigrams per liter of HA was obtained using *Bacillus subtilis* by overexpressing both *Streptococcus equisimilis hasA* and endogenous *tuaD* (*hasB* homologue). Chien and Lee (2007) constructed HA-producing *Lactococcus lactis* by coexpressing *hasA* and *ugd* from *Streptococcus equi* subsp. *zooepidemicus* in a nisin-controlled expression system and attained 0.65 g/l HA production by batch cultivation. In these studies, HA production genes were derived from pathogenic microorganisms. We attempted the construction of high-HA-producing recombinant *S. thermophilus* YIT 2084 strains by overexpressing endogenous *hasA*, *hasB*, and *glmU*. The construction scheme is shown in Fig. 20.4.

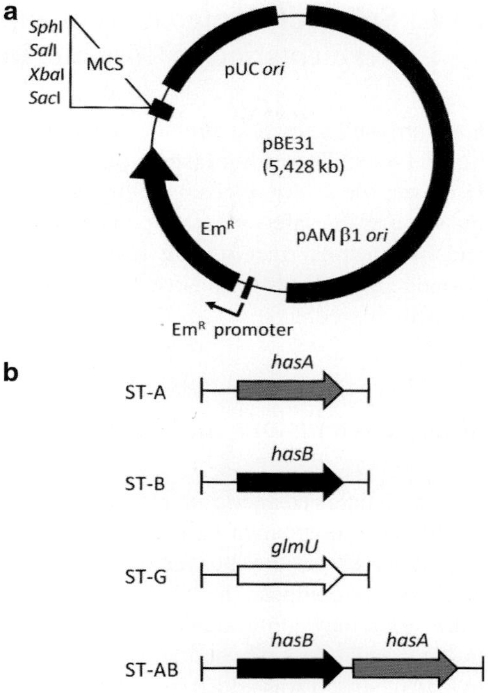

Fig. 20.4 Construction scheme for pBE vectors for ST-A, ST-B, ST-G, and ST-AB. Genetic map of pBE31 (**a**) and structure of inserted genes (**b**) (*pUC ori* replication region for *Escherichia coli*, *pAM β1 ori* replication region for *Streptococcus* spp., *EmR* erythromycin resistance, *MCS* multicloning site. (Modified from Izawa et al. 2011)

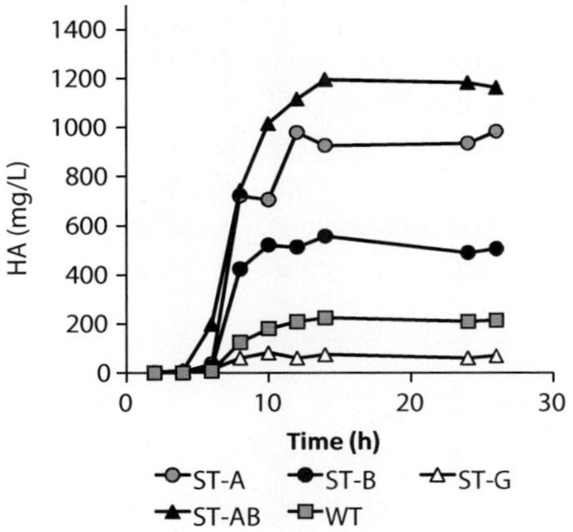

Fig. 20.5 Time-course of hyaluronic acid production. (Modified from Izawa et al. 2011)

The time-course of HA production by each recombinant strain is shown in Fig. 20.5. pBE31, which was designed as the *Enterococcus–Escherichia coli* shuttle vector, worked well in *S. thermophilus* YIT 2084. HA was produced at 1,200 mg/l by the strain overexpressing both *hasA* and *hasB* derived from *S. thermophilus* YIT 2084 (Izawa et al. 2011).

20.4 Safety of Components in Supernatant of LAB Fermentation as Cosmetic Ingredients

LAB are well known as probiotics and have been used in the preparation of traditional foods. Their long history proves the safety of foods prepared using LAB. However, the safety of cosmetic ingredients produced by LAB fermentation is not the same as the safety of such foods because there are differences between the skin and the intestinal tract in many respects. It is important to validate the safety of the cosmetic ingredients from LAB-fermented supernatants in the development of such cosmetic ingredients.

20.5 Conclusions and Perspectives

In this chapter, cosmetic ingredients produced by LAB fermentation were introduced. The supernatant of LAB fermentation has a hydration effect on the skin; thus, it is useful as the source of cosmetic ingredients, namely, lactic acid and amino acids, which compose the NMF. The combination of a substrate or medium and a LAB strain may confer other cosmetic effects such as antioxidative effect, pH control, and cell stress prevention. As the first example, a novel effect (barrier against oxidative stress) was added to the preexisting ingredient, SE, by only changing the LAB strain. As the second example, the skin hydration effect of the *A. vera* substrate fermented by selected LAB was enhanced. As the third example, HA production of the skin was enhanced by applying isoflavone-aglycones, which were converted by *B. breve* Yakult. As the fourth example, HA has the added value of being safe because of its production using a novel GRAS LAB.

Thus, determining the adequate combination of a substrate and a LAB strain or *Bifidobacterium* could add new value to cosmetic ingredients. There are uncountable combinations of a substrate and a LAB strain; therefore, an infinite number of new ingredients could be developed.

References

Axelsson L (1998) Lactic acid bacteria: classification and physiology. In: Salminen S, von Wright A (eds) Lactic acid bacteria. Microbiology and functional aspects. Dekker, New York, pp 1–71

Chiba K (2007) Development of functional cosmetic ingredients using lactic acid bacteria in Japan. Jpn J Lactic Acid Bacteria 18:105–112

Chien LJ, Lee CK (2007) Hyaluronic acid production by recombinant *Lactococcus lactis*. Appl Microbiol Biotechnol 77:339–346

Chong BF, Blank LM, Mclaughlin R et al (2005) Microbial hyaluronic acid production. Appl Microbiol Biotechnol 66:341–351

Eshun K, He Q (2004) *Aloe vera*: a valuable ingredient for the food, pharmaceutical and cosmetic industries: a review. Crit Rev Food Sci Nutr 44:91–96

Fuller R (1989) Probiotics in man and animals. J Appl Bacteriol 66:365–378

Gleeson M, Bishop NC, Oliveira M et al (2011) Daily probiotic's (*Lactobacillus casei* Shirota) reduction of infection incidence in athletes. Int J Sport Nutr Exerc Metab 21:55–64

Guarner F, Schaafsma GJ (1998) Probiotics. Int J Food Microbiol 39:237–238

Hierro E, de la Hoz L, Ordóñez JA (1999) Contribution of the microbial and meat endogenous enzymes to the free amino acid and amine contents of dry fermented sausages. J Agric Food Chem 47:1156–1161

Hiraki Y (1975) Natural and moisturizing tendency on cosmetics. Fragrance J 13:18–20

Ishikawa F, Mizobuchi T, Aiyama R et al (2003) Gann yobo shokuhin (Foods for cancer prevention). Jpn Patent 3489930:7

Iwano K, Takahashi K, Ito T et al (2004) Search for amino acids affecting the taste of Japanese sake. J Brew Soc Jpn 99:659–664

Izawa N (2012) Development of cosmetic ingredients using lactic acid bacteria. In: Anazawa S (ed) Advanced technology for microbiological production in Japan. CMC, Japan, pp 177–185

Izawa N, Hanamizu T, Iizuka R et al (2009) *Streptococcus thermophilus* produces exopolysaccharides including hyaluronic acid. J Biosci Bioeng 107:119–123

Izawa N, Hanamizu T, Sone T et al (2010) Effects of fermentation conditions and soybean peptide supplementation on hyaluronic acid production by *Streptococcus thermophilus* strain YIT 2084 in milk. J Biosci Bioeng 109:356–360

Izawa N, Serata M, Sone T et al (2011) Hyaluronic acid production by recombinant *Streptococcus thermophilus*. J Biosci Bioeng 111:665–670

Izawa N, Sone T, Chiba K (2012) Development of cosmetic ingredient using the culture supernatant of *Streptococcus thermophilus*. J Jpn Cosmet Sci Soc 36:87–92

Kogan G, Šoltés L, Stern R et al (2007) Hyaluronic acid: a natural biopolymer with a broad range of biomedical and industrial applications. Biotechnol Lett 29:17–25

Kunji ER, Mierau I, Hagting A et al (1996) The proteolytic systems of lactic acid bacteria. Antonie van Leeuwenhoek 70:187–221

Lambers H, Piessens S, Bloem H et al (2006) Natural skin surface pH is on average below 5, which is beneficial for its resident flora. Int J Cosmet Sci 28:359–370

Laurent TC, Fraser JR (1992) Hyaluronan. FASEB J 6:2397–2404

Lee K, Lee J, Kim YH et al (2001) Unique properties of four lactobacilli in amino acid production and symbiotic mixed culture for lactic acid biosynthesis. Curr Microbiol 43:383–390

Miyazaki K, Hanamizu T, Iizuka R et al (2002) Genistein and daizein stimulate hyaluronic acid production in transformed human keratinocyte culture and hairless mouse skin. Skin Pharmacol Appl Skin Physiol 15:175–183

Miyazaki K, Hanamizu T, Iizuka R et al (2003) *Bifidobacterium*-fermented soy milk extract stimulates hyaluronic acid production in human skin cells and hairless mouse skin. Skin Pharmacol Appl Skin Physiol 16:108–116

Miyazaki K, Hanamizu T, Sone T et al (2004) Topical application of *Bifidobacterium*-fermented soy milk extract containing genistein and daizein improves rheological and physiological properties of skin. J Cosmet Sci 55:473–479

Nomoto K (2005) Prevention of infections by probiotics. J Biosci Bioeng 100:583–592

Savijoki K, Ingmer H, Varmanen P (2006) Proteolytic systems of lactic acid bacteria. Appl Microbiol Biotechnol 71:394–406

Sone T, Hanamizu T, Chiba K (2005) Plant ingredients fermented by lactic acid bacteria. Fragrance J 33:65–70

Spier HW, Pascher G (1956) Zur analytischen und funktionellen Physiologie der Hautoberfläche (Analytical and functional physiology of the skin surface). Hautarzt 7:55–60

Sur D, Mnna B, Niyogi SK et al (2011) Role of probiotic in preventing acute diarrhoea in children: a community-based, randomized, double-blind placebo-controlled field trial in an urban slum. Epidemiol Infect 139:919–926

Vaningelgem F, Zamfir M, Mozzi F et al (2004) Biodiversity of exopolysaccharides produced by *Streptococcus thermophilus* strains is reflected in their production and their molecular and functional characteristics. Appl Environ Microbiol 70:900–912

Visser S (1993) Proteolytic enzymes and their relation to cheese ripening and flavor: an overview. J Dairy Sci 76:329–350

Weindl G, Schäller M, Schafer-Korting M et al (2004) Hyaluronic acid in the treatment and prevention of skin diseases: molecular biological, pharmaceutical and clinical aspects. Skin Pharmacol Physiol 17:207–213

Widner B, Behr R, Dollen SV et al (2005) Hyaluronic acid production in *Bacillus subtilis*. Appl Environ Microbiol 71:3747–3752

Yamada H, Sone T, Kimura M et al (1982) Application of a lactic acid bacteria culture filtrate to the skin care. (1) Moisturizing effect on the horny layer. J Jpn Cosmet Sci Soc 6:238–243

Chapter 21
Structure of Tyrosinase and Its Inhibitor from Sake Lees

Yasuyuki Matoba and Masanori Sugiyama

Abstract Tyrosinase (EC 1.14.18.1) is a copper-containing oxidase that catalyzes the first two steps in mammalian melanogenesis and is responsible for the enzymatic browning reactions in damaged fruit during postharvest handling and processing. The accumulation of an abnormal amount of melanin in the skin is an esthetic problem. On the other hand, the undesirable browning of fruits causes a significant decrease in nutritional and market value. Neither hyperpigmentation in human skin nor enzymatic browning in fruits is desirable. These phenomena have promoted research to seek and develop potent tyrosinase inhibitors for use in cosmetics and foods. Structural information for tyrosinase would be useful to develop an inhibitor. We determined the tertiary structure of the *Streptomyces castaneoglobisporus* tyrosinase at a very high resolution in complex with a metallochaperone that we call a "caddie." In addition, we found some tyrosinase-inhibiting factors in the *n*-hexane extract of sake lees and showed that they are a mixture of triacylglycerols.

Keywords Anti-browning agent • Crystal structure • Sake lees • Tyrosinase

21.1 Introduction

Melanogenesis has been defined as the entire process leading to the formation of dark macromolecular pigments, that is, melanin. In mammals, melanin, which is formed in melanocytes, is synthesized from the amino acid L-tyrosine by a combination of enzymatically catalyzed and chemical reactions. The melanin-synthesizing

Y. Matoba • M. Sugiyama (✉)
Department of Molecular Microbiology and Biotechnology, Graduate School of Biomedical and Health Sciences, Hiroshima University, Kasumi 1-2-3, Minami-ku, Hiroshima 734-8551, Japan
e-mail: sugi@hiroshima-u.ac.jp

H. Anazawa and S. Shimizu (eds.), *Microbial Production: From Genome Design to Cell Engineering*, DOI 10.1007/978-4-431-54607-8_21,

ability is widely distributed in animals, plants, and microorganisms. Melanin-like pigments are synthesized in vegetables, fruits, and mushrooms by the enzymatic browning that occurs upon bruising or long-term storage, although the origin of the building blocks of the pigments is different from L-tyrosine. This undesirable browning causes a significant decrease in nutritional and market value. On the other hand, the role of melanin in mammals is to protect the skin from UV-induced injury. However, the accumulation of an abnormal amount of melanin in the skin, which causes melasma, freckles, ephelides, and age spots, is an esthetic problem. Such food and skin hyperpigmentations have brought on research to seek and develop potent melanogenesis inhibitors for use in anti-browning of foods and skin whitening, and a large number of inhibitors have been identified. Approaches to skin whitening include inhibition of melanin synthesis, interference with melanosome maturation and transfer, inhibition of inflammation-induced melanogenic response, and acceleration of skin turnover.

Melanogenesis is initiated with the oxidation of L-tyrosine to L-dopaquinone catalyzed by tyrosinase. This step is rate limiting in melanin synthesis because the remainder of the reaction sequence can proceed spontaneously at physiological pH. Tyrosinase (EC 1.14.18.1) harboring a catalytic center formed by dinuclear copper catalyzes the two distinct reactions, *ortho*-hydroxylation of L-tyrosine and the subsequent oxidation of L-DOPA to L-dopaquinone (Solomon et al. 1996). A series of reactions occurs under the concomitant reduction of molecular oxygen to water. Kojic acid and arbutin, which have been used as whitening agents in cosmetics, inhibit tyrosinase catalytic activity. The recent discovery with regard to the life cycle of tyrosinase has opened new paths to treatment for skin hyperpigmentation. In addition to the inhibition of tyrosinase catalytic activity, other approaches to treat hyperpigmentation include inhibition of tyrosinase mRNA transcription, aberration of tyrosinase glycosylation and maturation, and acceleration of tyrosinase degradation. Accordingly, various whitening agents, which inhibit melanogenesis in different ways, have been developed.

Because tyrosinase is an indispensable enzyme for melanogenesis, structural information about tyrosinase may be useful for the development of tyrosinase inhibitors. We determined the tertiary structure of the *Streptomyces castaneoglobisporus* tyrosinase at a very high resolution in complex with the metallochaperone that we call a "caddie" (Matoba et al. 2006). This finding represents the first crystal structure determination of a tyrosinase from prokaryotes or eukaryotes. On the other hand, we searched for safe and potent tyrosinase inhibitors from natural sources. Sake, made from fermented rice, is an alcoholic beverage of Japanese origin. Sake lees are generated as a by-product of sake fermentation. Because sake brewers (*Tohji* in Japanese) frequently have light and smooth skin, we hypothesized that whitening agents are present in sake lees. We found that the *n*-hexane extract of freeze-dried sake lees inhibits tyrosinase activity and showed that the constituents exhibiting inhibitory activity are a mixture of triacylglycerols (Jeon et al. 2006). In this chapter, we describe the structure of bacterial tyrosinase and its inhibitor, which is found in sake lees.

21.2 Overview of the Tyrosinase Structure

Many strains classified in the genus *Streptomyces* produce a melanin-like pigment (Claus and Decker 2006). The molecular mechanism underlying the activation processes of *Streptomyces* tyrosinase is noteworthy. The melanin-synthesizing operon of *Streptomyces antibioticus* is composed of two genes, which encode MelC1 and MelC2 proteins (Lee et al. 1988). Apotyrosinase (MelC2) forms a stable complex with MelC1 (Chen et al. 1992). Although apotyrosinase was not activated by copper added from the outside, the addition of copper ions to the purified complex gave rise to the incorporation of two copper ions. Furthermore, during the in vitro activation of the MelC1/MelC2 complex, Cu(II)-bound MelC2 was discharged from the complex, but no trace of the released MelC1 protein was detectable; this suggests that the released MelC1 protein might form an aggregate to enable its separation from the protein complex. We previously cloned a melanin-synthesizing gene from the chromosomal DNA of *S. castaneoglobisporus* HUT 6202, which is a bacterial strain that produces a melanin pigment in large amounts (Ikeda et al. 1996). This gene forms an operon consisting of two cistrons: one is an open-reading frame consisting of 378 nucleotides designated *orf378*, and the other is a tyrosinase-encoding gene, designated *tyrC*, that is located just downstream of *orf378*. The *orf378* gene encodes a MelC1-like protein. We call the gene product "caddie" because the protein may carry copper ions for tyrosinase.

Tyrosinase, which contains the di-copper center, is classified as a type 3 copper protein family, similarly to catechol oxidase and the respiratory pigment hemocyanin. The structures of tyrosinase were unknown until our determination was done, whereas the structures of catechol oxidase (Klabunde et al. 1998) and hemocyanin (Volbeda and Hol 1989; Hazes et al. 1993; Magnus et al. 1994; Cuff et al. 1998) had already been identified. We crystallized tyrosinase in a complex with the caddie and determined the crystal structure by the multiple isomorphous replacement method (Matoba et al. 2006). Furthermore, we obtained Cu(II)-bound tyrosinase complexed with a caddie by soaking the native crystal in a solution containing $CuSO_4$ (Fig. 21.1).

Tyrosinase takes α-helical structures, and a four-helix bundle is formed in its core region. The catalytic dinuclear copper center is lodged in the helical bundle. Each of the two copper ions in an active site is coordinated by three His residues, which are derived from the bundle, except His^{54}. One copper ion (designated Cu^{A}) is coordinated by His^{38}, His^{54}, and His^{63}. The second copper ion (Cu^{B}) is coordinated by His^{190}, His^{194}, and His^{216}. This di-copper center is located at the bottom of the large concavity as a putative substrate-binding pocket, which is formed by hydrophobic residues. Although the amino-acid sequence of tyrosinase has only 25.3 % and 26.0 % identities with those of the *Ipomoea batatas* catechol oxidase (Klabunde et al. 1998) and the odg domain of the *Octopus dofleini* hemocyanin (Cuff et al. 1998), respectively, their overall structures are quite similar.

The caddie protein, which covers the molecular surface of tyrosinase, seems to prevent substrate binding to the active site of tyrosinase. The caddie has one six-stranded β-sheet and one α-helix. The protein has no sequence similarity with

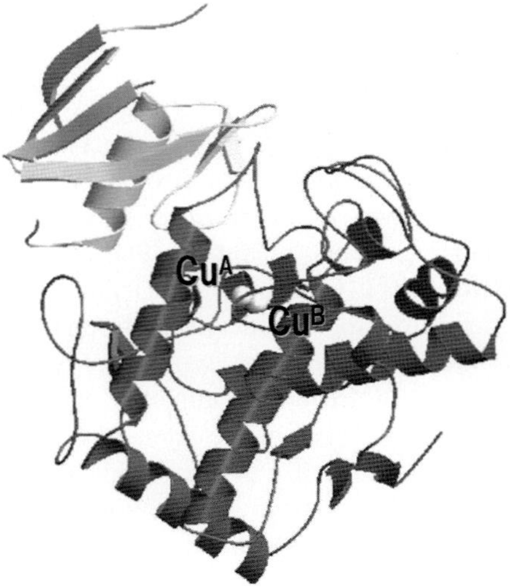

Fig. 21.1 Ribbon view of Cu(II)-bound form of tyrosinase complexed with a caddie. Tyrosinase and the caddie are shown in *dark gray* and *light gray*, respectively. Two copper ions at the active site are indicated by *spheres*

proteins structurally determined. According to the results obtained using the DALI program, however, the caddie has a significant structural similarity with the SH2 domain, which is contained in the proteins for the purposes of the signal transduction. Very recently, we performed kinetic and crystallographic studies to clarify the transfer mechanism of Cu(II) to tyrosinase, which is assisted by a caddie (Matoba et al. 2011). As a result, additional copper-binding sites in the caddie protein and hydrogen-bonding network around the catalytic center of tyrosinase were determined to be important for the effective transfer of Cu(II).

21.3 Structure of the Active Site of Tyrosinase

During catalysis, the type 3 copper center of tyrosinase adopts three redox forms (Solomon et al. 1996). The *deoxy* form [Cu(I)–Cu(I)] is a reduced species that binds dioxygen to yield the *oxy* form. In the *oxy* form, molecular oxygen binds in the form of peroxide in a μ–η^2:η^2 side-on bridging [Cu(II)–O_2^{2-}–Cu(II)] mode, which destabilizes the O–O bond and activates it. The *met* form [Cu(II)–Cu(II)] is recognized as a resting enzymatic form, in which Cu(II) ions are normally bridged with a small ligand, such as a water molecule or hydroxide ion. In these three redox forms of tyrosinase, the *oxy* form can catalyze both the monooxygenase and oxidase reactions, whereas the *met* form lacks monooxygenase activity. On the other hand, although the *oxy* form of catechol oxidase can oxidize catechols to the corresponding quinones, it lacks monooxygenase activity.

Interestingly, the side chain of Tyr^{98} in the caddie protein protrudes away from the core region of the protein and is accommodated in the substrate-binding pocket of tyrosinase (Fig. 21.2). The phenol ring is stacked with the imidazole ring of His^{194}

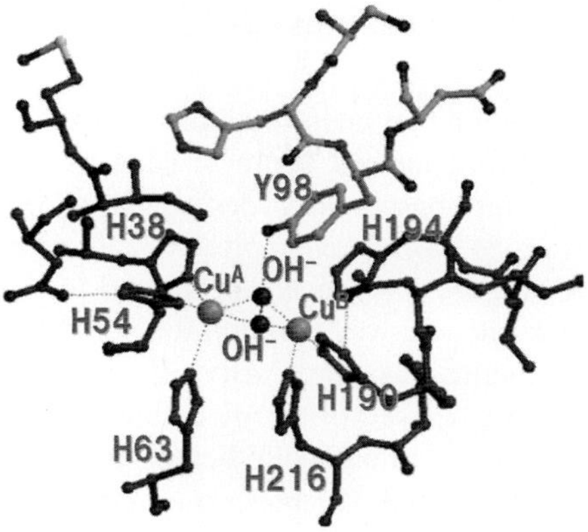

Fig. 21.2 Active center of tyrosinase found in the crystal structure of Cu(II)-bound form of tyrosinase complexed with caddie. Carbon atoms from the residues of tyrosinase and the caddie are shown in *dark gray* and *light gray*, respectively

of tyrosinase, which is one of the Cu^B ligands. In addition, its phenolic hydroxyl forms a hydrogen bond with a solvent atom (water molecule or peroxide ion), which forms a bridge with Cu^A and Cu^B in the active center of tyrosinase. Furthermore, when compared with the inhibitor (phenylthiourea) bound to catechol oxidase (Klabunde et al. 1998), the Tyr^{98} ring is perfectly aligned with the aromatic ring of the inhibitor. These results suggest that the Tyr^{98} of caddie functions as a competitive inhibitor to prevent the binding of the substrate L-tyrosine.

Tyrosinase belongs to the protein family having a type 3 copper center, as do catechol oxidase and hemocyanin. The reasons for the different functions exhibited by these proteins pose an interesting query. Hemocyanin plays a significant role in the transportation of oxygen in arthropods and mollusks. The protein has a domain that shields access to the di-copper center of the protein. Because the domain interferes with the binding of the substrate, hemocyanin can play only one role, namely, that of an oxygen transporter. On the other hand, the di-copper center of both catechol oxidase and tyrosinase is near the molecular surface, which is useful as a location to ensure that a substrate can access each catalytic site. Both tyrosinase and catechol oxidase catalyze the oxidase reaction, but the latter enzyme lacks the monooxygenase activity. One of the most significant differences between tyrosinase and catechol oxidase is the shape of the substrate-binding pocket. In the active site of catechol oxidase, the hydrophobic residue (Phe^{261}), which is located just above the Cu^A site, seems to partially prevent substrate binding. In contrast, the substrate-binding pocket of tyrosinase has a larger vacant space above the di-copper center if the caddie protein is liberated from tyrosinase. The catalytic specificity of tyrosinase and cathecol oxidase may depend on the shapes of the substrate-binding pockets.

Our report about the structure of tyrosinase, which was published in 2006, has been cited more than 300 times up to now, and proved to be of interest to many researchers. Based on the tertiary structures of tyrosinase, many studies have been performed to develop tyrosinase inhibitors or to clarify the catalytic mechanism of tyrosinase.

21.4 Tyrosinase Inhibitor from Sake Lees

There are two major methods to develop new drugs. One traditional way is discovery from natural sources, such as animals, plants, and bacteria. Another is a structure-based drug design based on the target protein structure. Kojic acid (Fig. 21.3a), which is produced by several species of filamentous fungi, especially *Aspergillus oryzae*, had been used as an anti-browning agent in cosmetics. However, as kojic acid was reported to exhibit cytotoxicity (Takizawa et al. 2003), its use in cosmetics was temporarily suspended in Japan. Therefore, safe tyrosinase inhibitors that are derived from safe edible materials are necessary.

Sake is an alcoholic beverage of Japanese origin made by fermenting rice with *A. oryzae* and *Saccharomyces cerevisiae*. Sake lees, which are generated as a by-product during sake production, are nutritious and useful for seasoning Japanese food. However, some of it is treated as industrial waste. Therefore, an effective use of sake lees may be valuable from an environmental point of view. Because the skin of sake brewers, who are known as *Tohji*, is frequently white, smooth, and healthy, whitening agents are assumed to be present in sake lees. We tried to identify the tyrosinase inhibitors that might be present in sake lees. We found that the *n*-hexane extract of freeze-dried sake lees inhibits tyrosinase catalytic activity (Jeon et al. 2006). The *n*-hexane extract of sake lees was used for further purification of the tyrosinase inhibitor. We finally obtained a mixture of triacylglycerols from the *n*-hexane extract, which exhibits the tyrosinase inhibitory effect. The ^{1}H-NMR spectra of a mixture of triacylglycerols indicated that one triacylglycerol molecule has an average of four olefinic double bonds. The profile of GC-MS showed a mixture containing some kinds of triacylglycerols. We examined the tyrosinase inhibitory effects of triolein and trilinolein (Fig. 21.3b, c), which are contained in the freeze-dried sake lees and commercially available, together with kojic acid as a positive control. Triolein and trilinolein have three and six olefinic double bonds, respectively.

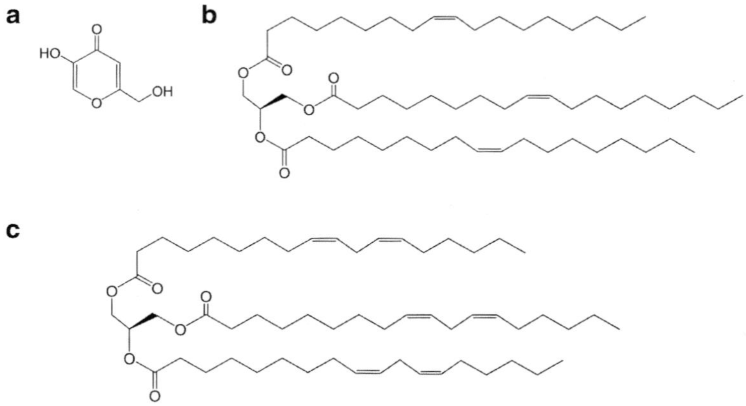

Fig. 21.3 Chemical structure of tyrosinase inhibitors: kojic acid (**a**); triolein (**b**); trilinolein (**c**)

The IC_{50} value of trilinolein for oxidase activity on mushroom tyrosinase was 8.4 μM. Because triolein was hardly soluble in a water-based solution, the IC_{50} value was not determined. The IC_{50} value of kojic acid was 14 μM, and the value was about 1.5 fold higher than that of trilinolein. On the other hand, the IC_{50} values of triolein, trilinolein, and kojic acid for oxidase activity on *Streptomyces* tyrosinase were 30, 1.0, and 7.8 μM, respectively. Thus, trilinolein inhibits tyrosinase activity more strongly than does triolein or kojic acid. The number of olefinic double bonds seems to be an important factor in the exhibition of tyrosinase inhibitory effect. The kinetic behaviors of the oxidation of L-DOPA catalyzed by the mushroom and *S. castaneoglobisporus* tyrosinases were investigated using various concentrations of trilinolein and L-DOPA. Judging from the results of Lineweaver–Burk plots, we conclude that trilinolein is a noncompetitive inhibitor. The inhibitory mode suggests that triacylglycerol may bind to some site other than the catalytic site of tyrosinase.

Through this study, we showed that unsaturated triacylglycerols, present in sake lees, inhibit tyrosinase catalytic activity. Therefore, sake lees may be used as a safe and potent anti-browning material. The origin of the unsaturated triacylglycerols found in sake lees is currently unknown, but they may be constituents in rice, molds, or yeast rather than products of the fermentation process.

21.5 Conclusion

This chapter contains the research fruits of two of our research projects. The objective was to develop anti-browning agents for applications in the cosmetic and food industries. The structure of tyrosinase, which we were the first to determine, will be useful for the development of tyrosinase inhibitors through in silico structure-based drug design. If the structure of human tyrosinase could be determined, it would be useful for drug design applications. However, the mammalian tyrosinase is known to be a glycosylated enzyme, which is difficult to crystallize. On the other hand, their bacterial counterparts are not glycosylated, because posttranslational glycosylation does not occur in bacteria. To understand the structure of mammalian proteins, which are difficult to crystallize because of glycosylation, the bacterial homologues should be crystallized first. In the near future, synthetic tyrosinase inhibitors will be developed based on our research results. However, the safety of any synthetic compound should be confirmed before commercialization. In this regard, the discovery of the anti-browning agents from natural sources, especially from food materials, would be important.

References

Chen LY, Leu WM, Wang KT et al (1992) Copper transfer and activation of the *Streptomyces* apotyrosinase are mediated through a complex formation between apotyrosinase and its trans-activator MelC1. J Biol Chem 267:20100–20107

Claus H, Decker H (2006) Bacterial tyrosinases. Syst Appl Microbiol 29:3–14

Cuff ME, Miller KI, van Holde KE et al (1998) Crystal structure of a functional unit from octopus hemocyanin. J Mol Biol 278:855–870

Hazes B, Magnus KA, Bonaventura C et al (1993) Crystal structure of deoxygenated *Limulus polyphemus* subunit II hemocyanin at 2.18 Å resolution: clues for a mechanism for allosteric regulation. Protein Sci 2:597–619

Ikeda K, Masujima T, Suzuki K et al (1996) Cloning and sequence analysis of the highly expressed melanin-synthesizing gene operon from *Streptomyces castaneoglobisporus*. Appl Microbiol Biotechnol 45:80–85

Jeon HJ, Noda M, Maruyama M et al (2006) Identification and kinetic study of tyrosinase inhibitors found in sake lees. J Agric Food Chem 54:9827–9833

Klabunde T, Eicken C, Sacchettini JC et al (1998) Crystal structure of a plant catechol oxidase containing a dicopper center. Nat Struct Biol 5:1084–1090

Lee YH, Chen BF, Wu SY et al (1988) A trans-acting gene is required for the phenotypic expression of a tyrosinase gene in *Streptomyces*. Gene (Amst) 65:71–81

Magnus KA, Hazes B, Ton-That H et al (1994) Crystallographic analysis of oxygenated and deoxygenated states of arthropod hemocyanin shows unusual differences. Proteins 19:302–309

Matoba Y, Kumagai T, Yamamoto A et al (2006) Crystallographic evidence that the dinuclear copper center of tyrosinase is flexible during catalysis. J Biol Chem 281:8981–8990

Matoba Y, Bando N, Oda K et al (2011) A molecular mechanism for copper transportation to tyrosinase that is assisted by a metallochaperone, caddie protein. J Biol Chem 286: 30219–30231

Solomon EI, Sundaram UM, Machonkin TE (1996) Multicopper oxidases and oxygenases. Chem Rev 96:2563–2606

Takizawa T, Mitsumori K, Tamura T et al (2003) Hepatocellular tumor induction in heterozygous p53-deficient CAB mice by 26-week dietary administration of kojic acid. Toxicol Sci 73:287–293

Volbeda A, Hol WG (1989) Crystal structure of hexameric haemocyanin from *Panulirus interruptus* refined at 3.2 Å resolution. J Mol Biol 209:249–279

Part VIII
Energy and Chemicals

Chapter 22
Toward Realization of New Biorefinery Industries Using *Corynebacterium glutamicum*

Haruhiko Teramoto, Masayuki Inui, and Hideaki Yukawa

Abstract Biorefinery technologies for the sustainable production of fuels and chemicals by bioprocesses using biomass resources as a feedstock have been extensively studied worldwide, aiming at construction of a society in the twenty-first century based on non-fossil renewable resources. The leading commercial production of bioethanol uses starch biomass (grains) as a feedstock. Because of concern about competition with food resources, the use of non-food biomass such as agricultural residues and energy crops is required. However, the lignocellulosic biomass contains significant amounts of C5 sugars such as xylose and arabinose, which present utilization difficulties for the microorganisms being used. This chapter describes our research and development at the Research Institute of Innovative Technology for the Earth (RITE) of growth-arrested bioprocesses using *Corynebacterium glutamicum*, an industrially useful bacterium. We constructed a genetically engineered strain that can consume xylose and arabinose at the same rate as innately preferred glucose without cell growth under oxygen deprivation. The efficient and simultaneous utilization of C6 and C5 sugars is advantageous for solution of the key technological barriers to realization of biorefinery industries.

Keywords Biorefinery • *Corynebacterium glutamicum* • Non-food biomass

22.1 Introduction

Biorefinery is an industrial technology concept, proposed by the United States Department of Energy, for the sustainable production of fuels and chemicals using biomass resources as a feedstock. The bio-based products are expected to be environmentally friendly and an alternative to a variety of the oil-based products from today's

H. Teramoto • M. Inui • H. Yukawa (✉)
Research Institute of Innovative Technology for the Earth,
9-2, Kizugawadai, Kizugawa, Kyoto 619-0292, Japan
e-mail: mmg-lab@rite.or.jp

H. Anazawa and S. Shimizu (eds.), *Microbial Production: From Genome Design to Cell Engineering*, DOI 10.1007/978-4-431-54607-8_22, © Springer Japan 2014

petroleum refineries. In the United States, research and development of biorefinery technologies have been intensively promoted since the 1990s as a national strategy for construction of a society in the twenty-first century based on non-fossil renewable resources. At present, the importance of these technologies is becoming noticed world-wide, and much effort has been made for the establishment of biorefinery industries.

According to F.O. Licht, bioethanol production is markedly increasing in the world. The bioethanol production in the United States accounts for 60 % of that in the world, and the U.S. Environmental Protection Agency approved E15 (a blend of gasoline and up to 15 % ethanol) in 2010. Thus, it is expected that demand for bio-ethanol will increase further. However, the usage of corn as a feedstock for the production of bioethanol had now reached 40 % of the total usage in the United States. Because of concern about competition with food resources, bioethanol production from non-food biomass is required. In this context, the United States government strongly promotes the development of biofuels using non-food biomass resources, that is, agricultural residues and energy crops as a feedstock. Cellulosic ethanol is expected to be more environmentally friendly; a life-cycle assessment indicates that it is highly effective for reduction of CO_2 emission.

The bio-based production of chemicals is also a promising field. The relevant target products as green chemicals have expanded from fine chemicals, including propanediol, to commodity chemicals, that is, acrylic acid, isoprene, various carboxylic acids, amines, and aromatics. Many joint ventures to develop green chemicals have been established. The future market size is estimated to reach US$70 billion in 2015 (Markets and Markets 2011) and US$100 billion in 2020 (Pike Research 2011). The use of non-food biomass resources as a feedstock for bio-chemicals is as highly desirable as for biofuels.

This manuscript describes our research and development at Research Institute of Innovative Technology for the Earth (RITE) of key technologies for the production of biofuels and biochemicals from non-food biomass.

22.2 Growth-Arrested Bioprocess

In a conventional fermentation process, a target compound is produced during cultivation of the compound-producing microorganisms with sugars as a feedstock. In other words, the sugar feedstock is converted to microbial biomass and a target product. The growth of microorganisms is a major negative factor for the economic efficiency because of (1) reduction of the specific feedstock consumption, (2) reduction of the target compound production rate, and (3) an increase in formation of by-products and (4) the purification cost involved.

Corynebacterium glutamicum, a high-GC gram-positive bacterium, has a long history as an industrial workhorse for the aerobic fermentation processes to produce amino acids such as glutamate and lysine. We found that this microorganism stops growing under oxygen deprivation, but maintains its major carbohydrate metabolic capabilities. Based on this property, we developed a growth-arrested bioprocess that can overcome the disadvantages intrinsic to the conventional fermentation processes just described (Inui et al. 2004b). The growth-arrested bioprocess decouples cell growth from the product production process. The product is produced from a sugar feedstock by the cells densely packed in a reactor. This process using cells as a chemical catalyst can achieve much higher STY (*S*pace *T*ime *Y*ield) compared to the conventional fermentation processes (Inui et al. 2004a, b; Okino et al. 2005).

22.3 Development of Non-food Biomass Utilization Technology

As for non-food biomass resources as a feedstock for biofuels and biochemicals, availability of agricultural residues and energy crops has been discussed by many research groups. The agricultural residues studied as feedstocks include corn stover, rice straw, and wheat straw. Various terrestrial plants, such as *Miscanthus*, switchgrass, and energy cane, have been studied as energy crops. The yield of energy crops per unit arable area can be higher by one order of magnitude than that of the corn mainly used for current production of bioethanol. In addition, energy crops can be cultivated under a wide range of conditions of climate and soil quality, thus avoiding competition with food crops for arable land. Therefore, it is expected that the use of energy crops will have a positive geopolitical impact, that is, promotion of agriculture in poor areas and creation of new employment opportunities.

The production of biofuels and biochemicals from non-food biomass consists of the saccharification and bioconversion processes. In the saccharification process, sugars are generated from lignocellulosic biomass, and are subsequently converted to desired products in a bioconversion process. In this process, cellulases used are key to economic efficiency, and it is necessary to reduce the cost of enzyme production to the largest extent possible. On the other hand, in the bioconversion process, the establishment of the technologies described next is indispensable.

The three technological characteristics necessary for bioconversion in economically feasible lignocellulosic biorefinery processes are (1) simultaneous utilization of C6 and C5 sugars, (2) tolerance to fermentation inhibitors, and (3) high productivity (Dien et al. 2003). In contrast to starch biomass consisting of C6 sugars mainly used for current bioprocesses, cellulosic biomass contains significant amounts of C5 sugars such as xylose and arabinose in addition to major C6 sugar glucose. Typical lignocellulosic biomass hydrolysates mainly contain glucose and xylose in a ratio of 2:1 to 3:2 with other minor sugars, such as arabinose. Thus, the microorganism used for the bioconversion of sugars derived from the lignocellulosic biomass should utilize C6 and C5 sugars efficiently and simultaneously. Furthermore, physical and thermochemical pretreatment is necessary to facilitate enzyme access for the hydrolytic reactions in the saccharification process, but excessive degradation of the lignocellulosic biomass generates by-products such as phenolics, furans, and organic acids. These by-products, called fermentation inhibitors, are highly problematic because they reduce the productivity of downstream bioconversion processes.

Although the physical and chemical methods for removing the fermentation inhibitors have been studied, the additional process increases production cost. Therefore, a bioconversion process tolerant to the fermentation inhibitors is strongly required.

22.3.1 Simultaneous Utilization of C6 and C5 Sugars

According to the U.S. Department of Energy, the current feedstock cost accounts for about 35 % of the total cost for the production of bioethanol (U.S. Department of Energy 2012). Thus, efficient utilization of sugars derived from the feedstock biomass is very important to establish cost-competitive biorefinery processes. For cellulosic ethanol, the native ethanol-producing microorganisms used, *Saccharomyces cerevisiae* and *Zymomonas mobilis*, cannot utilize C5 sugars, xylose and arabinose, as a fermentation carbon source. Although these microorganisms are genetically engineered to utilize C5 sugars, the current consumption rate of C5 sugars is quite low and further studies for strain improvement are in progress.

Because a *C. glutamicum* wild-type strain we have used also cannot utilize C5 sugars, we constructed a recombinant strain transformed with the *Escherichia coli* xylose metabolism gene *xylAB* (Kawaguchi et al. 2006). Xylose is metabolized by two-step reactions catalyzed by xylose isomerase (encoded by *xylA*) and xylulokinase (encoded by *xylB*) to D-xylulose-5-phosphate, an intermediate metabolite of the pentose phosphate pathway (Fig. 22.1). The resulting *C. glutamicum* recombinant strain grew on xylose as a sole carbon source. When this strain was aerobically grown with a mixture of glucose and xylose, glucose was preferentially consumed. In contrast, xylose was consumed in the presence of glucose by the growth-arrested cells of this recombinant strain under oxygen deprivation. The simultaneous utilization of glucose and xylose in the growth-arrested bioprocess is advantageous for the industrial use of non-food biomass because the bioconversion reaction times are shortened. We also introduced the *Escherichia coli* arabinose metabolism genes *araBAD* into *C. glutamicum* (Kawaguchi et al. 2008). Arabinose is converted to the intermediate metabolite (D-xylulose-5-phosphate) of the pentose phosphate pathway through three-step reactions catalyzed by L-arabinose isomerase (encoded by *araA*), L-ribulokinase (encoded by *araB*), and L-ribulose-5-phosphate 4-epimerase (encoded by *araD*) (Fig. 22.1). This genetic engineering conferred the ability of arabinose utilization to *C. glutamicum* in a manner similar to the results of the genetic engineering for xylose utilization described earlier. At that time the consumption rate of xylose and arabinose was notably lower than that of glucose, and significantly declined at lower concentrations of these C5 sugars. In this context, we screened many strains of *C. glutamicum* for the ability of utilization of C5 sugars and found a strain (ATCC 31831) that can utilize arabinose efficiently. Subsequently, we identified a transporter gene (*araE*) in the gene cluster containing arabinose metabolism genes (*araBDA*) on the chromosome (Kawaguchi et al. 2009). The corresponding genes involved in arabinose utilization are not found in the published genomic sequences of the other *C. glutamicum* strains and closely related bacterial species. Thus, it is likely that the cluster of genes involved in arabinose

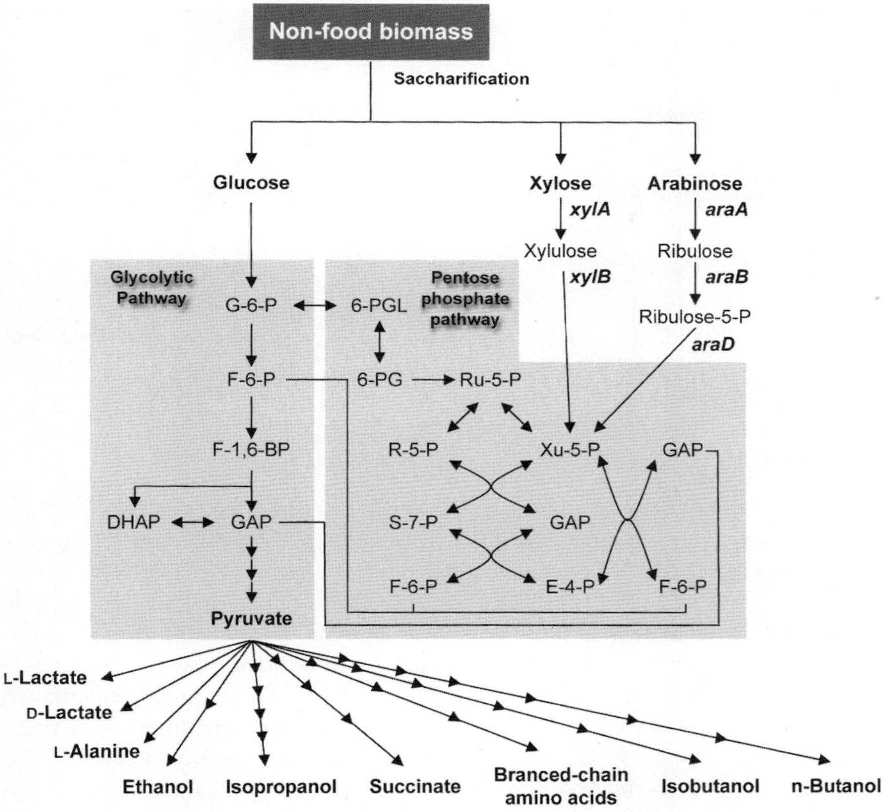

Fig. 22.1 Metabolic engineering of *Corynebacterium glutamicum* for the production of biofuels and biochemical from non-food biomass

utilization is acquired by *C. glutamicum* ATCC 31831 via a horizontal gene transfer. Introduction of the transporter gene (*araE*) into the respective xylose- and arabinose-utilizing recombinant strains resulted in a marked increase in the consumption rate of these C5 sugars, indicating that the *araE* gene product functions as an efficient transporter of not only arabinose but also xylose (Sasaki et al. 2009). Finally, we constructed a recombinant strain carrying both xylose and arabinose metabolism genes along with the pentose transporter gene. It is interesting to note that the resulting strain can simultaneously consume glucose, xylose, and arabinose at the same rates under oxygen deprivation (Fig. 22.2) (Sasaki et al. 2009). In addition, we have genetically engineered *C. glutamicum* for utilization of cellobiose (Sasaki et al. 2009) and mannose (Sasaki et al. 2011). Cellobiose is a partial degradation product of cellulose. Mannose is widely contained in lignocellulosic biomass, and especially abundant in the woody biomass hydrolysates. We think that the growth-arrested bioprocess based on these findings largely overcomes the technological barrier in the utilization of mixed sugars.

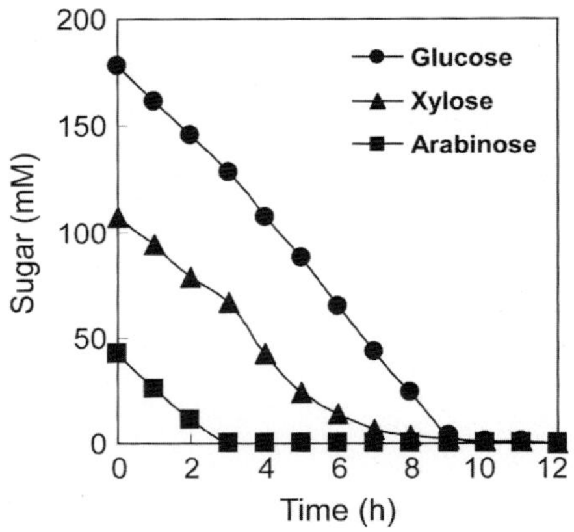

Fig. 22.2 Simultaneous consumption of a mixture of glucose, xylose, and arabinose by a recombinant *Corynebacterium glutamicum* strain

22.3.2 Tolerance to Fermentation Inhibitors

Hydrothermal treatment or acidic/alkaline treatment of lignocellulosic biomass before enzymatic hydrolysis generates phenolics, furans, and organic acids as by-products (Fig. 22.3). The phenolics derived from lignin degradation include 4-hydroxybenzaldehyde, vanillin, and syringaldehyde. Degradation of hemicellulose generates acetic acid, and also furans such as furfural and 5-hydroxymethyl-2-furaldehyde. Severe pretreatment, which can reduce the requisite amount of expensive saccharification enzymes, significantly increases the amounts of these fermentation inhibitors. In contrast, mild pretreatment decreases formation of the fermentation inhibitors, but more enzymes are required for saccharification. In these situations, we assessed effects of the fermentation inhibitors on the productivity of ethanol by the growth-arrested bioprocess using a recombinant ethanol-producing *C. glutamicum* strain (Fig. 22.3) (Sakai et al. 2007). As a result, the productivity was hardly affected by the representative inhibiting concentrations of the various inhibitors such as furfural, 5-hydroxymethyl-2-furaldehyde, 4-hydroxybenzaldehyde, vanillin, and syringaldehyde. Furthermore, a mixture of these inhibitors did not reduce ethanol productivity. It should be noted that these fermentation inhibitors inhibited the aerobic growth of *C. glutamicum*, indicating the advantage of the growth-arrested bioprocess over the conventional fermentation processes for solution of the technological barrier with the fermentation inhibitors.

22.3.3 Creation of High-Producing Strains

As described earlier, the growth-arrested bioprocess is based on the physiological property of *C. glutamicum*, which shows high metabolic capability without cell growth under oxygen deprivation. Under these conditions, sugars are converted to lactic acid, succinic acid, and a small amount of acetic acid (Inui et al. 2004b). The ratio of lactic

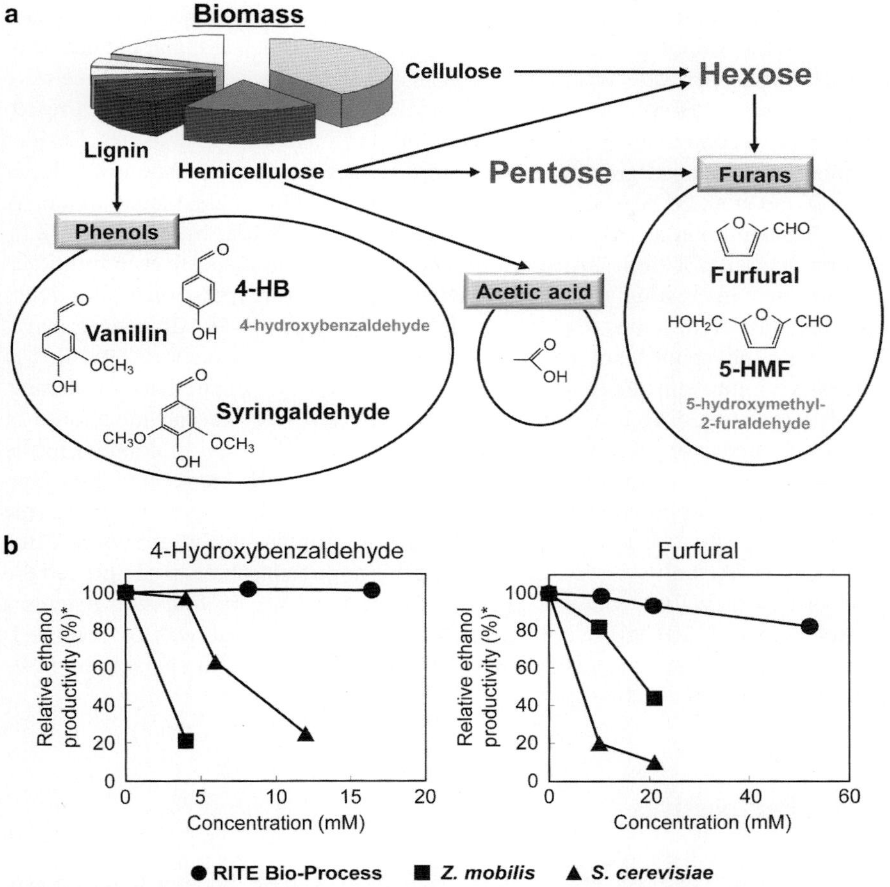

Fig. 22.3 Major fermentation inhibitors derived from non-food biomass (**a**) and their effects on ethanol productivity (**b**). *Relative to productivity without inhibitors

acid to succinic acid varies depending on the amount of bicarbonate supplemented. Bicarbonate supplementation enhances production of succinic acid because succinic acid is produced via a carbon dioxide fixation reaction catalyzed by phosphoenolpyruvate carboxylase followed by the reductive TCA pathway, indicating that carbon dioxide is possibly used as an auxiliary material in this bioprocess (Inui et al. 2004b). Based on the metabolic capability under oxygen deprivation, metabolic engineering of *C. glutamicum* enabled us to construct recombinant strains for high production of various valuable compounds by the growth-arrested bioprocesses. In this context, it should be noted that the genetic engineering tools of *C. glutamicum* have been sufficiently established. Furthermore, remarkable advances have been recently made in understanding the physiological function and regulation of genes in this bacterium by genome-wide analyses, such as transcriptomics and metabolomics, based on the whole-genomic sequences of multiple strains determined.

To construct an ethanol-producing strain, the *C. glutamicum* lactate dehydrogenase gene, which is responsible for lactic acid formation, was deleted, and the heterologous *Z. mobilis* pyruvate decarboxylase gene and alcohol dehydrogenase gene were highly expressed (Inui et al. 2004a). Ethanol productivity by the growth-arrested cells of this recombinant strain increased in proportion to the cell densities. Besides ethanol, we developed growth-arrested bioprocesses for the production of D-lactic acid (Okino et al. 2008b), succinic acid (Okino et al. 2008a), alanine (Jojima et al. 2010; Yamamoto et al. 2012), valine (Hasegawa et al. 2012a, b), etc. (Fig. 22.1). The productivity of these processes exceeded that of the respective conventional bioprocesses. Interestingly, we observed that stepwise overexpression of four genes encoding glycolytic enzymes (glyceraldehyde-3-phosphate dehydrogenase, pyruvate kinase, phosphofructokinase, and glucose-6-phosphate isomerase) resulted in successive improvements in glucose metabolism and the productivity of alanine, which is generated by a one-step reaction from pyruvate, an end metabolite of glycolysis, under oxygen deprivation (Yamamoto et al. 2012), in contrast to the minimal effect of overexpression of glycolytic genes on the conventional fermentation processes in previous studies. Therefore, an increase in glycolytic enzyme activity is an effective approach to make drastic progress in the growth-arrested bioprocesses. Actually, we also showed that this approach efficiently improved the productivity of valine, which is generated from pyruvate by four-step reactions (Hasegawa et al. 2012a). It is interesting to note that the pathway of branched-chain amino acids is a promising route for next-generation biofuels including isobutanol (Atsumi et al. 2008).

22.4 Perspectives

The projects for cellulosic ethanol in the United States have recently slowed down. The amount of cellulosic biofuel required to be used in 2011 by the Renewable Fuel Standard program (RFS2) was 6 million gallons (23,000 kl). However, the Environmental Protection Agency reported that no cellulosic biofuel was produced in 2011. A similar trend has been reported in 2012, implying that the technological barriers to industrialization are difficult to overcome. A breakthrough for the early solution of these issues is strongly required. The innovative technology for the efficient utilization of non-food biomass is fundamental to the industrialization of cellulosic biofuels, and it also enables the establishment of renewable green chemicals. Taking advantage of the growth-arrested bioprocess developed at RITE, we are collaborating domestically and internationally with both private and public sectors, aiming at early realization of biorefinery industries, that is, industrialization of cellulosic ethanol, butanol, amino acids, aromatic compounds, and other biochemicals.

Acknowledgments We thank Crispinus A. Omumasaba (RITE) for critical reading of the manuscript.

References

Atsumi S, Hanai T, Liao JC (2008) Non-fermentative pathways for synthesis of branched-chain higher alcohols as biofuels. Nature (Lond) 451:86–89

Dien BS, Cotta MA, Jeffries TW (2003) Bacteria engineered for fuel ethanol production: current status. Appl Microbiol Biotechnol 63:258–266

Hasegawa S, Suda M, Uematsu K, Natsuma Y, Hiraga K, Jojima T, Inui M, Yukawa H (2012a) Engineering of *Corynebacterium glutamicum* for high yield L-valine production under oxygen deprivation conditions. Appl Environ Microbiol 79:1250–1257

Hasegawa S, Uematsu K, Natsuma Y, Suda M, Hiraga K, Jojima T, Inui M, Yukawa H (2012b) Improvement of the redox balance increases L-valine production by *Corynebacterium glutamicum* under oxygen deprivation conditions. Appl Environ Microbiol 78:865–875

Inui M, Kawaguchi H, Murakami S, Vertès AA, Yukawa H (2004a) Metabolic engineering of *Corynebacterium glutamicum* for fuel ethanol production under oxygen-deprivation conditions. J Mol Microbiol Biotechnol 8:243–254

Inui M, Murakami S, Okino S, Kawaguchi H, Vertès AA, Yukawa H (2004b) Metabolic analysis of *Corynebacterium glutamicum* during lactate and succinate productions under oxygen deprivation conditions. J Mol Microbiol Biotechnol 7:182–196

Jojima T, Fujii M, Mori E, Inui M, Yukawa H (2010) Engineering of sugar metabolism of *Corynebacterium glutamicum* for production of amino acid L-alanine under oxygen deprivation. Appl Microbiol Biotechnol 87:159–165

Kawaguchi H, Vertès AA, Okino S, Inui M, Yukawa H (2006) Engineering of a xylose metabolic pathway in *Corynebacterium glutamicum*. Appl Environ Microbiol 72:3418–3428

Kawaguchi H, Sasaki M, Vertès AA, Inui M, Yukawa H (2008) Engineering of an L-arabinose metabolic pathway in *Corynebacterium glutamicum*. Appl Microbiol Biotechnol 77:1053–1062

Kawaguchi H, Sasaki M, Vertès AA, Inui M, Yukawa H (2009) Identification and functional analysis of the gene cluster for L-arabinose utilization in *Corynebacterium glutamicum*. Appl Environ Microbiol 75:3419–3429

Markets and Markets (2011) Global renewable chemicals market by geography, feedstocks, prices, applications trends and forecasts (2010–2015). http://www.marketresearch.com/MarketsandMarkets-v3719/Global-Renewable-Chemicals-Geography-Feedstocks-6111458/. Accessed 1 Feb 2011

Okino S, Inui M, Yukawa H (2005) Production of organic acids by *Corynebacterium glutamicum* under oxygen deprivation. Appl Microbiol Biotechnol 68:475–480

Okino S, Noburyu R, Suda M, Jojima T, Inui M, Yukawa H (2008a) An efficient succinic acid production process in a metabolically engineered *Corynebacterium glutamicum* strain. Appl Microbiol Biotechnol 81:459–464

Okino S, Suda M, Fujikura K, Inui M, Yukawa H (2008b) Production of D-lactic acid by *Corynebacterium glutamicum* under oxygen deprivation. Appl Microbiol Biotechnol 78:449–454

Pike Research (2011) Green chemical industry to soar to $98.5 billion by 2020. http://www.pikeresearch.com/newsroom/green-chemical-industry-to-soar-to-98-5-billion-by-2020. Accessed 20 Jun 2011

Sakai S, Tsuchida Y, Nakamoto H, Okino S, Ichihashi O, Kawaguchi H, Watanabe T, Inui M, Yukawa H (2007) Effect of lignocellulose-derived inhibitors on growth of and ethanol production by growth-arrested *Corynebacterium glutamicum* R. Appl Environ Microbiol 73:2349–2353

Sasaki M, Jojima T, Kawaguchi H, Inui M, Yukawa H (2009) Engineering of pentose transport in *Corynebacterium glutamicum* to improve simultaneous utilization of mixed sugars. Appl Microbiol Biotechnol 85:105–115

Sasaki M, Teramoto H, Inui M, Yukawa H (2011) Identification of mannose uptake and catabolism genes in *Corynebacterium glutamicum* and genetic engineering for simultaneous utilization of mannose and glucose. Appl Microbiol Biotechnol 89:1905–1916

U.S. Department of Energy (2012) Biomass multi-year program plan. http://www1.eere.energy.gov/biomass/pdfs/mypp_april_2012.pdf. Accessed Apr 2012

Yamamoto S, Gunji W, Suzuki H, Toda H, Suda M, Jojima T, Inui M, Yukawa H (2012) Overexpression of genes encoding glycolytic enzymes in *Corynebacterium glutamicum* enhances glucose metabolism and alanine production under oxygen deprivation conditions. Appl Environ Microbiol 78:4447–4457

Chapter 23
Hydrogen Production Using Photosynthetic Bacteria

Jun Miyake

Abstract Hydrogen production using photosynthetic bacteria is reviewed here with special reference to the contributions from Japan. Research has been focused on various aspects related to the subject, from basic to applied. The capability of biomass conversion to hydrogen was studied from the point of view of energetics. Application technologies as reactor designs have been studied extensively in Japan in the 1990s. Light penetration to the bacterial suspension was analyzed, and a method for maximizing hydrogen production from a vessel was proposed. After great problems with electricity supply, serious approaches are being carried out for realization of renewable energy systems.The combination of biologically prepared hydrogen with renewable energy such as solar cells and wind turbines is examined to stabilize the output of small-scale electric grids. The hydrogen production process is initiated by the conversion of light quanta to chemical energy. The early process of photosynthesis and the following chemical processes are still on the frontiers of science. A mysterious event of ATP synthesis at the mitochondrial membrane is also reviewed. Studies are also concerned with the structures and physical and biochemical functions of proteins in early photosynthesis. Applications of basic science on the early process of photosynthesis have been continuous in Japan with the development of photoelectric devices with Japanese cutting-edge technology of electron conductive materials as carbon nanotubes.

Keywords Biohydrogen • Energy conversion • Hydrogen • Photosynthesis • Solar energy

---

J. Miyake (✉)
Graduate School of Engineering Science, Osaka University,
1-3 Machikane-Yama, Toyonaka, Osaka 560-8531, Japan
e-mail: jun_miyake@bpe.es.osaka-u.ac.jp

H. Anazawa and S. Shimizu (eds.), *Microbial Production: From Genome Design to Cell Engineering*, DOI 10.1007/978-4-431-54607-8_23, © Springer Japan 2014

23.1 Introduction

Phototrophic microorganisms provide a simple method for photohydrogen generation. Much research has been done on photosynthetic bacterial hydrogen production. Photosynthetic bacteria produce hydrogen by light energy with organic molecules as electron donors. In Japan, the technology has been studied since the 1970s. All research subjects have been examined, namely, bacteriology, screening of bacteria, applied microbiology, energetics, and renewable energy development, reactors, waste treatments, and molecular studies. Recently, artificial photosynthesis has attracted growing interest. Government ministries such as METI/NEDO, MEXT/JST, local governments, and private companies support the research. In this chapter, the technology is detailed with special reference to Japanese studies. Several reviews have been published overviewing the technology in Japan and in the world (Asada and Miyake 1999; Miyake et al. 2001; Miyake et al. 2003; Miyake 2007; Miyake 1998; Kawasugi et al. 1998; Miyake et al. 1999a, b). Research on microbiology, such as screening, culture conditions, concomitant production of hydrogen with valuable chemicals, and wastewater treatment, has been carried out in various laboratories in Japan (Watanabe et al. 1998; Sasaki 1998; Watanabe et al. 1999; Noparatonaraporn et al. 2001; Ko and Noike 2002; Takabatake et al. 2004; Asada et al. 2008; Nara 2012).

23.2 The Importance of Photosynthetic Bacterial Hydrogen Production

Solar energy provides a huge total amount, 5.7×10^{24} J/year, irradiated to the Earth's surface. Carbon dioxide at 2×10^{11} t is fixed by photosynthesis per year (=3×10^{21} J/year). Humans use a large amount of energy, mostly from fossil resources, but their consumption is relatively small (3×10^{20} J/year) compared to the available natural energy (Hall 1998; Wakayama and Miyake 2001).

Compared to the total amount of solar energy, the effective energy concentration (energy/unit area) at any one point on the Earth's surface is limited, only about 1 kW/m^2, even at noon. The low effective energy concentrations limit the use of solar energy as a primary energy source and elevate the costs associated with its accumulation and transmission.

Utilization of low-density (high entropy state) energy sources requires new technology to facilitate the use of solar energy. Photosynthetic microorganisms could provide a very simple method with relatively minimal investment and resource requirements because the major part of the energy conversion is carried out within the cell without the need to use complicated plants.

Table 23.1 Classification of hydrogen-evolving bacteria

Enzyme	Class of bacteria		Genus of bacteria	Electron donor
Hydrogenase	Green algae	Heterocyst	*Chlamydomonas* *Chlorella* *Anabaena*	Water
	Blue-green algae	Non-heterocyst	*Oscillatoria*	
Nitrogenase	Photosynthetic bacteria	Non-sulfur	*Rhodopseudomonas* *Rhodospirillum*	Organic materials
		Sulfur	*Chromatium* *Thiocapsa*	Sulfate

23.2.1 New Technology of Solar Energy Conversion Using Photosynthesis

The use of renewable energy needs to overcome problems of low density and also their fluctuating nature. Before the industrial revolution, wood served as a major energy source. The industrial revolution gave rise to the widespread use of coal and petroleum, nonrenewable sources, as energy sources. This integrated system soon reached its limits, making it necessary to consider discrete social systems that utilize delocalized abundant energy sources.

Integration of resources is necessary for industrial use, but mechanical methods for this integration incorporate large energy expenditures. Existing problems cannot therefore be resolved without modification of the system. Microorganisms have the potential to simplify the conversion and accumulation of solar energy and energy utilization over large areas.

Biomass is produced by the actions of air, water, and soils using sunlight as the energy source; it is a renewable source of clean energy. Only a small portion of this biomass is presently used, and it is an unused resource that should be used effectively. Approximately 2 billion metric tons of biomass are present on the Earth, and 0.2 billion metric tons are added every year. Biomass produced per year corresponds to approximately ten times the yearly world energy use. Biomass can also be used as a substrate for energy production, particularly for hydrogen production.

By using microorganisms, hydrogen can be obtained from biomass. Diverse photosynthetic bacteria are able to produce hydrogen, ranging from photosynthetic microorganisms that depend on light energy for acquiring the necessary energy for growth to nonphotosynthetic microorganisms that depend on organic and inorganic compounds (Table 23.1). Microorganisms capable of producing hydrogen are classified into anaerobic bacteria, fermentation bacteria, aerobic bacteria, photosynthetic bacteria, and algae. These microorganisms can be used alone or as mixtures of multiple microorganisms, depending on the biomass to be utilized.

23.2.2 Photosynthesis Mechanisms for Evaluation of Efficiency

Biological energy conversions can be categorized into two groups: (1) photosynthesis (the process whereby solar energy is fixed to yield energy useful to organisms and industry), and (2) biomass conversion (the product of photosynthesis) into energy. Photosynthesis occurs in plants, algae, and photosynthetic bacteria, whereas biomass conversion reactions often occur in nonphotosynthetic microorganisms.

Photosynthesis is often regarded as a CO_2 anabolic reaction, whereby glucose is formed from CO_2 and water. CO_2 anabolism is an energy-consuming reaction in that it utilizes chemical energy produced by photosynthesis. In its narrowest sense, photosynthesis can be regarded as a process whereby energy is supplied for CO_2 anabolism. In a broader sense, photosynthesis, including CO_2 anabolism, can be divided into several steps: (1) photoelectric charge isolation using photon energy (conversion to electrical energy), (2) fixation of electrical energy in the form of chemical energy (ATP synthesis), and (3) chemical reactions involving ATP (fixation of CO_2, and hydrogen production).

The supply of energy for CO_2 anabolism is common to all organisms that exhibit photosynthesis. Energy conversion, ATP synthesis, and the production of both CO_2 and hydrogen, on the other hand, are not unique to photosynthetic organisms, but occur in all types of microorganisms, and are in fact similar to the respiratory processes that occur in the mitochondria of higher organisms.

Plant photosynthesis takes place only in the presence of visible light (400–700 nm). Solar light contains both visible and infrared components. Because visible light accounts for about 45 % of all solar energy, the maximum achievable energy efficiency for CO_2 fixation using solar radiation is approximately 13 %.

Bacterial photosynthesis is thought to be a relatively old form of photosynthesis. It incorporates the use of either organic or sulfur compounds as electron donors in the photosystem I (Fig. 23.1). In contrast to plant photosynthesis, cyclic photophosphorylation takes place in bacterial photosynthesis, that is, electrons are repeatedly excited in a cyclic manner, with ATP being generated in each cycle. Photosynthetic bacteria are also capable of reducing electron carriers such as NAD via a linear reaction similar to the electron transmission that occurs during plant photosynthesis (Fig. 23.1).

ATP is synthesized by using the energy of photons. Stoichiometry is not established. Excited electron energy is utilized in ATP synthesis. In contrast to ordinary chemical reactions, ATP synthesis does not give an integer (Wakayama and Miyake 2001). The ratio of excited photons to ATP produced is still a somewhat debatable issue: two or three photons give rise to the formation of two ATP molecules (Schlodder et al. 1982). Ohsawa and Hayashi have suggested a loose coupling between proton transport and ATP synthesis (Oosawa and Hayashi 1984, 1986; Oosawa 2000). We adopt the relationship as the range of 2–3 ATPs per 4 photons following the loose coupling hypothesis:

$$2\,\text{photons} + 1\ (\text{or more})\ \text{ADP} = 1\ (\text{or more})\ \text{ATP}. \tag{23.1}$$

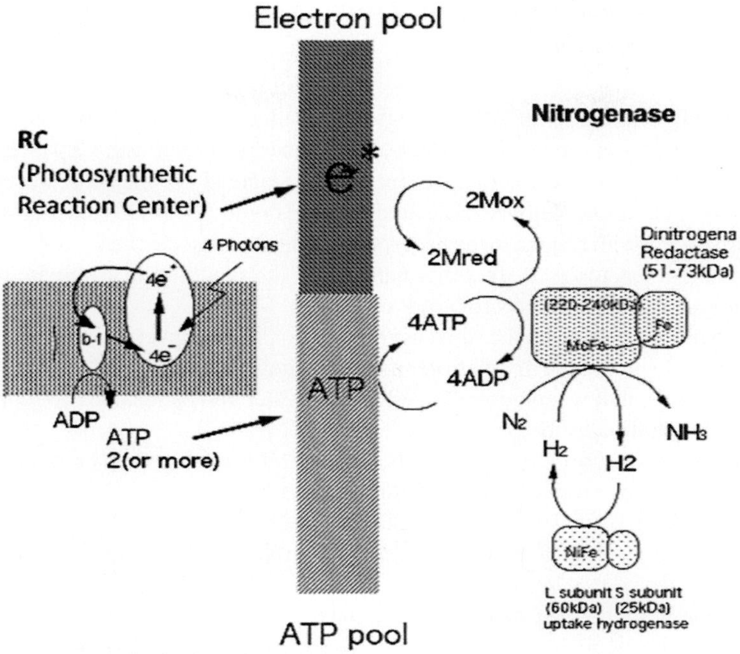

Fig. 23.1 Mechanism for hydrogen production by photosynthetic reaction systems

Electrons are extracted from organic substances as follows:

$$CH_2O + H_2O = 4e^- + 4H^+ + CO_2 \tag{23.2}$$

The structure of the photosynthetic reaction center (RC), involved in the early steps of photosynthesis, has been elucidated for certain photosynthetic bacteria (Fig. 23.1). Such chlorophyll-containing bacteria, which include *Rhodopseudomonas viridis* and *Rhodobacter sphaeroides*, show similarities with respect to the arrangement of chlorophyll and the three-dimensional structures of major portions of the proteins possessing that pigment. Such structural similarities between photosynthetic bacteria seem to suggest the acquisition of an optimal structure by these bacteria over a long evolutionary period.

Pigments such as bacteriochlorophyll are also present within the RC. Photoelectric charge isolation takes place within dimers of these bacteriochlorophyll pigments, resulting in the release of high-energy electrons, via the action of bacteriochlorophyll monomers such as bacteriopheophytin, quinone A, and quinone B. These high-energy electrons are subsequently conjugated with proton transportation in the cytochrome b/c_1 complex.

23.2.3 *Hydrogen Production Through Solar Energy Conversions*

Although large amounts of solar energy are irradiated to the Earth's surface, the actual concentration (energy/unit area) of solar energy at the Earth's surface is relatively low. The accumulation of solar energy for practical use therefore necessitates collection of solar energy over large areas, which could potentially be a costly venture, particularly with respect to the use of solar batteries for solar energy concentration. Photosynthetic bacteria, on the other hand, have the potential to eliminate requirements for such large batteries for solar collection.

Hydrogen appears to be the most useful form of solar-converted energy, in that it can be easily substituted for petroleum-based fuels. The use of hydrogen is also advantageous in that when burned it yields only water and hence does not contribute to environmental pollution.

Hydrogen production by photosynthetic bacteria is mostly catalyzed by nitrogenase. Nitrogenases require ATP for catalytic action:

$$2H^+ + 2e^- * (Fd) + 4ATP = H_2 + 4ADP \tag{23.3}$$

where e^-* are high-energy electrons in NADH, Fd_{red}, etc.

Bacterial mechanisms for photosynthetic hydrogen production are summarized in Figs. 23.2, 23.3, and 23.4. Upon exposure of ammonia-free media containing photosynthetic bacteria to light, nitrogenase activity is induced, resulting in hydrogen production. Organic substances such as lactic acid (Eq. 23.4) serve as electron donors in photosynthetic bacteria. In such reactions, ΔG is positive, indicating that

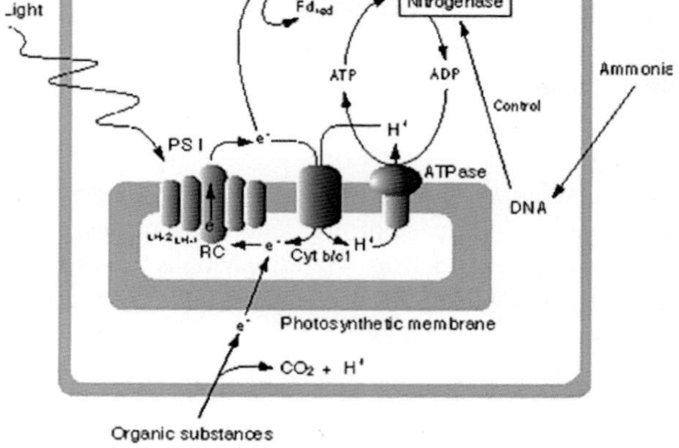

Fig. 23.2 Mechanism for hydrogen production by photosynthetic bacteria

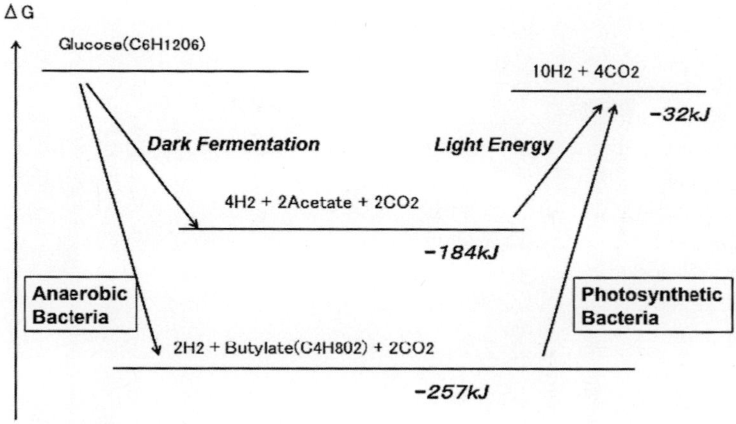

Fig. 23.3 Energy diagram of anaerobic and photosynthetic hydrogen production. The two methods are complementary (Miyake 1990; Miyake et al. 1985)

the use of solar energy allows photosynthetic bacteria to produce hydrogen through complete decomposition of organic substances. Anaerobes such as *Clostridium* also produce hydrogen but are incapable of completely utilizing energy or decomposing organic substances (Eq. 23.5):

$$C_3H_6O_3 + 3H_2O = 12H^+ + 3CO_2 + 12e^- = 6H_2 + 3CO_2 \quad \Delta G = 51.2\text{kJ} \quad (23.4)$$

$$C_6H_{12}O_6 + 2H_2O = 4H_2 + 2CH_3COOH + 2CO_2 \quad \Delta G = -184.2\text{kJ} \quad (23.5)$$

For detailed evaluation of hydrogen production of this type, we have to consider the decomposition of organic substances. Studies on hydrogen production through the exposure of organic wastes (waste fluids from food factories, pulp factories, etc.) to photosynthetic bacteria and light have been conducted in a number of countries. Research into hydrogen production using a combination of photosynthetic bacteria and anaerobes has also been conducted.

A key factor in determining the commercial applicability of hydrogen production processes is the rate at which hydrogen is produced. Bacteria have been widely investigated for their rates of hydrogen production. To date, *R. sphaeroides* has been identified as the bacterium having the highest hydrogen-producing rate (260 ml/mg/h) (Miyake and Kawamura 1987), with a photoenergy conversion efficiency (energy yielded by combustion of produced hydrogen/incident solar energy) of 7 %, determined using a solar simulator. Further strain development will potentially elevate the energy conversion efficiency of photosynthetic bacteria to levels comparable to those of solar batteries.

The rather low efficiency of biological energy conversion is balanced by stable productivity. Biological hydrogen production could not be the champion in renewable energy technologies, but complementary use with wind turbines and solar cells could open the way. Biological systems have profound capability to make the biosystem function against low energy density and time fluctuation (both are entropic problems).

Fig. 23.4 **(a)** Ocean float (volume 800 l). **(b)** Ocean float (tubular type). **(c)** Ocean float (plastic bag type). **(d)** Solar chaser (Himawari). Collected light is transferred to the reactor to illuminate the inside of the bacterial suspension

Solar batteries react to light energy immediately, but clouds and rains change the output. For actual application, we need to smooth the output using batteries. Photosynthetic bacterial hydrogen production does not correspond to the changes directly because the process is complicated. This complexity is the nature of photosynthesis, which enables plants to accumulate the energy to survive. The value of nature for practical applications is discussed later.

X-ray crystal analyses of hydrogenase and light-harvesting proteins have been intensively studied for improving oxygen tolerance and energy collection (Colbeau et al. 1998; Nagashima et al. 2012; Shomura et al. 2011).

23.3 Light Energy and Hydrogen Production in Reactors

23.3.1 Reactors for Hydrogen Production

The greatest advantage of photobiological hydrogen production by photosynthetic microorganisms is that sunlight is used as an energy source adding to substrates. Practical places for photobiological hydrogen production using sunlight are in the

Fig. 23.4 (continued)

so-called sunbelt zone, where strong sunlight intensity and long daylight hours are present. A METI-NEDO project has opened large-scale reactor design and operations (NEDO 1998). Various types of photobioreactors have been developed (Asada and Miyake 1999; Miyake et al. 2001; Miyake et al. 2003; Miyake 2007; Miyake 1998; Kawasugi et al. 1998; Okamoto et al. 2000; Ishikawa et al. 2008; Sakurai et al. 2010; Kitashima et al. 2012). Some types of the reactors developed in the METI-NEDO project are shown in Fig. 23.4.

Factors interfering with the use of sunlight for photobiological hydrogen production include those dependent on such properties as the frequent fluctuation of sunlight according to weather and the day/night cycle. Photobiological hydrogen production is saturated under light intensity at the meridian transit; the light is absorbed and scattered by photosynthetic bacteria and does not penetrate deep in the photobioreactor. Hydrogen production also depends on temperature.

Photosynthetic bacteria absorb visible light, including the near-infrared spectrum, and reactor temperature increases. Therefore, it is desirable to use heat-resistant photosynthetic bacteria. Because a bacterial strain with hydrogen-producing ability at 45 °C or higher has not yet been obtained, investigation and development of bacterial strains capable of producing hydrogen at a high temperature is desirable to make photobiological hydrogen production more efficient.

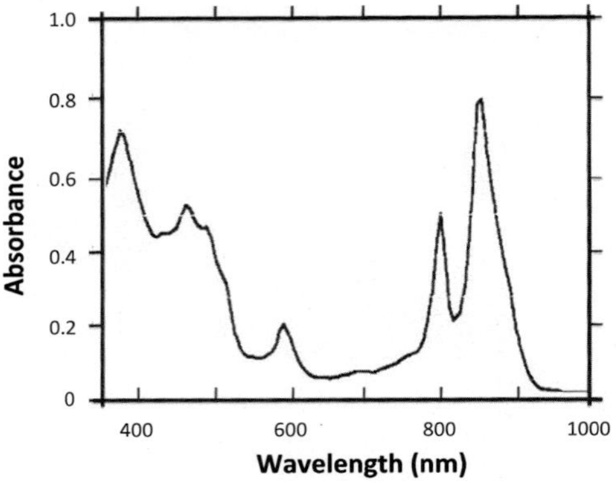

Fig. 23.5 Absorption spectrum of *Rhodobacter sphaeroides* RV: spectra of light at various depths in the bacterial culture. (Modified from Nakada et al. 1995)

Sunlight at noon is too great for efficient hydrogen production by photosynthetic bacteria, and the conversion efficiency from light to hydrogen is markedly decreased. To overcome this problem, diffuse irradiation of excess sunlight is being investigated. With *R. sphaeroides* RV strain, spatial diffusion of excess light was investigated using a reactor equipped with shading bands (Fig. 23.5). By setting shading bands on the surface of the photobioreactor, the conversion efficiency was improved 1.4 fold. Shading bands with solar cells gives additional energy capture. Time dispersion of excess sunlight is also a way to use sunlight efficiently.

23.3.2 Photoenergy Conversion Process in the Reactor

23.3.2.1 Light Penetration and Effective Wavelength Region

Reactors for photosynthetic bacterial hydrogen production have special difficulties. The reactor has to have a three-dimensional structure, but light illumination is possible only from one surface. Light penetration is an essential factor for massive photosynthetic reactions. Light is trapped at the shallow part of the suspension when the concentration is high. Lowering the concentration does not solve the problem as it also lowers the biochemical activity of hydrogen production. Balancing the biochemical function and photosynthesis have been studied (Nakada et al. 1995, 1996; Vasilyeva et al. 1999; Kondo et al. 2002, 2006; Katsuda et al. 2004; Katsuda 2008; Wakayama et al. 2000; Miyake et al. 1999a, b).

Nakada et al. (1995) examined light penetration with special interest in the alteration of the spectrum of incident light. Photosynthetic bacterial reaction proteins have peak absorption at around 800 nm. Absorption in the range of 500–700 nm is rather low, so light for the absorption maximum mostly does not contribute to hydrogen production. Light of 500–700 nm also used effectively: the light penetrates deeply into the cell suspension.

The foregoing observations provide the principal rule to enhance hydrogen production: reduction of light-harvesting pigments gives a more efficient system of energy capture.

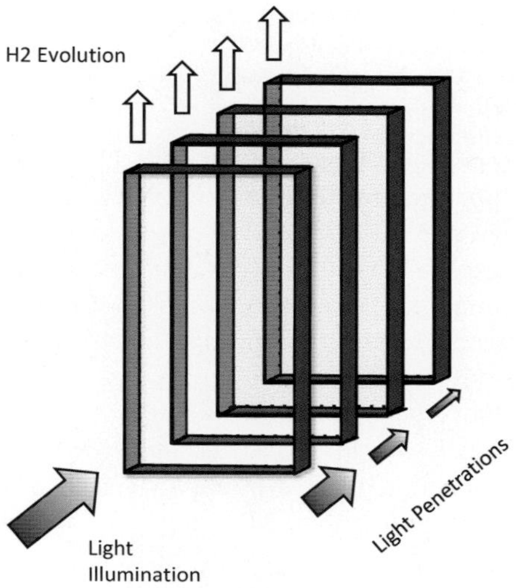

Fig. 23.6 Schematic view of photobioreactor with four compartments, prepared with polyacryl resin 0.5 cm thick (inner dimensions of each compartment were 0.5×5×12 cm). *Upper part* of window was covered with aluminum foil to give an irradiation area of 37.5 cm². Light path of each compartment was 0.5 cm. (Modified from Nakada et al. 1995)

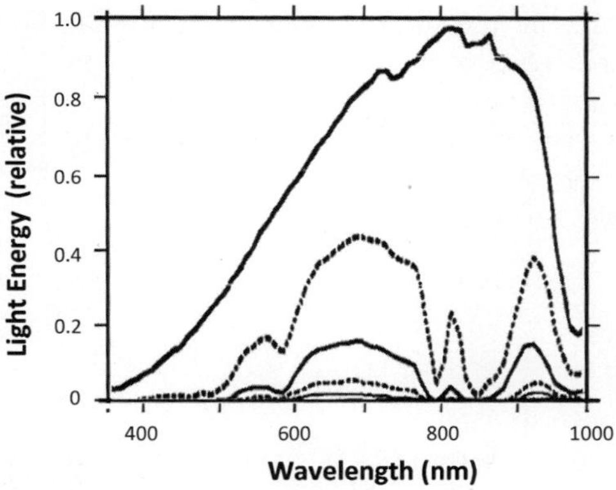

Fig. 23.7 Light spectrum in a reactor of *R. sphaeroides* RV: spectra of light at various depths in the bacterial culture. Depths were —, 0 cm (at the surface of the reactor); ....., 0.5 cm (at the surface of second compartment); —, 1.0 cm (third compartment); ....., 1.5 cm (fourth compartment); —, 2.0 cm (after passing through the fourth compartment). (Modified from Nakada et al. 1995)

Total hydrogen production by a reactor with a realistic depth should be maximized by the appropriate balance of pigments and the hydrogen-producing enzymes (Fig. 23.6).

23.3.2.2 Mutant with Reduced Pigments

Miyake and coworkers established the concept of making mutants with a reduced light-harvesting bacteriochlorophyll–protein complex (Vasilyeva et al. 1999; Kondo et al. 2002, 2006). By a combination of the mutant with the wild strain, hydrogen production efficiencies were enhanced up to 140 %.

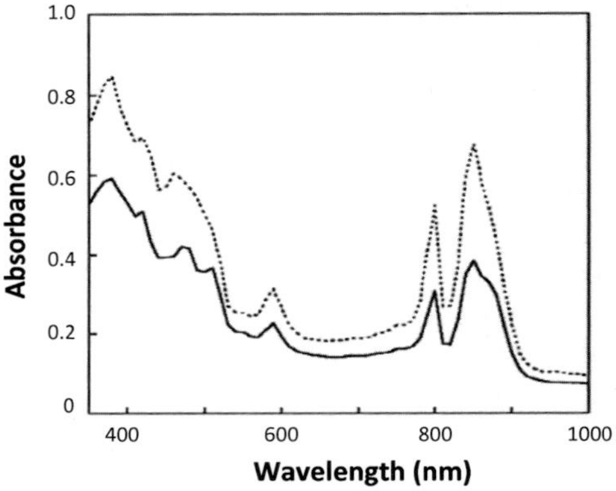

Fig. 23.8 Absorption spectra of two types of cells at cell concentration of 0.3 mg dry cell weight = ml. *Dashed* and *solid lines* indicate RV and MTP4, respectively. (Modified from Kondo et al. 2006)

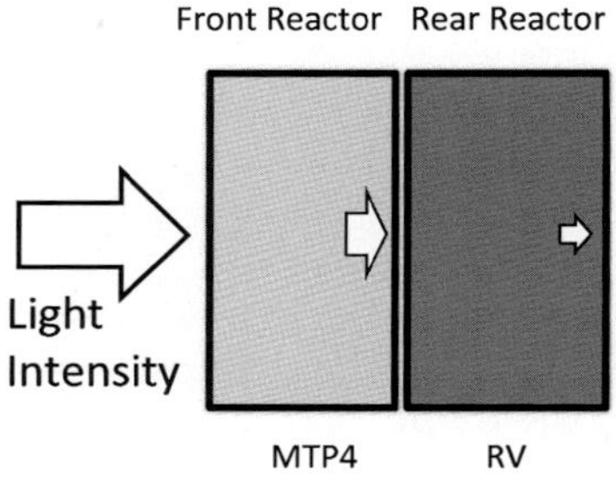

Fig. 23.9 Proposed combination of RV and MTP4. (Modified from Nakada et al. (1996)

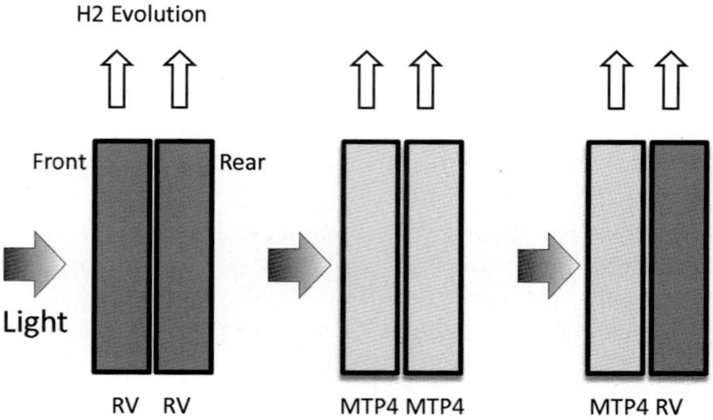

Fig. 23.10 Three arrangements in a double-layer reactor. (Modified from Kondo et al. 2006)

Mutants have been prepared for analyzing the effectiveness of light penetration and hydrogen production. Kondo et al. has prepared one mutant by ultraviolet irradiation, from a *R. sphaeroides* RV strain, that has reduced bacteriochlorophyll content compared to the cell mass (Fig. 23.7). The mutant penetrates light into the suspension more effectively than the wild type.

The problem of light penetration is the shading of the light in the deep part of the reactor by the bacteria in front of the incident light. The mutant and wild-type bacteria are contained in two transparent flat reactors (Fig. 23.8) and placed by changing the location along the incident light axis (Fig. 23.9). The result clearly shows the effectiveness of the combination (Fig. 23.10).

In the front vessel, hydrogen production was the same in both mutant and wild type. Excess light energy did not show a difference. In contrast, hydrogen production in the rear vessel was much affected. The combination of mutant (front) and wild type (rear) gave higher hydrogen production. Light energy in the rear vessel was increased by reduced pigments. In limited light energy, as in the rear vessel, a small difference in light energy could greatly alter hydrogen production.

Because photosynthetic bacteria have absorption maxima at around 800–850 nm and 400 nm, light energy of that wavelength should be most effective. However, as shown in Fig. 23.6, light around 600–800 nm, where the absorbance of bacteriochlorophyll is very low, functions well in photosynthesis and hydrogen production. Light of 600–800 nm was not adsorbed at the surface region and penetrates deeply in the reactor. The contribution of the light of the wavelength region should be high.

Melis et al. (1999) and the group of Nakajima et al. (2001) reported the same kind of concepts using cyanobacteria. Lowering the amount of pigment allowed greater penetration of light into the deep part. As light of a wavelength with low absorption is useful for hydrogen production (in the reactor in Fig. 23.6) gives equivalent meaning or extension of the concept of lowering the absorption of individual bacteria. The special nature of the absorption spectra should be the naturally equipped light penetration system. Having a window region in the spectra of chlorophylls could be interpreted as a strategy of the biological system of photosynthesis. A photosynthetic system utilizes the complicated relationship of light intensity and wavelength. We have to carefully consider the complexity when the mechanism for the conversion of light to hydrogen is examined. Monochromatic light could not directly give results from which we can reconstruct the total reaction in a reactor.

Light energy conversion in a reactor is not so simple as the vision of the eyes. The eye pigments (rhodopsin) have an absorption maximum at 550 nm where sunlight has maximum energy. Because vision does not need a three-dimensional (3-D) reactor, the light is sensed by the thin layer of the retina. Although both could have the same maxima, photosynthesis occurs in a 3-D structure in nature. In a pond, algae on the top of the pond absorb the sunlight but photosynthetic bacteria live with the rest of the wavelength region in the bottom. Chlorophyll–bacteriochlorophyll structures and absorption spectra are presumed to be designed by nature to realize maximum growth of photosynthetic organisms. The engineering strategy of reduction of the pigments is an essential one for future development.

23.4 Application for Electricity

23.4.1 Combination with Fuel Cells for Small-Scale Electric Grid System

Biological solar energy conversion is useful as the function is the combination of various types of reactions. Diffence of the reaction rates in the region of seconds to hours provides a resistant system for the frequent changes of light intensity. We could say that photosynthetic hydrogen production resembles a box with various buffers, which is completely different from a physical energy conversion system such as solar cells and wind turbines in which the output changes directly following changes in the energy sources.

Renewable natural power sources, such as sunlight and wind, have low energy density with time fractionation. On the other hand, demand asks for energy to be homogeneous and obtained in large quantities. Accumulation of energy of a low concentration state to a condensed one and diversed to uniformity does not occur naturally.

Both these difficulties are regarded as entropic problems. The former is exactly a reversed process of thermodynamics. The latter, to normalize the diversity, should be be a statistic dynamic one. Entropy is the essential barrier of the untilization of renewable natural energies.

Photosynthesis is useful for the collection and accumulation of low energy sources, an offset of the rather low efficiency. Physical processes have many drawbacks in reducing efficiency if we consider the production, setting, operation, and noramalization of the system. As already described, photosynthesis equips the modulation system of the time constants of the reactions. Nature is useful to normalize the outputs.

To realize energy capture and output to the social system, the combination with electricity should be considered not only in burning gasoline. A small-sized electric grid for a community of a certain population could be a good model (Miyake 2012). Hydrogen is simply converted to electricity by using fuel cells (Nakada et al. 1999). A system design of a grid composed of several elements (electric grid, wind turbine, solar cell, biohydrogen) is shown in Figs. 23.11 and 23.12. The effect of the reduction of entropic loss is to be examined theoretically and experimentally. A fuel cell for biological use was studied (Yuhashi et al. 2005; Ishikawa et al. 2006; Okuda-Shimazaki et al. 2008).

23.4.2 Photoelectric Devices and Artificial Photosynthesis and Energy Conversion

A photosynthetic reaction center is a natural solar battery. Charge separation occurs in the bacteriochlorophyll dimer. An electron is emitted from an edge of the protein. The positive hole is neutralized by an electron from the other edge of the molecule. Excitation with a photon generates electron flow across the photosynthetic membrane.

Thus, nature could be applied to make a solar cell. The difficulty comes from the control of the orientation. Exact arrangement of protein molecules is a technologically difficult question. Modifying the protein with genetics or combining molecules, positive–negative or hydrophobic–hydrophilic (Yang et al. 2000; Nakamura et al. 2000; Ueno et al. 1998; Yasuda et al. 1998; Hirata and Miyake 1994; Yasuda et al. 1994; Hara et al. 1994; Miyake et al. 1994), are typical modification strategies. The Langmuir–Blodgett (LB) method is one of the most commonly used for this purpose. High surface pressure at the air–water interface is apt to cause high tension in the molecule, leading to collapse. Compared to the PS I or PS II systems of plants, photosynthetic bacterial reaction centers are rather strong.

Photosynthetic antenna proteins surrounding the photosynthetic reaction centers are also reconstructed with the reaction centers on electrodes for photoelectric responses (Miyake et al. 1994). Artificial photosynthesis could provide a naturally supplied, renewable solar cell. An antenna is useful because the content of RC, the charge separator, could be reduced. Antennae also support the reaction center. The whole structure of the photosynthetic reaction system is stable against heat denaturation, retaining the alpha-helical structures of the proteins (Sumino et al. 2011; Nagata et al. 2012; Amao et al. 2012; Kondo et al. 2012).

To prepare photoelectric devices, an electrode for hydrogen production is required. Thiocapsa hydrogenase was found to be very resistant against physical treatment (Noda et al. 1998; Nakamura et al. 2001; Qian et al. 2002; Wenk et al. 2002; Qian et al. 2006; Liu et al. 2007; Sun et al. 2010; Kihara et al. 2011). LB films of thermostable hydrogenase were successfully prepared on electrodes using poly-l-lysine. The hydrogenase activity was maintained in the film. Hydrogen production was demonstrated by applying bias to the ITO electrode covered with hydrogenase LB film using methyl viologen as a mediator. The results show that the hydrogenase LB film has potential for an effective electrode system for hydrogen production.

A remarkable improvement was made on the electrode to hydrogenase using hydrogenase combined with a single-walled carbon nanotube forest (SWNT forest). The SWNT forest has a unique structure comprising dense and vertically aligned SWNTs with millimeter-scale height. Hydrogenase was spontaneously incorporated within the SWNT forest and confined on the sidewall of the rearranged architecture (Kihara et al. 2011). With a hydrogenase SWNT forest we were able to electrochemically produce hydrogen with an electron transfer efficiency exceeding 30 % without the use of any chemical mediator. A reverse oxidative reaction of hydrogen was also successfully performed using the same device. The SWNT forest can work as an effective and direct electron mediator for the oxidation/reduction reaction of hydrogenase. It is possible to make a highly efficient biofuel cell device without the use of any mediators.

References

Amao Y, Shuto N, Furuno K, Obata A, Fuchino Y, Uemura K, Kajino T, Sekito T, Iwai S, Miyamoto Y, Matsuda M (2012) Artificial leaf device for solar fuel production. Faraday Discuss 155:289–296

Asada Y, Miyake J (1999) Photobiological hydrogen production. J Biosci Bioeng 88:1–6

Asada Y, Ohsawa M, Nagai Y, Ishimi K, Fukatsu M, Hideno A, Wakayama T, Miyake J (2008) Re-evaluation of hydrogen productivity from acetate by some photosynthetic bacteria. Int J Hydrogen Energy 33:5147–5150

Colbeau A, Elsen S, Tomiyama M, Zorin N, Dimon B, Vignais PM (1998) *Rhodobacter capsulatus* HypF is involved in regulation of hydrogenase synthesis through the HupUV proteins. Eur J Biochem 251:65–71

Hall DO (1998) Will biomass be the environmentally friendly fuel of the future. Biomass Bioenergy 15:357–367

Hara M, Shigeno M, Asada Y, Miyake J (1994) Sheet formation of membrane proteins from photosynthetic bacteria. Mater Sci Eng C 2:13–18

Hirata Y, Miyake J (1994) Molecular construction of photosynthetic reaction center at interfaces by affinity with quinonylphosholipid. Thin Solid Films 244:865–868

Ishikawa M, Yamamura S, Takamura Y, Sode K, Tamiya E, Tomiyama M (2006) Development of a compact high-density microbial hydrogen reactor for portable bio-fuel cell system. Int J Hydrogen Energy 31:1484–1489

Ishikawa M, Yamamura S, Ikeda R, Takamura Y, Sode K, Tamiya E, Tomiyama M (2008) Development of a compact stacked flatbed reactor with immobilized high-density bacteria for hydrogen production. Int J Hydrogen Energy 33:1593–1597

Katsuda T (2008) A biochemical engineering study of bioproductions using photobioreactor. Seibutukougaku (in Japanese) 86:117–122

Katsuda T, Yegani R, Fujii N, Igarashi K, Yoshimura S, Katoh S (2004) Effects of light intensity distribution on growth of *Rhodobacter capsulatus*. Biotechnol Prog 20:998–1000

Kawasugi T, Pedroni P, Miyake M, Toriyama A, Fukunaga S, Takasaki K, Kawamoto T (1998) RITE biological hydrogen program. In: Zaborsky OR, Benemann JR, Matsunaga T, Miyake J, Pietro AS (eds) Biohydrogen. Plenum, New York, pp 501–505

Kihara T, Liu X-Y, Nakamura C, Park KM, Han SW, Qian D-J, Kawasaki K, Zorin NA, Yasuda S, Hata K, Wakayama T, Miyake J (2011) Direct electron transfer to hydrogenase for catalytic hydrogen production using a single-walled carbon nanotube forest. Int J Hydrogen Energy 36:7523–7529

Kitashima S, Masukawa H, Sakurai H, Inoue K (2012) Flexible plastic bioreactors for photobiological hydrogen production by hydrogenase-deficient cyanobacteria. Biosci Biotechnol Biochem 76:831–833

Ko I-N, Noike T (2002) Use of blue optical filters for suppression of growth of algae in hydrogen producing non-axenic cultures of *Rhodobacter sphaeroides* RV. Int J Hydrogen Energy 27:1297–1302

Kondo T, Arakawa M, Hirai T, Wakayama T, Hara M, Miyake J (2002) Enhancement of hydrogen production by an photosynthetic bacterium mutant with reduced pigment. J Biosci Bioeng 93:145–150

Kondo T, Wakayama T, Miyake J (2006) Efficient hydrogen production using a multi-layered photobioreactor and a photosynthetic bacterium mutant with reduced pigment. Int J Hydrogen Energy 31:1522–1526

Kondo M, Iida K, Dewa T, Tanaka H, Ogawa T, Nagashima S, Nagashima KV, Shimada K, Hashimoto H, Gardiner AT, Cogdell RJ, Nango M (2012) Photocurrent and electronic activities of oriented-His-tagged photosynthetic light-harvesting/reaction centre core complexes assembled onto a gold electrode. Biomacromolecules 13:432–438

Liu A-R, Wakayama T, Nakamura C, Miyake J, Zorin NA, Qian D-J (2007) Electrochemical properties of carbon nanotubes-hydrogenase conjugates Langmuir-Blodgett films. Electrochim Acta 52:3222–3228

Melis A, Neidhardt J, Benemann JR (1999) *Dunaliella salina* (Chlorophyta) with small chlorophyll antenna sizes exhibit higher photosynthetic productivities and photon use efficiencies than normally pigmented cells. J Appl Phycol 10:515–525

Miyake J (1990) Application of photosynthetic systems for energy conversion. In: Veziroglu TN, Takahashi PK (eds) Hydrogen energy progress. VIII. Proceedings 8th WHEC. Elsevier, New York, pp 755–764

Miyake J (1998) The science of biohydrogen: an energetic view. In: Zaborsky OR, Benemann JR, Matsunaga T, Miyake J, Pietro AS (eds) Biohydrogen. Plenum, New York, pp 7–18

Miyake J (2007) Task 21 report: biohydrogen. In: Hydrogen implementing agreement 2007 annual report: IEA agreement on the production and utilization of hydrogen. OECD/IEA, Paris and Bethesda

Miyake J (2012) Renewable energy path with biohydrogen. In: Proceedings of 2012 international symposium on advanced biological engineering (ISABE2012). Guilin, China

Miyake J, Kawamura S (1987) Efficiency of light energy conversion to hydrogen by the photosynthetic bacterium *Rhodobacter sphaeroides*. Int J Hydrogen Energy 12:147–149

Miyake J, Mao X-Y, Kawamura S (1985) Hydrogen production using an immobilized co-culture of anaerobic and photosynthetic bacteria. In: Hydrogen systems, vol 1. China Academic Publishers/Pergamon Press, Beijing/Oxford, pp 277–281

Miyake J, Majima T, Namba K, Hara M, Asada Y, Sugino H, Ajiki S, Toyotama H (1994) Thermal stability of dried photosynthetic membrane film for photoelectrodes. Mater Sci Eng C 1:63–67

Miyake J, Miyake M, Asada Y (1999a) Biotechnological hydrogen production: research for efficient light energy conversion. J Biotechnol 70:89–101

Miyake J, Wakayama T, Schnackenberg J, Arai T, Asada Y (1999b) Simulation of the daily illumination pattern of sunlight for bacterial photo-hydrogen production. J Biosci Bioeng 88:659–663

Miyake J, Matsunaga T, Pietro AS (eds) (2001) Biohydrogen II. Elsevier, London

Miyake J, Igarashi Y, Roegner M (eds) (2003) Biohydrogen III. Elsevier, London

Nagashima S, Kamimura A, Shimizu T, Nakamura-Isaki S, Aono E, Sakamoto K, Ichikawa N, Nakazawa H, Sekine M, Yamazaki S, Fujita N, Shimada K, Hanada S, Nagashima KV (2012) Complete genome sequence of phototrophic betaproteobacterium *Rubrivivax gelatinosus* IL144. J Bacteriol 194:3541–3542

Nagata M, Amano M, Joke T, Fujii K, Okuda A, Kondo M, Ishigure S, Dewa T, Iida K, Secundo F, Amao Y, Hashimoto H, Nango M (2012) Immobilization and photocurrent activity of a light-harvesting antenna complex II, LHCII, isolated from a plant on electrodes. ACS Macro Lett 1:296–299

Nakada E, Asada Y, Arai T, Miyake J (1995) Light penetration into cell suspensions of photosynthetic bacteria, relation to hydrogen production. J Ferment Bioeng 80:53–57

Nakada E, Nishikata S, Asada Y, Miyake J (1996) Hydrogen production by gel-immobilized cells of *Rhodobacter sphaeroides*: distribution of cells, pigments and hydrogen evolution. J Mar Biotechnol 4:38–42

Nakada E, Nishikata S, Asada Y, Miyake J (1999) Photosynthetic bacterial hydrogen production combined with a fuel cell. Int J Hydrogen Energy 24:1053–1057

Nakajima Y, Tsuzuki M, Ueda R (2001) Improved productivity by reduction of the content of light-harvesting pigment in *Chlamydomonas perigranulata*. J Appl Phycol 13:95–101

Nakamura C, Hasegawa M, Yasuda Y, Miyake J (2000) Self-assembling photosynthetic reaction centers on electrodes for current generation. Appl Biochem Biotechnol 84–86:401–408

Nakamura C, Noda K, Zorin NA, Akutsu H, Miyake J (2001) Cytochrome c3–Langmuir-Blodgett film for hydrogen evolving device. Synth Met 117:285–288

Nara M (2012) Study on production of bio-hydrogen gas as energy. Trans Jpn Soc Mech Eng B 78:1053–1057

NEDO (New Energy and Industrial Technology Development Organization) (1998) Report of research and development on environmentally friendly technology for the production of hydrogen

Noda K, Zorin NA, Nakamura C, Miyake M, Gogotov IN, Asada Y, Akutsu H, Miyake J (1998) Langmuir-Blodgett film of hydrogenase for electrochemical hydrogen production. Thin Solid Films 327(329):639–642

Noparatonaraporn N, Takeno K, Sasaki K (2001) Hydrogen and poly-(hydroxy)alkanet production from organic acid by photosynthetic bacteria. In: Miyake J, Matsunaga T, Pietro AS (eds) Biohydrogen II. Pergamon/Elsevier, Amsterdam, pp 33–40

Okamoto M, Miyahara T, Mizuno O, Noike T (2000) Biological hydrogen potential of materials characteristic of the organic fraction of municipal solid wastes. Water Sci Technol 41:25–32

Okuda-Shimazaki J, Kakehi N, Yamazaki T, Tomiyama M, Sode K (2008) Biofuel cell system employing thermostable glucose dehydrogenase. Biotechnol Lett 30:1753–1758

Oosawa F (2000) The loose coupling mechanism in molecular machines of living cells. Genes Cells 5:9–16

Oosawa F, Hayashi S (1984) A loose coupling mechanism of synthesis of ATP by proton flux in the molecular machine of living cells. J Phys Soc Jpn 53:1575–1579

Oosawa F, Hayashi S (1986) The loose coupling mechanism in molecular machines of living cells. Adv Biophys 22:151–183

Qian D-J, Wenk S-V, Nakamura C, Wakayama T, Zorin N, Miyake J (2002) Photo-induced hydrogen evolution by use of porphyrin, EDTA, viologens and hydrogenase in solutions and Langmuir-Blodgett films. Int J Hydrogen Energy 27:1481–1487

Qian D-J, Chen H-T, Liu B, Xiang X-M, Wakayama T, Nakamura C, Miyake J (2006) Monolayers and Langmuir-Blodgett films of palladium-mediated multiporphyrin arrays with different central ions. Colloids Surf A Physicochem Eng Asp 284(285):180–186

Sakurai H, Masukawa H, Kitashima M, Inoue K (2010) A feaxibility study of large-scale photobiological hydrogen production utilizing mariculture-raised cyanobacteria. In: Hallenbeck PC (ed) Recent advances in phototrophic prokaryotes. Springer, New York, pp 291–304

Sasaki K (1998) Hydrogen and 5-amino-levulinic acid production by photosynthetic bacteria. In: Zaborsky OR, Benemann JR, Matsunaga T, Miyake J, Pietro AS (eds) Biohydrogen. Plenum Press, New York, pp 133–392

Schlodder E, Gräber P, Witt HT (1982) Mechanism of phosphorylation in chloroplasts. In: Barber J (ed) Electron transport and photophosphorylation. Elsevier, New York, pp 105–176

Shomura Y, Yoon K, Nishihara H, Higuchi Y (2011) Structural basis for a [4Fe-3S] cluster in the oxygen-tolerant membrane-bound [NiFe]-hydrogenase. Nature (Lond) 479:253–257

Sumino A, Dewa T, Kondo M, Morii T, Hashimoto H, Gardiner AT, Cogdell RJ, Nango M (2011) Selective assembly of photosynthetic antenna proteins into a domain-structured lipid bilayer for the construction of artificial photosynthetic antenna systems: structural analysis of the assembly using surface plasmon resonance and atomic force microscopy. Langmuir 27:1092–1099

Sun Q, Zorin NA, Chen D, Chen M, Liu TX, Miyake J, Qian DJ (2010) Langmuir-Blodgett films of pyridyldithio-modified multiwalled carbon nanotubes as a support to immobilize hydrogenase. Langmuir 26:10259–10265

Takabatake H, Suzuki K, Ko I-N, Noike T (2004) Characteristics of anaerobic ammonia removal by a mixed culture of hydrogen producing photosynthetic bacteria. Bioresour Technol 95:151–158

Ueno T, Hara M, Kamo N, Fujii T, Miyake J (1998) Control of unidirectional topological orientation of cross-linked complex composed of bacterial photosynthetic reaction center and horse heart cytochrome *c* reconstituted in proteoliposomes. J Biochem (Tokyo) 124:485–490

Vasilyeva LG, Miyake M, Khatipov E, Wakayama T, Sekine M, Hara M, Nakada E, Asada Y, Miyake J (1999) Enhanced hydrogen production by a mutant of *Rhodobacter sphaeroides* having an altered light-harvesting system. J Biosci Bioeng 87:619–624

Wakayama T, Miyake J (2001) Hydrogen from biomass. In: Miyake J, Matsunaga T, Pietro AS (eds) Biohydrogen II. Elsevier, London, pp 41–52

Wakayama T, Nakada E, Asada Y, Miyake J (2000) Effect of light/dark cycle on bacterial hydrogen production by *Rhodobacter sphaeroides* RV—from hour to second range. Appl Biochem Biotechnol 84–86:431–440

Watanabe M, Sasaki K, Nakashimada Y, Nishio N (1998) High density cell culture of a marine photosynthetic bacterium *Rhodovulum* sp. with self-flocculated cells. Biotechnol Lett 20(12):1113–1117

Watanabe M, Suzuki Y, Sasaki K, Nakashimada Y, Nishino N (1999) Flocculating property of extracellular polymeric substance derived from a marine photosynthetic bacterium, *Rhodovulum* sp. J Biosci Bioeng 87:625–629

Wenk S-O, Qian D-J, Wakayama T, Nakamura C, Zorin N, Rogner M, Miyake J (2002) Biomolecular device for photoinduced hydrogen production. Int J Hydrogen Energy 27:1489–1493

Yang Q, Liu X-Y, Hara M, Lundahl P, Miyake J (2000) Quantitative affinity chromatographic studies of mitochondrial cytochrome *c* binding to bacterial photosynthetic reaction center, reconstituted in liposome membranes and immobilized by detergent dialysis and avidin-biotin binding. Anal Biochem 280:94–102

Yasuda Y, Sugino H, Toyotama H, Hiratsu Y, Hara M, Miyake J (1994) Control of protein orientation in molecular photoelectric devices using Langmuir-Blodgett films of photosynthetic reaction centers from Rhodopseudomonas viridis. Bioelectrochem Bioenerg 34:135–139

Yasuda Y, Toyotama H, Hara M, Miyake J (1998) Effects of the counter ions on the orientation control of photosynthetic reaction center proteins by an applied bias-voltage Langmuir-Blodgett method. Thin Solid Films 327(329):800–803

Yuhashi N, Tomiyama M, Okuda J, Igarashi S, Ikebukuro K, Sode K (2005) Development of a novel glucose enzyme fuel cell system employing protein engineered PQQ glucose dehydrogenase. Biosens Bioelectron 20:2145–2150

Chapter 24
Production of Biofuels and Useful Materials by Anaerobic Organisms in Ecosystem of Methane Fermentation

Yutaka Nakashimada and Naomichi Nishio

Abstract The demand for saving used energy and its recovery from unused matter has been intensified by fears of fossil fuel depletion and global warming. In this context, the methane fermentation process to treat organic wastes with energy recovery has been widely used. However, improvement and modification of methane fermentation have been needed to expand availability to various types of organic matter. In this chapter, we introduce our attempts to treat some unusual organic matter with recovery of methane as a sustainable energy. Furthermore, it was also described that not only methane but also useful biofuels and biomaterials such as hydrogen, ethanol, optically active materials, and a physiologically active substance, vitamin B_{12}, can be produced using microorganisms participating in the ecosystem of methane fermentation.

Keywords Ethanol • Fermentation • Hydrogen • Methane • Optically active substance • Solid wastes • Vitamin B_{12} • Wastewater

24.1 Introduction

Methane fermentation has been commonly used for the biological treatment of high-strength wastewater and energy recovery as methane. The process has been practiced in Japan for wastewaters from food-processing factories such as beer, sugars, soft drinks, potatoes, chemicals, and so on. Methane fermentation is a highly sophisticated ecosystem in which a variety of microorganisms play a different role for decomposition of organic material and production of methane and CO_2. Methane fermentation can be subdivided into the following three phases. A schematic drawing of methane

Y. Nakashimada (✉) • N. Nishio
Department of Molecular Biotechnology, Graduate School of Advanced Sciences of Matter, Hiroshima University, 1-3-1 Kagamiyama, Higashi-Hiroshima 739-8530, Japan
e-mail: nyutaka@hiroshima-u.ac.jp

H. Anazawa and S. Shimizu (eds.), *Microbial Production: From Genome Design to Cell Engineering*, DOI 10.1007/978-4-431-54607-8_24, © Springer Japan 2014

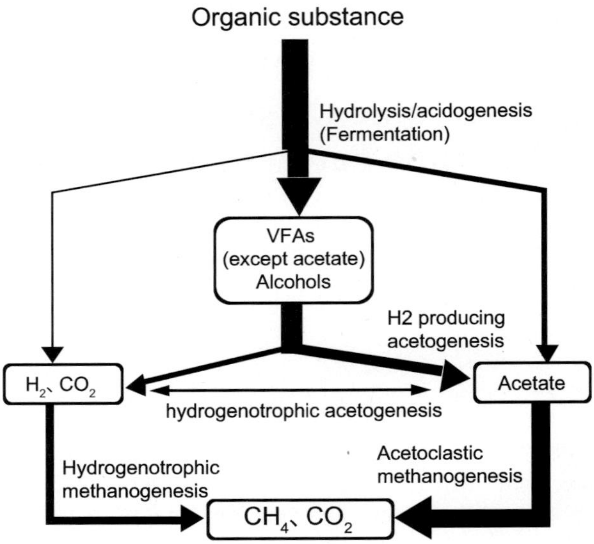

Fig. 24.1 Simplified schematic representation of methane fermentation

fermentation of organic matter is presented as Fig. 24.1. In the phase of hydrolysis and acidogenesis, after composite particulate materials are disrupted, as microbes or hydrolyzing enzymes can attack biopolymers such as proteins, polysaccharides, and lipids, biopolymers are enzymatically hydrolyzed into soluble organic compounds such as amino acids, sugars, and free long-chain fatty acids and glycerol. Then, many kinds of fermentative microorganisms convert soluble organic compounds to volatile fatty acids (VFAs), alcohols, H_2, and CO_2. The produced metabolites such as propionate, butyrate, and ethanol are oxidized to acetate, H_2, and CO_2 by acetogenic bacteria (H_2-producing acetogenesis). Because substrates that methanogens can utilize are very limited, in the industrial process of anaerobic digestion, the usual substrates for methane formation are limited to H_2 and CO_2, or formate by hydrogenotrophic methanogen, or acetate by acetoclastic methanogen produced via acidogenesis and acetogenesis as follows, respectively (methanogenesis). The removal of H_2 by hydrogenotrophic methanogens is indispensable to go on with syntrophic acetogenesis of higher organic acids because the total reaction becomes exergonic by coupling with acetogenesis and methanogenesis.

Recently, we developed some novel methane fermentation processes to expand the availability to recalcitrant organic wastes such as high-solid wastes and toxic compounds. Furthermore, because a variety of microorganisms that have different metabolic functions for decomposition of organic material and production of methane participate in methane fermentation, the ecosystem is a mine of worthy microorganisms to produce useful materials in addition to methane. In this chapter, we describe not only the novel methane fermentation process we developed but also useful material production with anaerobic microorganisms in the ecosystem of methane fermentation.

24.2 Advanced Methane Fermentation for Treatment of Different Organic Wastes with Energy Recovery

To achieve rapid and effective anaerobic digestion, some processes have been developed so far, such as upflow anaerobic filter process (UAFP) (Young and McCarty 1969), upflow anaerobic sludge blanket (UASB) (Lettinga et al. 1983), and anaerobic attached film expanded-bed reactor (AAFEB) (Jowell et al. 1981) to increase cell retention, and a two-phase digestion process to optimize both acidogenesis and methanogenesis. However, to enhance these processes, it is necessary to determine their applicability to recalcitrant organic wastes and wastewaters containing toxic compounds, high-solid organic materials, or extremely low strength organic matter.

24.2.1 Advanced UASB Reactors

24.2.1.1 Alternative Use of UASB Reactor for Sewage Treatment

In sewage treatment, chemical oxygen demand (COD, approximately 400 ppm) and biochemical oxygen demand (BOD) in influent have been decreased to less than 20 ppm under an aerobic condition using the activated sludge process. However, the disadvantages of this process are that (1) it requires aeration energy, that is, electricity is needed to supply air and that (2) excess activated sludge is produced, for example, 50 g sludge from 100 g organic carbon. The UASB method has been developed as an efficient anaerobic wastewater treatment process. However, it is difficult to remove organic carbon to less than 20 ppm by this anaerobic process. To overcome this problem, a novel system consisting of a UASB anaerobic pretreatment unit and a DHS (down-flow hanging sponge cube) aerobic posttreatment unit was proposed by Machdar et al. (2000) for a low-cost and easy maintenance process; the system requires neither external aeration input nor excess sludge withdrawal. The process performance of the aforementioned novel sewage treatment system using the sulfur reduction-oxidation reaction of microbes in the UASB pretreatment unit was investigated in a pilot-scale reactor with actual sewage for more than 600 days at the Higashi-Hiroshima Sewage Treatment Center (Takahashi et al. 2005). The reactor system consisted of a denitrification reactor (1.4 m^3), a UASB reactor (8.4 m^3), and a DHS reactor (13.9 m^3) with a recirculation line. The total BOD in the influent (128 ppm) decreased to 39 ppm in the UASB effluent, and 7 ppm in the DHS effluent with a low degree of suspended solid (SS) accumulation, namely, 2–3 % of the inlet SS level, under the conditions of overall hydraulic retention times (HRTs) of 12–24 h and a recirculation ratio of 0.3–2 throughout the experimental period. These results revealed that sulfate-reducing bacteria, not methanogens, contributed to the degradation of organic matter in the UASB reactor even at low temperatures (below 10 °C).

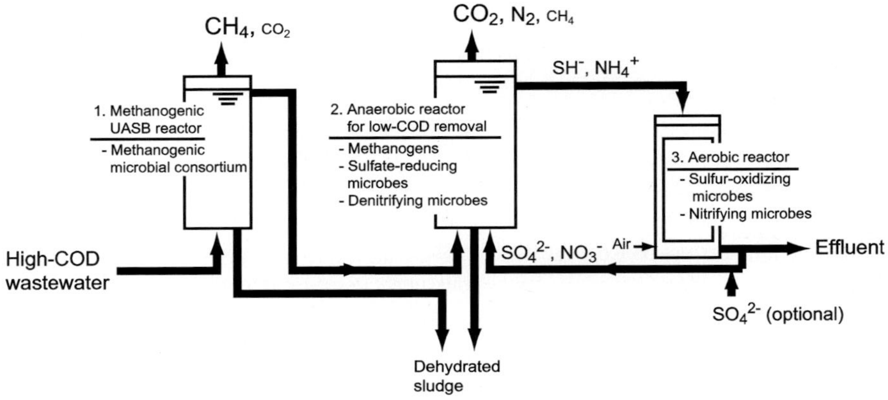

Fig. 24.2 Schematic diagram of complete anaerobic organic matter removal process (CARP)

24.2.1.2 Complete Anaerobic Treatment of High-Strength Wastewater (CARP Process)

In conventional wastewater treatment processes, methane fermentation processes such as UASB are commonly used for the biological treatment of high-strength wastewater (e.g., COD = 4,000 ppm) and for energy recovery by recovering methane. Such processes have been used widely in Japan for wastewaters from food-processing factories that produce commodities such as beer, sugar, soft drinks, and potato products. However, because of the low COD and BOD removal efficiencies of such processes (less than 90 %), aerobic treatment (the activated sludge process) has been generally used as a posttreatment to further decrease the COD and BOD level to less than 20 ppm. This activated sludge process requires high energy consumption for aeration, as already mentioned. If anaerobic treatment is possible as a posttreatment, the methane fermentation process will become more useful. Therefore, methane fermentation following the anaerobic posttreatment process (complete anaerobic organic matter removal process, CARP) was proposed as shown in Fig. 24.2. By using CARP, the COD in wastewater (initially 4,000 ppm) can be decreased to 20 ppm after these anaerobic treatments.

In the CARP process, three types of microorganism should be functioning. Methane-producing microbes can convert COD and BOD to methane and carbon dioxide in a methanogenic reactor (1). Denitrifying microbes that convert nitrate to nitrogen gas (USEPA 1993) and sulfate-reducing microbes that convert sulfate to sulfide (Fauque 1995) mainly act as BOD and COD eliminators in an anaerobic reactor (2). This process, however, requires an aerobic reactor (3) for posttreatment to recycle the sulfate from the sulfide produced by the sulfur-oxidizing microbes and to recycle the nitrate from the ammonia produced by nitrifying microbes: this posttreatment contributes to the removal of ammonia. The sulfate-reducing and sulfur-oxidizing microbes can function under not only eutrophic but also oligotrophic conditions

(Holmer and Storkholm 2001). They contribute to the removal of organic matter at low (less than 20 ppm) BOD and COD, and can also function at low temperatures (Yamaguchi et al. 2006).

CARP was used for wastewater treatment discharged from food-processing plants using three reactors sequentially. Reactor 1 is a mesophilic methanogenic UASB reactor (35 °C) in which methane can be produced. Reactor 2 is a second UASB reactor under ambient temperature in which nitrate or sulfate reduction mainly occurs but methane production is not expected. Reactor 3 is a DHS aerobic reactor used to convert ammonia to nitrate and sulfide to sulfate. A portion of the effluent from reactor 3 is recycled to reactor 2, in which COD and BOD are decreased as a result of nitrate being reduced to nitrogen. Sulfate can be supplemented into the stream for recycling sulfate to activate sulfate-reducing bacteria optionally. The influent BOD and SS level were adjusted to approximately 2,000 ppm and 500 ppm, respectively. The temperatures in reactors 2 (denitrifying/sulfate-reducing) and 3 (nitrification/sulfate-oxidizing) were below 10 °C, and sometimes close to 1 °C during operation days 150–350 because it was winter. However, the BOD decreased to within the range of 100–200 ppm in the methanogenic UASB effluent and to within the range of 10–20 ppm in the final effluent, respectively. Moreover, even during winter, the final BOD decreased to 1 ppm. The methane conversion efficiency was approximately 80 %.

In conclusion, CARP is useful for wastewater treatment without the need to conduct the activated sludge process as posttreatment.

24.2.2 Methane Production from Solid Materials

Methane fermentation has been used for anaerobic wastewater treatment and energy recovery. In particular, the UASB process has been developed for high-rate methane fermentation of high-strength wastewater (Pette and Versprille 1982), in which microbial granule formation including those of acidogens and methanogens is a key factor. However, this process is applicable only in wastewater containing a low SS level. It is, therefore, difficult to apply a UASB system to solid biomass directly, and pretreatment to liquefy the SSs is required.

24.2.2.1 Cellulosic Materials

Although cellulose and hemicellulose are the main components of biomass, it is difficult to treat them directly using a high-rate methane fermentation process. Nakashimada et al. (2000a) investigated the direct conversion of cellulose to methane using defined methanogens and the anaerobic fungus *Neocallimastix frontalis*, which is found in ruminants and produces cellulolytic and hemicellulolytic enzymes such as cellulase and xylanase (Kartikeyan et al. 2001). This fungus can degrade biomass and produce acetate, formate, H_2/CO_2, lactate, and a small amount of ethanol, all of which are suitable substrates for methanogens and methanogenic consortia (Fukuzaki et al. 1995, 1991; Jee et al. 1987; Nishio et al. 1985). Cocultures of *N. frontalis* with *Methanobacterium formicicum* and *Methanothaeta concilii* were successfully performed to produce methane. When cellulose solution was added five times to a bioreactor in which 24 g l^{-1} cellulose in total had been added, formate and hydrogen accumulations were not observed. After 24 days, 150 mM CH_4 was produced and 57 mM acetate remained in the medium. Then, the accumulated acetate was consumed while methane formation occurred, and after 55 days, 178 mM CH_4 was produced.

24.2.2.2 Organic Wastes with High Salinity

The recent environmental pollution of coastal seawater has resulted in the accumulation of large amounts of organic sediment in the mud in some areas at the bottom of Hiroshima Bay in Japan. In particular, mud layers 2–3 m thick have frequently been observed in oyster farming areas because of the large amounts of organic matter being continuously excreted by oysters.

Mud sediment can be treated by anaerobic acidogenic fermentation followed by the cultivation of photosynthetic bacteria (Takeno et al. 1999). In acidogenic fermentation, almost all the fatty acids produced were acetate molecules. The combined addition of vitamins such as nicotinic acid, thiamine, and biotin markedly induced the production of acetate from sea mud sediment. Furthermore, by using the UASB reactor, which included methanogenic sludge that was acclimated to the mud sediment, the post acidogenic fermentation liquor can be treated as previously mentioned (Takeno et al. 2001). When a two-stage UASB reactor system was used for the high-rate anaerobic treatment of mud sediment, acetic acid from the mud sediment was present in the effluent of the acidogenic reactor in each run, yielding approximately 110 mM methane from 278 g wet wt mud/l after 4 days for each run (Fig. 24.3).

Solid wastes are discharged during the processes of making traditional Japanese foods, for example, tofu and soy sauce, and traditional drinks such as sake and shochu. We aimed to treat these solid wastes by methane fermentation. Soy sauce refuse (SSR), similar to bean curd refuse, is a highly nutritious biomass, but its utilization is difficult because of its high salt content of 10 % (w/w). Some of this type of refuse is currently being used as animal feed for cattle, but demand for it may not continue. In Japan, more than 100,000 t of SSR is produced per year; therefore, the treatment of this solid waste is also an important goal and major challenge. Although anaerobic microbial digestion is often a successful treatment for wastewater in the food industry, no such method that can be applied to solid and high-salt wastes such as SSR has been reported. Nagai et al. (2002) reported that thermophilic methanogenic sludge obtained from a municipal wastewater treatment plant could be used as seed to decrease the SS level in SSR and produce methane successfully. At 25 g wet wt/l SSR, the production of 120 mM CH_4 and 50 % (w/v) decrease in SS level were observed after 35 days. The acclimation of sludge to the waste was effective for increasing methane production rate in the pH-controlled fed-batch culture using a stirred tank reactor.

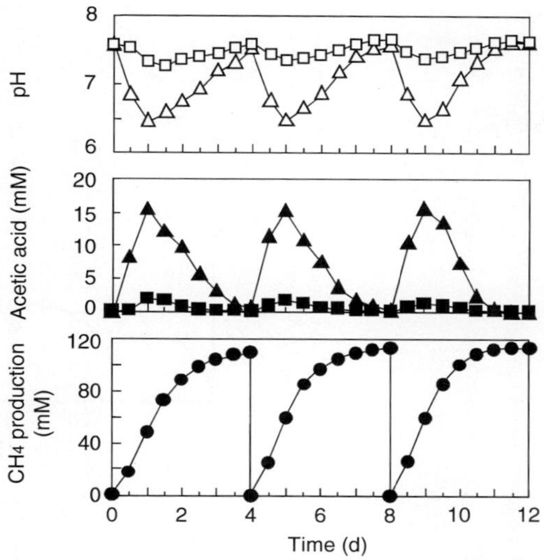

Fig. 24.3 Profiles of methane production, acetic acid concentration, and pH during methane fermentation of mud sediment in a two-stage UASB reactor system at 37 °C. *Open triangles*, pH in acidogenic reactor; *open squares*, pH in methanogenic reactor effluent; *closed triangles*, acetic acid concentration in acidogenic reactor effluent; *closed squares*, acetic acid concentration in methanogenic reactor effluent; *closed circles*, methane production in methanogenic reactor. (Adapted from Takeno et al. 2001)

24.3 Dry Methane Fermentation of Organic Solid Wastes (Am-Met Process)

Because the recycling of solid wastes is vigorously promoted in Japan and the need to recover energy from organic wastes is increasing, a dry digestion plant, which has been developed for a KOMPOGAS process (Wellinger et al. 1993), has been operating in Kyoto (http://takuma.co.jp/news/2001/20010921.html). Three types of waste (i.e., garbage and leftovers from hotels, yard waste, and used paper) mixed at various ratios are used to control the C/N ratio. The plant has maintained stable operations with the mixture, generating biogas by the decomposition of volatile solid (VS) at a rate of about 820 Nm^3 per ton of VS.

Because the aerobic-activated sludge process is common in Japan, an abundant amount of excess sludge (2,000,000 t as dry sludge) has been discharged annually. Although some of this sludge has been treated by methane fermentation to reduce the amounts of the excess sludge after condensation (approximately 4 % solid content), most of the sludge is dumped after dehydration (approximately 20 % solid content) or incineration. If the dehydrated excess sludge can be treated directly by methane fermentation, dry methane fermentation (water content, 80 %) will be more useful because the plant scale could be reduced by one fifth. Thus, the dehydrated excess sludge was fermented semicontinuously to methane under a dry

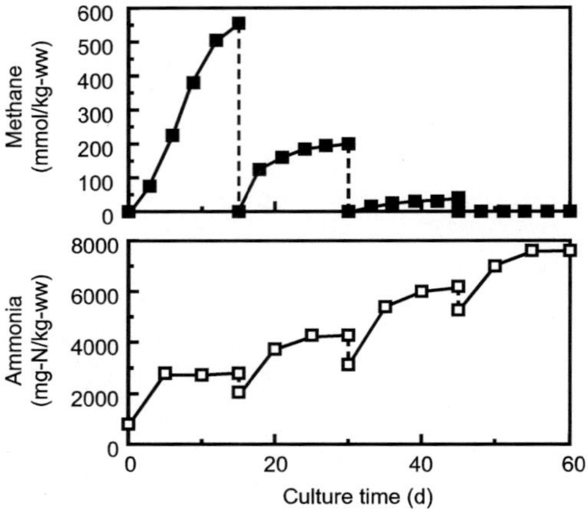

Fig. 24.4 Repeated batch methane fermentation of excess activated sludge under semi-dry condition (water content = 80 % w/w). *ww*, wet weight sludge

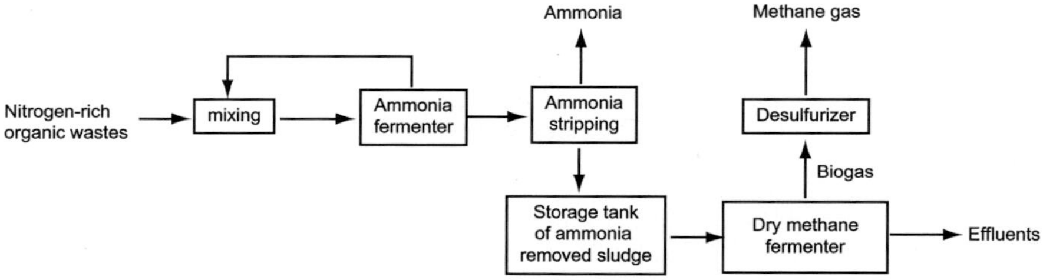

Fig. 24.5 Schematic drawings of concept of dry ammonia-methane fermentation (am-met process)

condition (water content = 80 %). Although methane was produced at a yield of approximately 610 mmol/kg wet wt, this methane production gradually decreased and eventually ceased (Fig. 24.4); this is the reason that the dry fermentation of only excess sludge has not yet succeeded. However, it is very important to note that ammonia fermentation almost constantly occurred, indicating that the cessation of methane fermentation might result form the accumulation of ammonia produced from protein in the excess sludge. Thus, the ammonia-removed dehydrated excess sludge was fermented to methane semicontinuously under the same dry condition (water content = 80 %) by changing the solid waste retention time (SRT). Methane production began even at an SRT of 20 days. The ammonia concentration was maintained at less than 3,000 mg N/kg wet wt, and VS removal and ammonia nitrogen removal efficiencies were 52 % and 59 %, respectively (Nakashimada et al. 2008).

These results demonstrated that the dry methane fermentation of dehydrated excess sludge is possible at a relatively short SRT, which corresponded to the HRT for the wet methane fermentation, if the ammonia produced is removed. If the ammonia produced can be completely removed, dry fermentation of sludge can be possible at shorter SRTs. Thus, we proposed dry ammonia-methane fermentation (Am-Met process) as a novel process to treat nitrogen-rich wastes (Fig. 24.5) and have demonstrated it was useful for the anaerobic treatment of nitrogen-rich wastes such as chicken feces (Abouelenien et al. 2010; Abouelenien et al. 2009), and food wastes (Yabu et al. 2011).

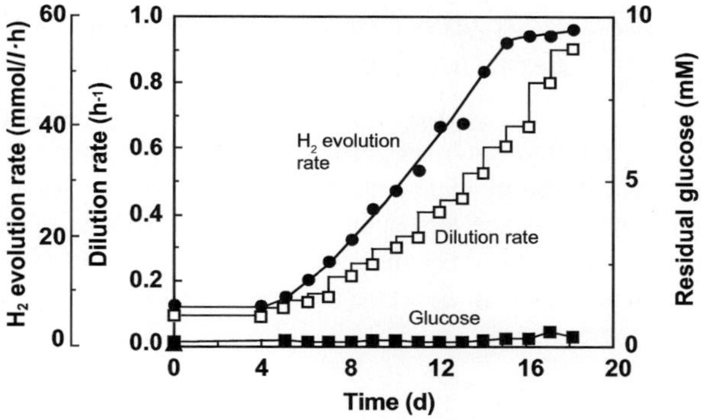

Fig. 24.6 Hydrogen fermentation of mutant AY-2 in packed-bed reactor. (Adapted from Rachman et al. 1998)

24.4 Useful Material Production with Anaerobic Microorganisms from the Ecosystem of Methane Fermentation

24.4.1 Hydrogen Production with Hydrolytic Microorganisms

Hydrogen can be used as clean energy and is generally produced by photosynthetic and fermentative microorganisms under anaerobic conditions in pure cultures. Fortunately, hydrogen can be produced in the hydrolysis and acidogenesis stages in the methanogenic ecosystem. If high-rate and high-yield hydrogen fermentation is achieved, it is possible that the hydrogen gas produced can be directly connected to a fuel cell system without any reforming being necessary. Furthermore, methane can be produced from fatty acids such as acetate, propionate, and butyrate that accumulate in liquid broth after hydrogen fermentation.

To gain an understanding of the potential of hydrogen production from carbohydrates, continuous hydrogen production using self-flocculating cells of *Enterobacter aerogenes* HU-101 isolated from methanogenic sludge as the high-rate hydrogen producer and its mutant AY-2 (Rachman et al. 1997, 1998) was carried out in a packed-bed reactor under a glucose-limiting condition. As shown in Fig. 24.6, the maximum hydrogen production rates were 31 and 58 mmol/l h for HU-101 and AY-2, respectively, at a dilution rate of 0.67 h^{-1}. Using *E. aerogenes* HU-101, we clarified the relationship between hydrogen production and carbon source with various redox states (Nakashimada et al. 2002b). It was clearly indicated that the redox state of carbon sources directly affected hydrogen production by *E. aerogenes*. High-rate hydrogen production from glycerol is now possible at an HRT of 1.5 h. In addition, a higher yield, approximately 2 mol/mol glucose, can also be obtained by mutation (Ito et al. 2004). The fermented broth contained 2,3-butandiol, ethanol, lactate, and acetate, which are good substrates for UASB methane fermentation.

24.4.2 Hydrogen Production Combined with Methane Fermentation (Hy-Met Process)

As it was found that hydrogen could be efficiently produced using hydrolytic/acidogenic bacteria in the methanogenic sludge, we have proposed this process as a hydrogen-methane two-stage fermentation (Hy-Met) process. In this process, the energy in the hydrogen produced is converted to electricity by the fuel cell system and the produced methane is used to generate heat energy to heat the two reactors and to satisfy heat requirements.

24.4.2.1 Bread Wastes

In Japan, a total of 100,000 t/year solid bread waste is discharged. For this waste, the Hy-Met process was applied (Nakashimada and Nishio 2003; Nishio et al. 2004). In a batch culture, in which 100 g wet wt/l bread waste (43 % water content, w/w) was treated at 55 °C with 10 % (w/v) of a thermophilic sludge collected from an anaerobic digester of sewage sludge at a sewage treatment plant in Hiroshima, Japan, the waste was fermented to hydrogen and VFAs under pH-uncontrolled (initial pH 7) and pH-controlled conditions at 7 or 5. Although the pH-uncontrolled condition yielded only 70 mM H_2 with an 80 % decrease in SS level, the pH 7-controlled condition yielded 240 mM H_2 with a 91 % decrease in SS level after 24 h. The culture broth contained 150 mM each of acetate and butyrate, and the TOC concentration was approximately 20,000 ppm. On the other hand, under the pH 5-controlled condition, only 100 mM H_2 was produced, and lactate (approximately 220 mM) was mainly produced in the culture broth. Next, culture broth of the hydrogen fermentation of the bread waste, which contained approximately 20,000 ppm of TOC concentration, was used for methane production. This culture broth was diluted to yield TOC concentration of 2,000–5,000 ppm and supplied continuously to a UASB methane reactor, in which acclimatized methanogenic granules were inoculated. When the organic loading rate was increased by increasing the dilution rate stepwise, the optimum loading rate was 9.5 g TOC/l day yielding 80 % TOC removal, a methane production rate of 400 mmol/l day, and a methane yield of approximately 0.6 as the carbon base. These results show that when reactor volumes for hydrogen and methane fermentations are set to a ratio of 1:2.1, SS level will be decreased by 91 % at a the loading rate of 29 g wet wt/l day, and the hydrogen and methane yields will be 2.4 mol/kg wet wt and 8.6 mol/kg wet wt, respectively.

The amount of energy recovered from the Hy-Met process using bread waste was estimated on the basis of these results. To treat the waste discharged from one factory at 2.67 t/day, a 26.7 m^3 hydrogen fermentation reactor in which 145 m^3 hydrogen/day will be produced, which corresponds to 214 kWh when the conversion efficiency of the fuel cell system is 50 %, and a 56-m^3 methane fermentation reactor in which 514 m^3 methane/day will be produced, which corresponds to 530 l oil/day.

24.4.2.2 Beer Waste

Recently, a UASB methane fermentation process has often been used in a beer manufacturing factory. However, because the pressed filtrate from the spent malt in the lauter tun at the beer factory contains a high density of suspended matter, such filtrate is difficult to apply to the UASB reactor. Therefore, only the filtrate obtained from the pressed filtrate after SS removal has been treated using the UASB reactor. The Hy-Met process was applied directly to this pressed filtrate (Mitani et al. 2005). When continuous culture for hydrogen fermentation and the following UASB methane fermentation was carried out at 50 °C and 37 °C, respectively, COD removal was more or less the same compared with the single UASB methane fermentation that is currently in use in this factory. However, the total amount of energy recovered as the sum of hydrogen and methane increased to 103 kJ/l from 90 kJ/l, which corresponds to the amount of the suspended matter solubilized during the hydrogen fermentation. Our results also demonstrated that waste treatment could be carried out without the removal of suspended matter from the pressed filtrate by connecting the hydrogen fermentor before UASB methane fermentation.

24.4.3 Hydrogen and Ethanol Production from Biodiesel Wastewater with a Hydrolytic Microorganism

Biodiesel fuels have attracted a great deal of attention recently because they are an alternative to petroleum-based fuel, renewable and nontoxic, contribute to a favorable energy balance, and produce fewer harmful emissions than gasoline. Although biodiesel fuels are produced chemically and enzymatically, glycerol is essentially generated as the by-product (Du et al. 2003; Vicente et al. 2004). If there is an increase in the production of biodiesel fuels in the world, then the problem of efficiently treating wastes containing glycerol will need to be faced. Because glycerol is the best substrate for hydrogen production by *E. aerogenes* (Nakashimada et al. 2002a), hydrogen production was examined from glycerol-containing wastes discharged after the biodiesel manufacturing process. In continuous culture with a porous, ceramic packed-bed reactor as a support material for fixing cells in the reactor using flock-formed *E. aerogenes*, the maximum hydrogen production rate reached was 63 mmol/l/h, giving an ethanol yield of 0.85 mol/mol glycerol in the culture broth (Fig. 24.7) (Ito et al. 2005). This result indicates that we can produce biodiesel, hydrogen, and ethanol from vegetable oils and animal fats or their wastes.

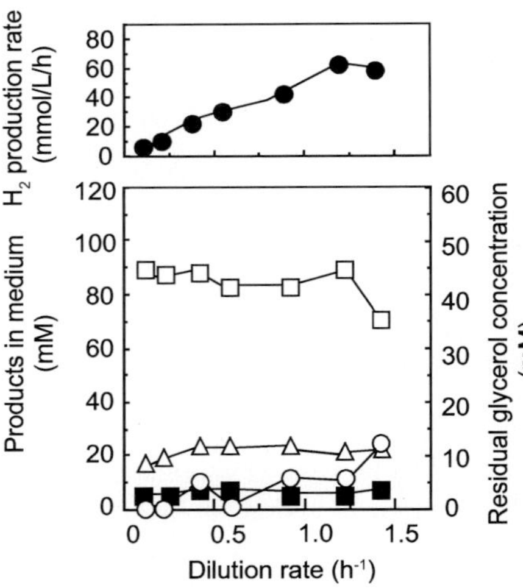

Fig. 24.7 Hydrogen and ethanol production by self-flocculated *Enterobacter aerogenes* HU-101 from glycerol-containing wastes discharged from biodiesel manufacturing factory. *Closed circles*, hydrogen; *open squares*, ethanol; *open triangles*, 1,3-propanediol; *closed squares*, lactate; *open circles*, glucose in the medium. (Adapted from Ito et al. 2005)

24.4.4 Useful Material Production with CO_2 Fixation

Anaerobic microorganisms participating in methane fermentation can fix CO_2 in catabolism. For example, hydrogenotrophic (hydrogen-consuming) methanogens produce methane with CO_2 as carbon source and hydrogen as energy source. Using this property, mutual conversion of H_2–CO_2 to formate that is liquid at room temperature is possible (Nishio et al. 1983; Eguchi et al. 1985). Although hydrogen is an attractive clean fuel, transportation of hydrogen gas is costly because of the necessity of high pressure and special storage bottles. If hydrogen is converted to formate at the site where hydrogen is produced and transported formate is reconverted to hydrogen, it would be beneficial for improvement of economical efficiency.

Syngas fermentation is a biological process to produce useful metabolites using chemolithotrophs that catabolize the mixed gas of H_2, CO, and CO_2 generated by gasification of various organic substances. The biological production of fuels and chemicals through syngas fermentation offers several advantages over conventional sugar fermentation technology because in syngas fermentation whole biomass including nondegradable components such as lignin via gasification is used.

A group of anaerobic bacteria known as acetogens that can grow autotrophically (Drake et al. 2008) has been investigated for syngas fermentation. In mesophilic acetogens such as *Clostridium ljungdahlii* (Klasson et al. 1993), ethanol production from syngas has been reported. Also, thermophilic acetogens such as *Moorella* sp. HUC22-1 (Sakai et al. 2004, 2005), which is closely related to *Moorella thermoacetica* ATCC39073, have been investigated for ethanol production from H_2–CO_2. The use of thermophilic bacteria for syngas fermentation will facilitate the recovery of ethanol because aqueous ethanol will readily vaporize at temperatures above 50 °C, so that it will enable continuous distillation of ethanol. Furthermore, thermophilic bacteria have higher growth and metabolic rates than mesophilic bacteria, and there is low risk of microbial contamination (Payton 1984; Taylor et al. 2009). In this context, thermophilic acetogens such as *Moorella* spp. should be more promising candidates for syngas fermentation than mesophilic bacteria.

To improve the production of biofuel and renewable materials from syngas by *Moorella* spp., molecular breeding will be indispensable. In previous research, therefore, we have developed genetic transformation and heterologous expression system for the type strain of *Moorella thermoacetica* ATCC39073 (Kita et al. 2013). The transformation system consists of an orotate monophosphate decarboxylase gene (*pyrF*) deletion mutant, strain dpyrF, as a transformation host and a *pyrF* gene as the positive marker to recover uracil auxotrophy. Using the developed transformation system, we successfully constructed a lactate-producing mutant of strain dpyrF that expressed the lactate dehydrogenase gene from *Thermoanaerobacter pseudethanolicus* ATCC33233 (Kita et al. 2013). This result clearly demonstrated that the developed system could be used as a genetic tool for improved production by thermophilic acetogens of target metabolites such as alcohols or fatty acids. Such attempts are now in progress.

24.4.5 Production of Optically Active Materials with Hydrolytic Microorganisms

Hydrolytic/acidogenic anaerobes such as *Enterobacter*, *Klebsiella*, and *Bacillus* spp. produce 2, 3-butanediol that is used for a biofuel. Among them, *Paenibacillus polymyxa* ATCC 12321 (formally *Bacillus polymyxa*) produces optically active (*R*,*R*)-2,3-butanediol that is an ingredient of liquid crystal and pharmaceutical products at more than 98 % e.e. of optical purity. Under anaerobic conditions, however, the maximum yield of (*R*,*R*)-2,3-butanediol is 0.67 mol/mol glucose because excess reducing power has to be reoxidized to keep the redox balance in catabolism, resulting in by-product formation such as ethanol and acetate. To increase the yield of (*R*,*R*)-2,3-butanediol, therefore, it is needed to scavenge the excess reducing power (mainly NADH) without formation of by-products. For this purpose, microaerobic culture was popular, in which a small amount of oxygen was supplied to re-oxidize NADH (Magee and Kosaric 1987). However, we found the microaerobic culture decreased the optical purity of (*R*,*R*)-2,3-butanediol (Nakashimada et al. 1998). Instead of microaerobic culture, we investigated addition of fatty acids such as acetic or propionic acid as the electron acceptor together with glucose (Fig. 24.8) (Nakashimada et al. 2000b). In the study, addition of acetic acid significantly increased the yield of (*R*,*R*)-2,3-butanediol from 0.59 to 0.88 mol/mol glucose without the decrease of the optical purity. It is noted that acetic acid is reduced to ethanol during the fermentation. Because ethanol is the most popular biofuel, this is a method for simultaneous production of a worthy biochemical and biofuel. Furthermore, *Paenibacillus polymyxa* can utilize xylose together with glucose at 39 °C (Marwoto et al. 2002, 2004); this is a great advantage for production of the biomaterial and biofuels from cellulosic biomass.

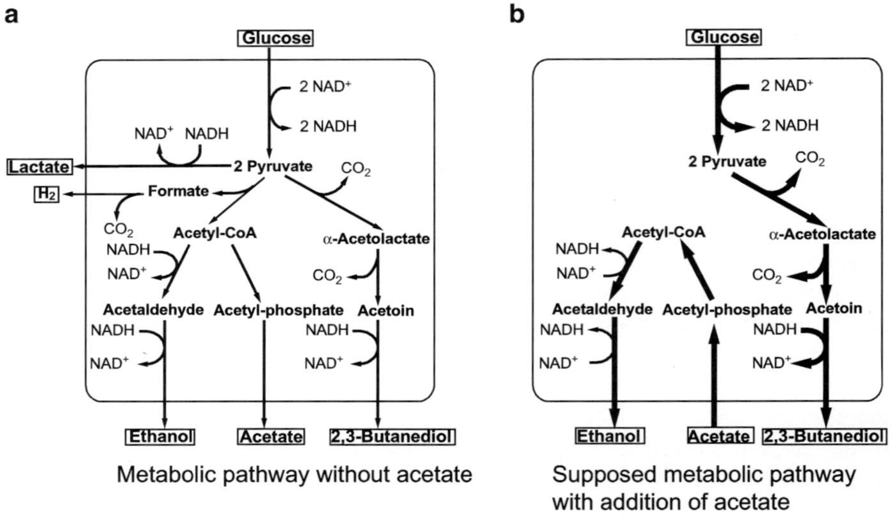

Fig. 24.8 Supposed metabolic pathway by *Paenibacillus polymyxa* of glucose (**a**) and glucose with addition of acetate (**b**)

In the presence of nitrate, some acidogenic microorganisms reduce nitrate to nitrogen gas with carbohydrates as the electron donor, which is named "denitrification." In such a denitrifying bacteria, we found that *Paracoccus denitrificans* converted ethyl acetoacetate to ethyl (*R*)-β-hydroxybutyrate coupled with nitrate reduction at the yield of 64 % and optical purity of 99 % e.e. (Nakashimada et al. 2001). Because nitrate can supply the reactor with the conventional fed-batch strategy unlikely for oxygen, more attention should be given to the use of electron acceptors besides oxygen.

24.4.6 Vitamin Production with Methanogen and Acetogen

An acetoclastic methanogen, *Methanobacterium barkeri*, may be of interest for corrinoids production because of participation of corrinoids (B_{12}) in methanogenesis, a high content of corrinoids when grown on methanol medium and non-inhibition of the cellular growth by the main fermentation product (CH_4), and use of methanol as an inexpensive, water-soluble, and neutral substrate. Thus, we investigated efficient corrinoid production with methanol as the carbon source by *M. barkeri*. Addition of cobalt (the essential metal for corrinoid) was effective to increase corrinoids content, reaching a maximum value of 5.8 mg B_{12}/g dry cell at 9.6 mg/l $CoCl_2$ $6H_2O$ in the batch culture. To avoid the inhibitory effects of methanol on cell growth, a repeated batch culture was carried out for attaining a high cell mass for corrinoids production (Fig. 24.9). In each batch, the added

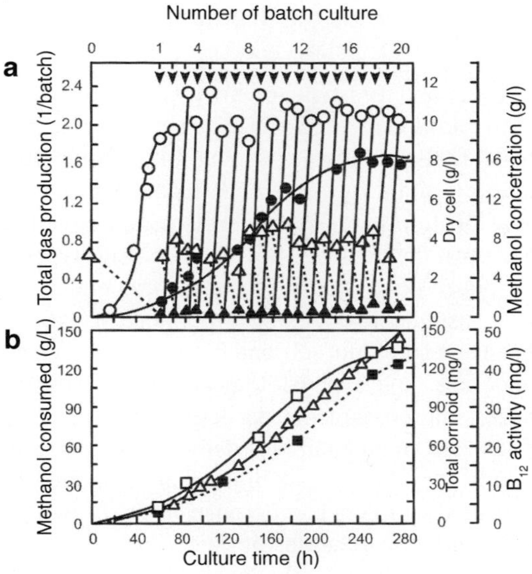

Fig. 24.9 Repeated batch culture of *Methanosarcina barkeri* for B_{12} production. **(a)** *Open circles*, total gas production per batch; *closed circles*, cell growth (dry cell); *broken line* shows methanol addition (*open triangles*) and methanol consumption (*closed triangles*); *arrows* indicate nutrient feed to the culture. **(b)** *Open squares*, total corrinoid as dicyano form; *closed squares*, B_{12} activity of the corrinoid; *open triangles*, total methanol consumed. (Adapted from Mazumder et al. 1986)

methanol was almost completely utilized and cells continued to grow. After 20 intermittent additions during 280 h of culture, total corrinoids concentration was as high as 135 mg/l, of which more than 70 % was excreted into the culture supernatant and the remaining was in the cells (Mazumder et al. 1986). However, because separation and identification of corrinoids produced in the culture revealed that 33 % of total corrinoids was factor III (5-hydroxybenzimidazolyl cobamide) and the remaining corrinoids were base-free corrinoids such as cobinamide (factor B) and cobyrinic acid and/or cobinic acid, containing one or more carboxylic groups on the corrin nucleus, the development of methods to convert such incomplete forms to B_{12} is expected.

On the other hand, an acetogen, *Acetobacterium* sp., produces a complete form of vitamin B_{12} and had 5 mg/g dry cell of high intracellular content of B_{12} in the medium supplemented with glucose (Bainotti et al. 1996) or methanol-formate (Bainotti and Nishio 2000). However, high-density culture was difficult because of inhibition by produced acetate. Thus, the culture system equipped with a removal system of acetate was developed. In the system, once acetate concentration reached the inhibitory level, the medium in the reactor for *Acetobacterium* sp. was recycled into the fixed-bed reactor containing an acetoclastic methanogen, *Methanosaeta concilii*, to consume produced acetate. When the acetate removal system was combined with the continuous feeding of the medium by pH stat in which the fresh medium was fed into the reactor once the culture pH increased to the set value with consumption of acetate, the productivity of B_{12} was 20 fold higher than that of the batch culture (Bainotti et al. 1997).

24.5 Concluding Remarks

The demand for saving of energy used and its recovery from unused matter has been high traditionally in Japan. The tendency has been intensified by fear of depletion of fossil fuel and global warming. In the foregoing context, efficient methane fermentation typified by the UASB reactor became widely used, and improvement and modification of methane fermentation have been carried out to expand the availability of the high-rate process to various types of organic matter. In this chapter, we introduced our attempts to treat some unusual organic matter and recover methane as a sustainable energy. Furthermore, it was also described that not only methane but also useful biofuels and biomaterials such as hydrogen, ethanol, optically active materials, and a physiologically active substance, vitamin B_{12}, can be produced using microorganisms participating in the ecosystem of methane fermentation. These are small examples demonstrating that the ecosystem of methane fermentation is a gold mine as a biological resource. The physiology of anaerobic microorganisms is still unclear because of the difficulty of the culture and the complexity of the symbiotic relationship such as hydrogen-producing acetogenic bacteria and hydrogenotrophic methanogens. We believe that deep understanding of the physiology and metabolism of anaerobic bacteria will give more beneficial properties to contribute to environmental protection and the welfare of human beings.

References

Abouelenien F, Kitamura Y, Nishio N, Nakashimada Y (2009) Dry anaerobic ammonia-methane production from chicken manure. Appl Microbiol Biotechnol 82(4):757–764

Abouelenien F, Fujiwara W, Namba Y, Kosseva M, Nishio N, Nakashimada Y (2010) Improved methane fermentation of chicken manure via ammonia removal by biogas recycle. Bioresour Technol 101(16):6368–6373

Bainotti AE, Nishio N (2000) Growth kinetics of *Acetobacterium* sp. on methanol-formate in continuous culture. J Appl Microbiol 88(2):191–201

Bainotti A, Setogaichi M, Nishio N (1996) Kinetic studies and medium improvement of growth and vitamin B_{12} production by *Acetobacterium* sp. in batch culture. J Ferment Bioeng 81:324–328

Bainotti AB, Futagami K, Nakashimada Y, Chang Y-J, Nagai S, Nishio N (1997) pH auxostat continuous culture of *Acetobacterium* sp. coupled with removal of acetate by *Methanosaeta concilii*. Biotechnol Lett 19:989–993

Drake HL, Gossner AS, Daniel SL (2008) Old acetogens, new light. Ann N Y Acad Sci 1125:100–128

Du W, Xu YY, Liu DH (2003) Lipase-catalysed transesterification of soya bean oil for biodiesel production during continuous batch operation. Biotechnol Appl Biochem 38(2):103–106

Eguchi SY, Nishio N, Nagai S (1985) Formic-acid production from H_2 and bicarbonate by a formate-utilizing methanogen. Appl Microbiol Biotechnol 22(2):148–151

Fauque GD (1995) Ecology of sulfate-reducing bacteria. In: Barton LL (ed) Sulfate reducing bacteria. Kluwer Academic/Plenum, New York, pp 217–242

Fukuzaki S, Nishio N, Sakurai H, Nagai S (1991) Characteristics of methanogenic granules grown on propionate in an upflow anaerobic sludge blanket (UASB) reactor. J Ferment Bioeng 71:50–57

Fukuzaki S, Nishio N, Nagai S (1995) High rate performance and characterization of granular methanogenic sludges in upflow anaerobic sludge blanket reactors fed with various defined substrates. J Ferment Bioeng 79:354–359

Holmer M, Storkholm P (2001) Sulphate reduction and sulphur cycling in lake sediments: a review. Freshw Biol 46:431–451

Ito T, Nakashimada Y, Kakizono T, Nishio N (2004) High yield production of hydrogen by *Enterobacter aerogenes* mutants with decreased α-acetolactate synthase activity. J Biosci Bioeng 97(4):227–232

Ito T, Nakashimada Y, Senba K, Matsui T, Nishio N (2005) Hydrogen and ethanol production from glycerol-containing wastes discharged from biodiesel manufacturing process. J Biosci Bioeng 100(3):260–265

Jee HS, Ohashi T, Nishizawa Y, Nagai S (1987) Limiting factor of nitrogenase system mediating hydrogen-production of *Rhodobacter sphaeroides* S. J Ferment Technol 65(2):153–158

Jowell WJ, Switzenbaum MS, Morris JW (1981) J Wat Pollut Control Fed 53:482

Kartikeyan S, Murakami M, Nakashimada Y, Nishio N (2001) Efficient production of cellulolytic and xylanolytic enzymes by the rumen anaerobic fungus, *Neocallimastix frontalis*, in a repeated-batch culture. J Biosci Bioeng 91(2):153–158

Kita A, Iwasaki Y, Sakai S, Okuto S, Takaoka K, Suzuki T, Yano S, Sawayama S, Tajima T, Kato J, Nishio N, Murakami K, Nakashimada Y (2013) Development of genetic transformation and heterologous expression system in carboxydotrophic thermophilic acetogen *Moorella thermoacetica*. J Biosci Bioeng (in press). doi:10.1016/j.jbiosc.2012.1010.1013

Klasson KT, Ackerson MD, Clausen EC, Gaddy JL (1993) Biological conversion of coal and coal-derived synthesis gas. Fuel 72(12):1673–1678

Lettinga G, Roersma R, Grin P (1983) Biotechnol Bioeng 25:1701

Machdar I, Sekiguchi Y, Sumino H, Ohashi A, Harada H (2000) Combination of a UASB reactor and a curtain type DHS (downflow hanging sponge) reactor as a cost-effective sewage treatment system for developing countries. Water Sci Technol 42(3–4):83–88

Magee RJ, Kosaric N (1987) The microbial production of 2,3-butanediol. Adv Appl Microbiol 32:89–159

Marwoto B, Nakashimada Y, Kakizono T, Nishio N (2002) Enhancement of (R, R)-2,3-butanediol production from xylose by *Paenibacillus polymyxa* at elevated temperatures. Biotechnol Lett 24(2):109–114

Marwoto B, Nakashimada Y, Kakizono T, Nishio N (2004) Metabolic analysis of acetate accumulation during xylose consumption by *Paenibacillus polymyxa*. Appl Microbiol Biotechnol 64(1):112–119

Mazumder TK, Nishio N, Hayashi M, Nagai S (1986) Production of corrinoids including vitamin B_{12} by *Methanosarcina barkeri* growing on methanol. Biotechnol Lett 8(12):843–848

Mitani Y, Takamoto Y, Atsumi R, Hiraga T, Nishio N (2005) Hydrogen and methane two-stage production directly from brewery effluent by anaerobic fermentation. Master Brewers Assoc Americas TQ 42:283–289

Nagai H, Kobayashi M, Tsuji Y, Nakashimada Y, Kakizono T, Nishio N (2002) Biological and chemical treatment of solid waste from soy sauce manufacture. Water Sci Technol 45(12):335–338

Nakashimada Y, Nishio N (2003) Hydrogen and methane fermentation of solid wastes from food industry. Foods Food Ingredients J Jpn 208(9):703–708 (in Japanese)

Nakashimada Y, Kanai K, Nishio N (1998) Optimization of dilution rate, pH and oxygen supply on optical purity of 2, 3-butanediol produced by *Paenibacillus polymyxa* in chemostat culture. Biotechnol Lett 20(12):1133–1138

Nakashimada Y, Kartikeyan S, Murakami M, Nishio N (2000a) Direct conversion of cellulose to methane by anaerobic fungus *Neocallimastix frontalis* and defined methanogens. Biotechnol Lett 22(3):223–227

Nakashimada Y, Marwoto B, Kashiwamura T, Kakizono T, Nishio N (2000b) Enhanced 2,3-butanediol production by addition of acetic acid in *Paenibacillus polymyxa*. J Biosci Bioeng 89(6):661–664

Nakashimada Y, Kubota H, Takayose A, Kakizono T, Nishio N (2001) Asymmetric reduction of ethyl acetoacetate to ethyl (R)-β-hydroxybutyrate coupled with nitrate reduction by *Paracoccus denitrificans*. J Biosci Bioeng 91(4):368–372

Nakashimada Y, Rachman MA, Kakizono T, Nishio N (2002) H_2 production of *Enterobacter aerogenes* altered by extracellular and intracellular redox states. Int J Hydrogen Energy 27(11–12):1399–1405

Nakashimada Y, Ohshima Y, Minami H, Yabu H, Namba Y, Nishio N (2008) Ammonia-methane two-stage anaerobic digestion of dehydrated waste-activated sludge. Appl Microbiol Biotechnol 79(6):1061–1069

Nishio N, Eguchi SY, Kawashima H, Nagai S (1983) Mutual conversion between H_2 plus CO_2 and formate by a formate-utilizing methanogen. J Ferment Technol 61:557–561

Nishio N, Kayawake E, Nagai S (1985) Rapid methane production from formate or acetate in fixed bed bioreactors. J Ferment Technol 63:205–209

Nishio N, Nakashimada Y, Mitani Y, Hiraga T (2004) Hydrogen-methane two-stage fermentation (Hy-Met Process) for anaerobic waste treatment. In: Proceedings of 15th world hydrogen energy conference, Yokohama, Japan, pp 28E-07.pdf (CD-ROM)

Payton MA (1984) Production of ethanol by thermophilic bacteria. Trends Biotechnol 2:153–158

Pette KC, Versprille AI (1982) Application of the UASB-concept for wastewater treatment. In: Hughes DE, Stafford DA, Wheatley BI et al (eds) Anaerobic digestion 1981. Elsevier/North-Holland Biochemical Press, Amsterdam, pp 121–137

Rachman MA, Furutani Y, Nakashimada Y, Kakizono T, Nishio N (1997) Enhanced hydrogen production in altered mixed acid fermentation of glucose by *Enterobacter aerogenes*. J Ferment Bioeng 83(4):358–363

Rachman MA, Furutani Y, Nakashimada Y, Kakizono T, Nishio N (1998) Hydrogen production with high yield and high evolution rate by self-flocculated cells of *Enterobacter aerogenes* in a packed-bed reactor. Appl Microbiol Biotechnol 49(4):450–454

Sakai S, Nakashimada Y, Yoshimoto H, Watanabe S, Okada H, Nishio N (2004) Ethanol production from H_2 and CO_2 by a newly isolated thermophilic bacterium, *Moorella* sp. HUC22-1. Biotechnol Lett 26(20):1607–1612

Sakai S, Nakashimada Y, Inokuma K, Kita M, Okada H, Nishio N (2005) Acetate and ethanol production from H_2 and CO_2 by *Moorella* sp. using a repeated batch culture. J Biosci Bioeng 99(3):252–258

Takahashi M, Yamaguchi T, Abe K, Araki N, Sumino H, Yamazaki S, Nishio N (2005) Process performance and ecological significance of a pilot-scale sewage treatment system by combining of UASB and DHS reactor system enhancing a sulfur-redox cycle action. In: 1st IWA-ASPIRE 2005 conference, Singapore, pp 7D-2.pdf (CD-ROM)

Takeno K, Sasaki K, Watanabe M, Kaneyasu T, Nishio N (1999) Removal of phosphorus from oyster farm mud sediment using a photosynthetic bacterium, *Rhodobacter sphaeroides* IL106. J Biosci Bioeng 88:410–415

Takeno K, Nakashimada Y, Kakizono T, Nishio N (2001) Methane fermentation of coastal mud sediment by a two-stage upflow anaerobic sludge blanket (UASB) reactor system. Appl Microbiol Biotechnol 56(1/2):280–285

Taylor MP, Eley KL, Martin S, Tuffin MI, Burton SG, Cowan DA (2009) Thermophilic ethanologenesis: future prospects for second- generation bioethanol production. Trends Biotechnol 27:398–405

USEPA (1993) Manual for nitrogen control. In: U.S. Environmental Protection Agency technical report. Office of Research and Development and Office of Water, Washington, DC

Vicente G, Martinez M, Aracil J (2004) Integrated biodiesel production: a comparison of different homogeneous catalysts systems. Bioresour Technol 92(3):297–305

Wellinger A, Wyder K, Metzler AE (1993) KOMPOGAS: a new system for the anaerobic treatment of source separated waste. Water Sci Technol 27:153–158

Yabu H, Sakai C, Fujiwara T, Nishio N, Nakashimada Y (2011) Thermophilic two-stage dry anaerobic digestion of model garbage with ammonia stripping. J Biosci Bioeng 111(3):312–319. doi:10.1016/J.Jbiosc.2010.10.011

Yamaguchi T, Bungo Y, Takahashi M, Sumino H, Nagano A, Araki N, Imai T, Yamazaki S, Harada H (2006) Low strength wastewater treatment under low temperature conditions by a novel sulfur redox action process. Water Sci Technol 53:99–105

Young JC, McCarty PL (1969) The anaerobic filter for waste treatment. J Water Pollut Control Fed 41(5):R160–R173

Index

H. Anazawa and S. Shimizu (eds.), *Microbial Production: From Genome Design to Cell Engineering*, DOI 10.1007/978-4-431-54607-8, © Springer Japan 2014

V

W

X

Y

Z

Printed by Printforce, the Netherlands